ARTICULATED LADIES

GENDER AND THE MALE COMMUNITY IN EARLY CHINESE TEXTS

HARVARD-YENCHING INSTITUTE

MONOGRAPH SERIES, 53

ARTICULATED LADIES

GENDER AND THE MALE COMMUNITY IN EARLY CHINESE TEXTS

Paul Rouzer

Published by the Harvard University Asia Center
for the Harvard-Yenching Institute
Distributed by Harvard University Press
Cambridge, Massachusetts, and London, England 2001

Printed in the United States of America

The Harvard-Yenching Institute, founded in 1928 and headquartered at Harvard University, is a foundation dedicated to the advancement of higher education in the humanities and social sciences in East and Southeast Asia. The Institute supports advanced research at Harvard by faculty members of certain Asian universities and doctoral studies at Harvard and other universities by junior faculty at the same universities. It also supports East Asian studies at Harvard through contributions to the Harvard-Yenching Library and publication of the *Harvard Journal of Asiatic Studies* and books on premodern East Asian history and literature.

Library of Congress Cataloging-in-Publication Data

Rouzer, Paul F.

Articulated ladies : gender and the male community in early Chinese texts / Paul Rouzer.

p. cm. -- (Harvard-Yenching Institute monograph series ; 53)

Includes bibliographical references and index.

ISBN 0-674-00527-9 (cloth : alk. paper)

1. Chinese literature--History and criticism. 2. Gender identity in literature. I. Title. II. Series.

PL2261.R68 2001

895.1'09353--dc21 00-053984

Index by the author

Printed on acid-free paper

Last figure below indicates year of this printing

10 09 08 07 06 05 04 03 02 01

for Jenny

Preface and Acknowledgments

Since gender and its representation have become subjects of interest to students of cultural, social, and literary history from the ancient period to the present, I have written these essays in the hope that they would find readers both inside and outside the field of Chinese literature studies. Consequently, I have tried to supply translations (or at least partial translations) of all the texts I discuss and have even supplied as an appendix a complete translation of the *You xianku*. I have also supplied an introductory table of important Chinese historical periods. I hope that this book will thus prove readable and interesting for both the specialist and nonspecialist.

My first debt of gratitude is owed to the talented scholars who over the past several decades have already translated and written on many of the texts I discuss here: people such as Richard Mather, David Knechtges, Burton Watson, Kwong Ying Foon, and Georges des Rotours, to name just a few. Their groundbreaking and erudite work has made my own discussions of Chinese literature possible. My policy to use my own translations throughout resulted more from a desire for consistency in style than to any thought on my part to improve on their renderings.

Chapters 1 through 4 and Chapter 7 were finished earlier than the other parts of the book and have had many readers and critics. I owe special thanks to Gustav Heldt, whose valuable discussions in the early days of writing inspired me in my work on Chapter 1 and my point of view throughout. Robert Hymes and Joseph Roe Allen took great pains to make content and stylistic suggestions on this part of the book, and I have gratefully incorporated almost all their suggestions. Others, too, have lent me their assistance: Cynthia Chennault generously gave me a copy of her essay on palace poetry, which has incited me to further thoughts on literary composition during the Six Dynasties; Pei-i Wu made some suggestions on the world of the examination candidate during the Tang; and Stephen Owen helped me over some rough passages in the *You xianku* translation. The anonymous referees of my manuscript also made important suggestions, many of which I have tried to incorporate, or at least to answer. Finally, I owe much to the suggestions and good sense of John Ziemer, who has always been for me an ideal editor. As always, however, I alone am responsible for any errors.

I would also like to extend special thanks to the faculty of the Department of East Asian Languages and Civilizations at Harvard, Chairman Peter Bol, department administrator Frankie Hoff, and the staff of the Yenching Library for supporting me materially and spiritually during a rough period in my career. Above all, however, I owe the greatest debt to my wife, Jennifer Carpenter, who continues to be a judicious critic as well as a source of inspiration.

P.R.

Contents

Table of Dynasties

Zhou 周, ca. 1027–256 B.C.
Warring States 戰國, 403–221 B.C.
Qin 秦, 221–207 B.C.
Western (Former) Han 西漢, 206 B.C.–A.D. 8; royal family: Liu 劉
Eastern (Latter) Han 東漢, 25–220; royal family: Liu 劉
Wei 魏, 220–65; royal family: Cao 曹
Western Jin 西晉, 265–317; royal family: Sima 司馬
Six Dynasties 六朝 (Southern Dynasties 南朝)
  Wu 吳, 222–80; royal family: Sun 孫
  Eastern Jin 東晉, 317–420; royal family: Sima 司馬
  Song 宋, 420–79; royal family: Liu 劉
  Qi 齊, 479–502; royal family: Xiao 蕭
  Liang 梁, 502–57; royal family: Xiao 蕭
  Chen 陳, 557–89; royal family: Chen 陳
Sui 隋, 581–618; royal family: Yang 楊
Tang 唐, 618–907; royal family: Li 李
Song 宋, 960–1279
Yuan 元, 1260–1368
Ming 明, 1368–1644
Qing, 清, 1644–1911

ARTICULATED LADIES

GENDER AND THE MALE COMMUNITY IN EARLY CHINESE TEXTS

Introduction

No matter whether a woman is beautiful or ugly, once she enters the palace she is the subject of jealousy; and no matter whether a gentleman is worthy or useless, once he enters the court he is despised.

—Zou Yang 鄒陽 (206–129 B.C.)

This book endeavors to show how gender and desire are represented in elite, male-authored literary texts during an early phase of Chinese culture (200 B.C.–1000 A.D.). Above all, it discusses the intimate relations between the representation of gender and the political and social self-representations of elite men: it shows where gender and social hierarchies cross paths. By way of introduction, we may cite one of the earliest texts dealing with public life, the writings of the philosopher Mencius 孟子 (fl. fourth c. B.C.).

Jing Chun 景春 said, "Were not Gongsun Yan 公孫衍 and Zhang Yi 張儀 truly great men? Once they grew angry, all the feudal lords trembled; but when they remained at peace, then the fires [of war] were extinguished throughout the world."

Mencius replied, "How could these have succeeded in becoming great men? Have you not studied the Rites? When a man is first capped, his father gives him direction; but when a daughter marries, her mother gives her direction: she sees her off at the gate and cautions her, 'Since you are going to your home, you must be respectful and cautious. Do not oppose your husband: for to take compliance as your standard is the Way of concubines and wives' (*qiefu zhi dao* 妾婦之道).

"If you dwell in the broad dwelling of the world; if you stand in the proper position in the world; if you carry out the great Way in the world; if, when you obtain your ambitions, you share them with the people; if, when you do not obtain them, you go solitary on your chosen way; if you are incapable of excess in spite of wealth and rank; if you are incapable of being moved from your goal in spite of poverty and low position; if you do not bend to power and military might—then you may be called a great man." (*Mencius* III.B.2)[1]

Mencius has often been considered one of the most famous supports of the "Confucian" social system in China; yet his brand of "Confucianism" here makes an oppositional gesture *against* social conventions. His unbending sense of moral correctness compels him to evoke a vision of a hero, the Great Man—an elite, scholarly image of the ideal male. Mencius arrives at this description by first demonstrating his contempt for two prominent men of his time and then constructing an analogy from rituals of adulthood and marriage, of men and of women. The process of becoming a "great man" (*zhangfu* 丈夫) begins with the ceremony of "capping"—for *zhangfu* also means "adult." From the admonitions of the father, a man grows in wisdom and knowledge, learning to take the correct course or Way (*dao* 道) as the exigencies of fate and the demands of politics direct him. The Great Man is the master of circumstances, expecting both failure and defeat and remaining undisturbed in the face of earthly vicissitudes. The key traits here are defined negatively: the inability to go to excess (*bu neng yin* 不能淫) and the inability to be moved from one's goals (*bu neng yi* 不能移). The daughter, on the other hand, does not pass through an initiation ceremony in Mencius' description—she only gets married. The lesson she learns from her mother is one of obedience, not merely to fate or to a moral code but also to her husband. The endurance demanded of the Great Man becomes in her case simple compliance; stoicism of a far different kind.

It is clear in this passage that Mencius is linking Gongsun Yan and Zhang Yi with concubines and wives; none of these four is capable of the ideal of the Great Man (although Mencius himself may be capable of it). Gongsun and Zhang were orators, men who made their living by traveling from state to state and persuading rulers to take action through the power of their rhetoric. But wherein lies their feminine quality? Perhaps it lies in the fact that, as wandering orators, they leave one state for another, much as the wife or concubine leaves her family home and travels to her husband's house.[2] If this is true, then their pursuit of pre-eminence ("once they grew angry, all the feudal lords trembled") violates the subordinate position required of women. The philosopher thus grants that such orators *do* have a place in society, if a modest one. But they are still incapable of becoming Great Men, of gaining status on the basis of their moral conduct.

Most have seen Mencius' text as yet another manifestation of the Confucian's distrust of the artful production of words. In some later texts (beginning in the first century B.C.), literary rhetoric continues to be associated with "the way of concubines and wives." Perhaps the implication is that, granted the weak and subordinate position of women in society, words are one of the few ways they can obtain their desires. On the other hand, the Great Man supposedly does not speak; he provides an example. If he is virtuous but unheeded, he will simply "go solitary on his chosen way." Yet it could hardly escape most of Mencius' followers that a virtuous minister must make some attempt to convince his ruler of the correct course of action and thus potentially enter a "woman's" domain. In spite of anti-rhetorical arguments, one must ultimately distinguish between wrong and right rhetoric.

Another example of female conduct may provide insight into the male public world—in this case, the famous account of Mencius' mother in the *Lienü zhuan* 列女傳 (Biographies of eminent women), a Han dynasty moral guide. At one point, Mencius inadvertently walks in on his wife when she is still undressed; he is so revolted by her casual behavior that he refuses to be intimate with her.

> His wife then went to see his mother and asked to be sent home. She said, "I have heard that the Way of conduct between husband and wife does not reach the bedchamber. However, just now I was too informal, and when my husband saw me improperly dressed, he was unhappy; that is why he's treating me as if I were merely a guest. It is not right for a woman to live as a guest; therefore, let me return to my mother and father."
>
> Mencius' mother summoned Mencius and said to him, "You should ask first before you enter a room as a demonstration of respect—or make some loud noise to warn those inside. When you enter, you should lower your head, so as not to catch anyone off guard. You have not behaved properly, yet you blame others for their faults. Aren't you rather far off the mark in this?"
>
> Mencius apologized and decided to keep his wife. A Gentleman said, "His mother knew propriety and was enlightened in the Way of the mother-in-law."[3]

Women are entitled (required) to protest in decorous ways if larger issues of propriety are involved. In fact, throughout the biography, Mencius' mother uses persuasive language (simple but eloquent) to keep her son on the right path. The woman restores harmony in the household through her persuasions, earning the acknowledgment that she knows "the Way of the mother-in-law."

Although the *Lienü zhuan* was ostensibly written for women, it addresses larger questions of power and subordination, silence and speech, that implicate the male "public" world as well. With the establishment of the Qin and Han empires in the third and second centuries B.C., the official found himself bound even more closely to his ruler through a complex ritual hierarchy of relations. For him, it was not always clear why the Way of the great man should be followed, rather than the Way of concubines and wives. Perhaps the general unpopularity of Mencius in the Han, in contrast to two other great Confucian texts, the *Xunzi* 荀子 and the *Li ji* 禮記 (Record of rites), with their emphasis on ritual control and etiquette, suggests less interest in courage and the isolated defense of ideals. Immovability is less admired; compliance more so. Yet if literati saw themselves increasingly taking the woman's part in their relationships with the ruler, the problem of remonstrance remained. Could they employ their "women's weapons" of persuasion for the forces of good, or would they have to suffer in silence? The Han philosopher Yang Xiong 揚雄 (53 B.C.–A.D. 18), adducing the female body as proof, suggested that the minister could speak out, if caution and judgment shaped his rhetoric. "Someone asked me, 'Since a woman has beauty, does writing also have beauty?' I replied, 'It does. But just as it is wrong for a woman to bring disorder to her charm by applying makeup, so it is wrong for the reasoning of a piece to be ruined by excessive verbiage.'"[4] Earlier, when Mencius had spoken of the Way of women, calling them "concubines and wives" (*qiefu* 妾婦), he may simply have meant married women in general; his odd usage in putting the more subordinate term first (instead of saying *fuqie*, "wives and concubines") may merely have emphasized women's lowly position in the marriage relation. For Yang, however, the female compound must be split: Is the "woman" in question a wife, like Mencius' wife and mother, with the skill and right to speak discreetly, or is she merely a plaything, a bought slave, whose artificial, manipulative beauty could undermine the harmony of the household?[5] For the skilled statesman and writer, this tension between the sense of propriety and the need for expressive rhetoric was strong. Furthermore, increased passion could induce a flood of language, leaving one open to charges of insubordination and dissatisfaction. Resentment of injustice at the hands of the ruler could threaten the special relationship, and remonstrance break into satire, jeremiad, or lament.[6]

For writers facing this dilemma, the semi-mythical figure of Qu Yuan 屈原 (fl. early third c. B.C.?) served as an ambiguous role model. It was believed that this virtuous statesman from the state of Chu had vented his frustration at his ruler's destructive policies in problematic ways—first, in a number of passionate and elaborate laments, most famously, the *Li sao* 離騷 (Encountering sorrow); second, through his own suicide. Did his passion and death represent excessive egoism or the selfless act of the unhesitatingly loyal? *Encountering Sorrow* itself seems to undermine the Confucian distrust of superficial language with a radical privileging of rhetoric/ornament. Note its beginning:

I am the distant descendant of the high god Gaoyang,	帝高陽之苗裔兮
My late father was named Bo Yong—	朕皇考曰伯庸
When the *sheti* stars faced the first month	攝提貞於孟陬兮
I descended on the day *gengyin*.	惟庚寅吾以降
My father, observing my first arrival,	皇覽揆余于初度兮
Presented me with a goodly name.	肇錫余以嘉名
He named me Proper Measure	名余曰正則兮
And titled me Divine Balance.	字余曰靈均
Already I had inward beauty in great measure,	紛吾既有此内美兮
And I coupled it with a refined appearance.	又重之以修能
I donned fragrant river grass and hidden angelica,	扈江離與辟芷兮
I twisted autumn orchids for my girdle.	紉秋蘭以爲佩
Swift time flowed; I could not catch it.	汨余若將不及兮
I feared the years would not wait for me.	恐年歲之不吾與
At dawn I culled the hill's wood hibiscus;	朝搴阰之木蘭兮
At evening plucked the river reeds of the isles.[7]	夕攬洲之宿莽

The narrator (whom I will identify as Qu Yuan, as the Han dynasty reader would) begins with a claim of inner "beauty" (virtue) by right of birth—both his line of descent and the auspiciousness of his nativity. This makes him eminently suitable to be a court advisor. Yet his first act is one of self-adornment. It is suggested that the beauty within gives him the special right to adorn himself *without*, to provide himself with decoration and thus make himself more attractive to his sovereign. After that, sensitive to the dangers of time passing (and the passing of his own attractiveness), Qu commits himself to seeking out other herbs and plants; he now becomes the desiring one, the one who will seek out "fragrant" rulers later in the poem. A Han reader would make the link between outer adornment and the power of "heightened" language almost automatically. Qu is proposing a striking countermodel to Yang's painted concubine: the person who manifests inner virtue with surface allure; the lovely hermaphrodite who desires and is desirable, who may be chosen for her/his beauty and who will seek beauty out. Although Qu's model may be dangerous to emulate (arousing suspicion of excessive emotion, of reckless moral individualism, and of self-destruction), it also promises freedom for the scholar/literatus/official, who can create through the power of language a zone in which he can adopt poses of subordination and domination and create fantasies of power and victory or console himself with laments of failure. He sometimes articulates himself as the female in a conscious strategy, and the articulation of women in texts will inevitably be influenced by this act of identification. Yet articulated women will always be located within a non-existent border space between ostensible object and ostensible subject—a focus of textual desire both through possession and through identification. Never in male-authored texts will this articulation fully resolve itself—the potential for multiple interpretations is continually present.

If we as modern readers are to engage with the literary products of this elite class (particularly when they speak the language of desire and sexual longing), we must engage as well with this powerful political model of expression and sympathize with its motives. We must see that the sexual moves made in literary texts are seductions of public power, in which court ladies, fellow officials, and the ruler himself become obsessions and objects of seduction. This book is an attempt at reinvigorating this reading and at demonstrating how much we lose if we either divorce the social from the sexual or suppress the social altogether.

Modern scholarship on traditional Chinese literature, despite the increasing sophistication of analysis and greater attention paid to social context, still has much to say on the way gender and sexuality are represented in texts. This absence is particularly obvious in discussions of the early period and of the fundamental early writings that helped to shape the language, conventions, and tropes of later erotic literature (and also helped shape conceptions of female authorship). Although various factors may account for this, the lack of social and cultural sources that are so richly available in Ming and Qing studies is perhaps most important. Also significant is the distaste shown at times by modern Western and Asian scholars for the texts' presumed immorality and their general sameness of tone: many of them involve male voyeuristic depictions of the female, often in vulnerable situations (abandoned by her husband, ignored by her emperor) and seem on the surface to have been composed mainly for male titillation.[8] Yet if the rapidly expanding fields of Ming and Qing cultural history and criticism are to deal intelligently with issues of gender, they must come to grips with these literary roots. I attempt here not a comprehensive history of literary eroticism but, rather, a selective examination of some of the more significant works, in the hope that this will suggest certain strategies of analysis and certain recurring themes that will prove fruitful as more scholarship is done in this complicated field.

This is not, however, a simple enterprise. The practitioners of gender criticism are becoming increasingly aware of the pitfalls of attempting to define or analyze representations of "male" and "female" in literature, particularly in a premodern literature. Above all, any attempt to relate Chinese literati-authored texts to some female "reality" is impossible when discussing the pre-Song era, a time period when so little of social history is known and when what is known is invariably filtered through the class interests of literati and (sometimes) aristocratic women. To understand the social significance of such texts for actual, historical gender relations is a hypothetical possibility, but it cannot be carried out with the richness of understanding that someone working on a much later period (the Ming or the Qing, for example) could manage.

A problem also emerges when we attempt to link male writing about women with the literary output of women themselves. Although in the late Ming a specific discourse arose that attempted to define the distinctive qualities of "women's writing," such a discourse does not really exist for earlier periods. In the poetry of the female authors Xue Tao 薛濤 (768–831) and Yu Xuanji 魚玄機 (ca. 844–68), for example, class, social context, and literary conventions are arguably more significant than any effort on the part of these poets to write *as* women; if we demand a female voice for them, we fall victim to a process whereby the differences between men and women are essentialized and rendered ahistorical. We consequently create a vision of early literature that is merely a pale imitation of modern and modernist concerns, in which a false teleology leading to modern progressive political interests forces a misreading of the texts themselves.[9]

We can discover a (partial) way out through a consideration of desire and sexuality and their relationship to power. Here, if we admit that the male literati world was the only world that can be known well through literary texts, we can locate the dynamics of desire squarely within that world's social and political concerns. Above all, we must recognize that the high literary language *is* male—even to the point that literacy is reproduced in the individual largely with the understanding that he will use it to participate in the public male structures of a nominally "Confucian" polity (one wrote so as to govern). This world necessarily attempts, through its control of literacy, to exclude women, although its constructions of heterosexual desire, combined with the permeability of gender, suggest that "women" could enter into the text, either through male adoption of gendered behavior (writing *as* a woman in the process of writing *about* her)[10] or (in reality) through actual mastery of the language that articulated the political concerns of the male literati class. This last option, potentially, might allow for women to participate by writing as literati or by writing as literati writing as women. Whether women (and by this I mean to emphasize literate women, which in itself creates some complex social relationships among text, author, and audience) end by confirming the male game or by subverting it is a difficult point to determine, if determinable at all. But at least a more complex model is created.[11]

Of course, even while operating within the world of the educated male literatus, I am still constrained by the lack of detailed social history for the Han through the Tang. I am not saying that social historiography of sufficient sophistication does not exist for these periods. Rather, the simplest issues of social construction (class composition, power relations, geographical distribution of elites, economic factors) are still heatedly debated by historians. If this were not the case, I might ask more complex and history-specific questions: for example, does the representation of female gender in male-authored texts during the late seventh century implicate political and social ideologies at Empress Wu's court? Although guesses have been made at the social constituents of political factions during this period, it must be admitted that we cannot examine Tang China with the certainty and critical sophistication of those who write about Elizabethan England. Perhaps one day we will know enough to make more detailed conclusions, but the limited survival of Tang textual material may be an insurmountable barrier.

Such methodological problems are not restricted to early and middle-period China. When scholars or critics set out to draw conclusions of a historical nature from texts, they inevitably raise the issue of how such texts may be said to "represent" a historical reality. Those who rely too heavily on the anecdotal or the coincident (for example, by drawing conclusions or analogies from interesting juxtapositions) open themselves to charges of indulging in the sort of "homologous" thinking recently critiqued by Fredric Jameson in his discussion of "new historicist" writing. To state Jameson's argument simply (perhaps oversimply), the free linkage of textual motifs with certain social constructions in the world outside the text is merely a return to the ultimately "ahistorical" tendencies of structuralism. Without a theoretically sophisticated model that can prove the appropriateness of such linkages, such writing cannot make valid claims to be representing that exterior world.[12]

On the other hand, when dealing with a premodern society, texts may provide the only insight into the culture, and even then there is much of the culture that simply is unrepresented. We can be modest in making claims about how much of the world we are really reconstructing, but this does not mean that the task is completely futile as long as we do not make all-encompassing assertions about what we can accomplish. In responding to Jameson's critique and the paradoxes of new historicism in general, James Chandler has argued for the synecdochal success of the anecdote in representing a culture to a degree. He suggests that texts both produce and reflect the world experienced by a particular age.[13] He quotes in this respect Kenneth Burke's *A Grammar of Motives*: "Men seek for vocabularies that will be faithful *reflections* of reality. To this end, they must develop vocabularies that are *selections* of reality."[14]

From my own perspective, I see the discussions in this book as giving a piece of a bigger picture that we will never completely know. They produce one version of the truth, not some absolute solution to the "problem of the past." I believe I can demonstrate certain recurring patterns in the way gender is represented in early texts and show that these patterns bespeak certain recurring interests, obsessions, and self-contradictions among many literati writers. I am not claiming to tell the whole story of gender or of the dynamics of the early Chinese male community. But I do hope that my readings will lead others to confront the complexities of texts that were in the past read too simply or with too many modern presuppositions.

My arguments here also engage with certain recent explorations of gender theory. Although many of its ultimate conclusions are no longer accepted, Michel Foucault's *The History of Sexuality* (1976) articulated the issue of the construction of gender and the historical analysis of sexuality most forcefully. His insights (if not his conclusions) have guided the cultural criticism of gender in the past fifteen years and have transformed the field of feminist studies in particular, which now tends to avoid an "essentialist" position in defining the differences between "female" and "male." Most significantly, Foucault led scholars to question assumptions about sexuality that had been taken for granted within Western bourgeois culture at least from the time of Freud; as a result, the focus has shifted to the role that social structures (rather than psychoanalytic development) play in defining desire and its objects. This tendency was reinforced by the feminist anthropologist Gayle Rubin, who in her 1975 essay "The Traffic in Women" put forth the idea that gender construction in many if not most cultures is rooted in a mandatory heterosexuality employed by kinship structures: heterosexual congress (especially within marriage) is not innately a "proper" or "typical" form of sexual conduct but is learned through society's enforcement of a family system in which women serve as tokens of exchange between men.

Foucault and Rubin have been particularly influential on other scholars in the critical investigation of sexuality and its relation to forms of cultural expression. In *Between Men: English Literature and Male Homosocial Desire* (1985), Eve Kosofsky Sedgwick argued through her readings of English literary texts that the figuration of male heterosexual desire often screens other forms of amity and enmity that act to define male social interaction; she terms these forms "homosocial," arguing that they often function independent of modern, bourgeois definitions of "homosexual" activity. More recently, Judith Butler (in *Gender Trouble* [1990] and *Bodies That Matter* [1993]) has proposed that sexuality is a kind of performance governed and prescribed by society and its ideologies and that the individual is capable of perceiving sexuality in ways radically different from those of modern heterosexuality, depending on the societal pressures to which he/she is subject. The potential of this open-ended reading of gender for an examination of early Chinese literature is great—not because it establishes a broader, more inclusive model for sexuality into which Chinese texts can be inserted, but because its largely critical and skeptical nature encourages the careful reader against making rash "modern" conclusions about the way desire functions in the text. In the Chinese case, with the absence of a modern, Western biological analysis of sex (a form of thinking that has helped define our own assumptions at least since the nineteenth century),[15] the gendered body is created by many other "languages," of which the Taoist medico-religious manual is only the most familiar to the modern student.[16]

All the texts discussed in this book are the works of an elite that engages in acts of cultural production either to validate a literati society they have constructed (Six Dynasties) or to demonstrate their own right to participate in public life, often through an imperial bureaucracy (Tang). Since none of these texts could have been created wholly in the context of a "private life" unexamined by peers and superiors, we must disabuse ourselves of the notion that some supposedly uncomplicated heterosexuality is present in these works (the male desire for the female object) and read these texts instead in the light of "homosocial" modes of desire—whether of the literatus' longing for the ruler's attention, his concern for his own position within the social order, or his desire to bond with or triumph over one's peers. A post-Foucauldian perspective allows us to dispense with the belief that desire is a "metaphor" for political relations, as if the "sexual" could be differentiated from the "non-sexual"; instead, desire, passion, and eroticism emerge not merely in the surface relationships described in the texts but in the social site of production and the purpose of writing as well.[17]

Chapter 1, "The Give and Take of Gender," discusses and expands many of the issues introduced above. Each of the later chapters takes up a text or texts for closer discussion. Chapter 2, "The Traffic in Goddesses," investigates one of the earliest figurations of a desirable woman in poetry, the sexually available deity found in the Han-era pseudo–Song Yu 宋玉 rhapsodies *Gaotang* 高唐 and *Shennü* 神女 (The goddess). Some attention has been focused heretofore on the goddess figure as the primary representation of female idealization in male sexuality, and thus as an emblem of the mystery and power of womankind;[18] I see her more as a male creation meant to clarify the courtier's subordination to, or (less often) subversion of, the ruler's authority: as the poet controls the narratives of the goddess's sexual invitation and seduction, he also controls the voyeuristic satisfaction or disappointment of his master. This is reflected even more clearly in the tongue-in-cheek "Master Dengtu" 登徒子好色賦 (also attributed to Song Yu), in which the ruler is rhetorically cuckolded by his loyal ministers.

As the Han ends and a literary and cultural elite defines itself more distinctly (particularly within the defensive community of great families that regrouped near Nanjing following the conquest of the North by non-Han peoples in 317), the literatus/ruler relation is complicated by an increase in aristocratic tensions that express themselves through competitive philosophical and artistic discourses. Chapter 3, "The Competitive Community," examines this society as described in the *Shishuo xinyu* 世說新語 (Worldly tales, new series), a fifth-century book of anecdotes that details the actions of the cultural elite. In particular, I illustrate how expression comes into existence through the necessity of finding a mode of ritualized, nonviolent competition between males—which nonetheless reroutes itself into new kinds of violence (symbolic and real). This leads in turn to Chapter 4, "Spectator Sports," an investigation of the *Yutai xinyong* 玉臺新詠 (New verses from a jade terrace), a volume of erotic verse edited at the Liang court in the early sixth century. Although the *Yutai xinyong* helped create the "palace poem," a recognized early locus for the tropes that later defined female desirability in traditional China, I examine the text instead as a product of "male bonding," in which sexual role playing, cross-gender personae, homosocial competition for the female, and verse competition create a space for erotic and political play that nonetheless acknowledges serious consequences.

Chapter 5, "The Textual Life of Savages," investigates Chinese literati perspectives on the "savage Other," beginning with the classic texts from the Han histories that deal with the nomadic pastoral people, the Xiongnu 匈奴. Caught between two contradictory models (Xiongnu as "dark mirror" of China, a competing cultural center, and Xiongnu as peripheral, "uncivilized"), Chinese literati eventually settled on a way of portraying the Chinese/Xiongnu interaction that stresses the link between self-worth and cultural identity—those who travel to Xiongnu lands are in danger of losing their integrity as "Han." Some of the texts that work through this issue are pre-Tang and Tang poems that build upon the legend of Wang Zhaojun 王昭君, a minor Han-era court lady married to the Xiongnu chieftain in order to solidify diplomatic alliances. Her "court lady" identity enables writers to play with gender identification once again, routing political/international affairs through considerations of the represented alternations between male and female, emperor and court favorite, barbarization and sinicization.

Chapter 6, "From Ritual to Romance," observes how the Tang male community interacts with elements of the heterosexual love affair found in Tang narratives. It begins by tracing the movement of erotic traditions from the Six Dynasties into the Tang through a reading of the seventh-century short story *You xianku* 遊仙窟 (A dalliance in the immortals' den). By relating the adventures of a lone traveler who happens upon a community of sexually available women with whom he exchanges poems (and other things), this text alters the *Jade Terrace* dynamic from the community of males competing in the portrayal of a female object to a single male presenting male/female competition to a male audience. Although it liberally employs palace-poetry eroticism as the method of evoking desire, it elevates the individual male above the male community and suggests that sexual display will soon be seen as a version of the display of literary ability; women take on a more pervasive role in texts as validators of male talent. This trend continues in later narrative, although male-female interaction is displayed against a continuing anxiety on the part of the literatus about his place in society and in various male-constructed social worlds—the traditional family, the imperial bureaucracy, the world of the urban underground. This is demonstrated through a discussion of several romances and related stories. In some ways, the conception of romance itself and the invention of free sexual exchange between men and women can become comprehensible in these narratives only when they are viewed as variants on certain forms of friendship and alliance already existing within the community of men.

Chapter 7, "Honor Among the Roués," deals specifically with the phenomenon of the courtesan and her relations with the male elite, especially with civil service examinees visiting the capital. A reading of Sun Qi's (n.d.) 孫棨 collection of anecdotes concerning famous courtesans, the *Beili zhi* 北里志 (Account of the northern wards, ca. 880), observes how its rhetoric creates scenarios in which literati self-esteem depends on the role of the courtesan: as a "professional" trained and equipped with specialized literary talents, she serves as a mirror for the literatus himself and as a force that validates his own accomplishments (because she stands "outside" his own class but nonetheless functions as an organ of his society). As a result, courtesan culture acts as a distorted version of literati culture in general, while it also becomes a site within which examination literati attempt to free themselves from dependence on ruler and bureaucracy and to create the illusion of a world within which they can operate autonomously.

The Afterword, "Lost in a Sea of Coral," looks briefly at the poetry of the late Tang author Li Shangyin 李商隱 (813?–58) as well as the songs of the early Song poet Liu Yong 柳永 (987–1053). Some suggestions about how Li and Liu constructed their own readership lead to some thoughts on how the way of reading gender in literati texts was beginning to change and led the way to a more "late imperial" world of sexuality.

CHAPTER ONE

The Give and Take of Gender

Method is the *path, after* one has been along it.
—Marcel Granet

Reading Gender

If I were engaged in a history of gender in literature or a survey of how desire is manipulated through the inherent structures of socialized sexual behavior, then I could begin by tracing a genealogy of desire, demonstrating the continuity of conventions, of the establishment of genres, the *tradition* as it looks from our perspective. Such a survey would commence inevitably with the *Shi jing* 詩經 (Classic of odes), the collection of over three hundred songs, hymns, and poems that have their first beginnings most probably in the early Eastern Zhou period (eighth–fifth c. B.C.). For many, they represent the commencement of a literary and cultural tradition, as well as the canonical expression of poetry composition as an essential moral and sacred component of the Confucian project of gentlemanly self-cultivation. Coherence and identity for the collection of texts known as "Chinese literature" begin with the myth of identification and centralization manifested in this group of poems. I, too, will start with the *Odes*, but for different reasons: to illustrate instead the complexities involved in constructing gender from Chinese literary texts and to suggest the degree to which any concept of a "tradition" is inevitably created by a historically situated reader. Gender often cannot be determined here with any sense of epistemological certainty; it shifts and transforms itself into different representations with every fresh perspective. To illustrate this complex disorientation, to decenter this originating myth and tell instead of multiplicity, uncertainty, and (especially) the shifts of gender, we can commence with a conflict of narratives. Consider *Odes* 64, "Quince." I render it here in as clumsy and as literal a form as I can muster:

Tossed to me a quince, 投我以木瓜
Repaid with a precious *ju*-gem; 報之以瓊琚
It was not that I repaid— 匪報也
Forever by this means create amity. 永以爲好也

Tossed to me a peach, 投我以木桃
Repaid with a precious *yao*-gem; 報之以瓊瑤
It was not that I repaid— 匪報也
Forever by this means create amity. 永以爲好也

Tossed to me a plum, 投我以木李
Repaid with a precious *jiu*-gem; 報之以瓊玖
It was not that I repaid— 匪報也
Forever by this means create amity.[1] 永以爲好也

I present the bare minimum of meaning, to demonstrate the indeterminacy of the language of classical poetry, as well as the vagueness induced by the ignorance (on our part) of the context of composition. There are no subject pronouns here; from at least the Han era, no one was sure of the nature of the gems exchanged or what kind of fruit was meant. The word translated as "amity" (*hao* 好) allows for numerous interpretations, from "goodwill" (whatever that might be) to "erotic desire." Moreover, I have left one word in the second line of each stanza untranslated: a direct object pronoun of indeterminate gender and number. Repaid *her*? Repaid *him*? Repaid *them*? We can make a few generalizations about what the poem does communicate: some act of gift giving and its reciprocation; their invocation in a poem in incantatory style, with stanzas that differ only in the identity of the gifts; the oath-like declaration that the reciprocation is not a "true" return but merely a guarantee or assurance of *hao*. It may seem to the modern reader (a reader with particular interests and concerns) that the major unanswered questions relate to gender and desire.

The elite, cultural tradition that made of the *Odes* a canonical text was well aware of these indeterminacies; the poems enter hermeneutic analysis several centuries after their probable creation when the contexts of composition were already unknowable—not only the specific contexts of each text, but also the larger context of the society that produced them (early Zhou

China was very different from the empires of Qin and Han). The process by which such a problematic text became one of the sacred books of a religion has attracted a considerable scholarship in its own right.[2] I focus, however, on the interpretations (both ancient and modern) of "Quince"—for it is an axiom of Chinese literature that although the *Odes* cannot be proved to mean one thing, readers across the centuries have always claimed to know what the one thing is.

First reading. "Quince" is a courtship poem, written in the voice of a man, responding to gifts from a woman. This supposition may be traced at least as far back as the twelfth-century neo-Confucian philosopher Zhu Xi 朱熹 (1130–1200), whose moral disapproval of the poem will engage us later, but it has become an almost mandated assumption in this century as a result of the "deconfucianization" of the *Odes.* As both Western scholars and Chinese intellectuals sought to reject the moral scholasticism of distinctively "sacred" Confucian readings, they turned to a romantic, anthropological reading that conceived early Zhou culture as a pre-statist utopia—distinctive, above all, for the genial way in which courtship and marriage rituals were practiced. Confucian interpretations had sought to "repress" the erotic dimensions of many of the poems (now seen as analogous to European folk songs in theme and style) and to "conceal" the surprising autonomy women had possessed in early Chinese society. Of course (such proponents argue), this does not mean that women were not subject to severe societal constraints—the large number of wives' laments in the *Odes* suggests otherwise. Still, the outstanding presence of women in the *Odes* surely does suggest a greater degree of female independence. Predictably, many have seen these poems as revealing a framework on which a new feminist history might be constructed (although such a position is not without its problems).[3] Our barren, incomprehensible text can now be given a determinate meaning:

She tossed to me a quince;
In repayment I gave her a *ju*-gem.
But it wasn't really a repayment—
Just shows I will love her forever.[4]

I will not deny the possibilities of this reading; in other Odes, gender and content are explicit enough to suggest some sort of courtship ritual and to indicate that women and men had considerably more freedom to address each other directly in such rituals than they would (at least among the elites) just a few centuries later. However, the "agrarian utopia" that this reading implies can convince only when we are willing to accept a strictly hypothetical and not terribly detailed social context. Even at the most basic level, one may ask: If this is courtship, why must the woman throw the fruit? Although there may be scientific or anthropological analogies or reasons that lead to such a conclusion, I detect certain gender-assumptions at play here: woman as nature, bearing fruit, possessed of "earthiness"; man as culture, the carver and giver of gems. Behind the idealized, romanticized gender reversal (woman as initiator) lies a continuing male control: the decision to reciprocate, the regularization of the relationship, the promise to "love her forever." Above all, man has the power of speech in this reading; the woman is silent.[5]

We can perhaps dismiss these reservations and grant that many popular twentieth-century assumptions about gender roles in folk culture may indeed hold true for this hypothesized Zhou paradise. Yet there is always the danger of romanticizing the poem, of seeing "Quince" as a pure expression of "love," or seeing the courtship ritual as a charming product of an essentially unproblematic and apolitical society.[6] Such a reading might perversely lead us to create equally unjustified stories that might be hidden under the assertions of the song. What is concealed under this poem-as-performative-utterance, this oath of betrothal or courtship? Does the act of repetitive exchange, of continual trade, conceal tensions and anxieties? What makes the promise of eternal love *necessary*? Why should the (male) speaker find it important to assure his readers that the gift of gems (practically speaking, far more valuable than the fruit) is not a fair recompense? A sentimental gesture, of course, this elevation of *hao* over and above the economics of the gift; one could hypothesize the sublimation of a need to control through gradations of commodity value. Once gifts are unequal in economic terms, the purity of "love" is at risk.

The reader may object that I am overinterpreting, but this is surely a possible result when a social context is lacking. Brutal demystification here arises not from a desire to historicize the poem, but as an equally conceivable counterproduct of an original positing—the dark side of an impossible utopia projected into the irrecoverable past. The questions attendant on any real historicizing, on the other hand, remain unanswered because the context can only be invented. If this is a courtship poem, then what kinship relations stand behind the action? How might a song such as "Quince" actually be performed in a social setting? What allows such a poem to be copied and preserved? These questions, like our phantom demystification of utopia, involve issues of power: especially of gender and class. Marriage as an institution is seen by feminist historians as a form of female subordination that guarantees male relations and hierarchies,[7] and nothing in what we know of the history of early China (derived from archaeological or textual sources) would suggest a reading significantly at odds with this. Literacy and textual preservation are an issue here as well, for without the easy reproduction of texts that allows for mass literacy, the power to transcribe a folk poem and disseminate it to literate members of the society inevitably belongs to the elite classes. What use could such a class have for "Quince"? Are they reifying the lower classes, projecting the poem as an idealized product of the common folk? Or are they appropriating the text for their own lives, rewriting it as the manifestation of an elite truth? Or, very easily, both? In the light of such questions, our "heterosexual" reading of "Quince"—espoused as the self-evident manifestation of a simpler, pre-Confucian society—fragments into a multitude of possibilities.

Second reading. Perhaps we should retreat then and consider what is at stake for the early Confucian reading of the poem, the one found in the commentaries attributed to the Mao 毛 family beginning in the second century B.C. and elevated to orthodoxy by the end of the Han. Here, the ideological motivations for interpretation are easy enough to trace; in keeping with the ethical management of society envisioned by an elite Confucian project, stabilized readings of each of the Odes become necessary. The result, as is well known, is itself the construction of a mythistory—almost every poem is given a specific context intimately connected with the political and moral history of Zhou China. Since the first 160 poems are divided into groups based on the various feudal states from which they supposedly came, the Mao commentaries make them a sort of running narrative on the rise and fall of each state's rulers. Within this project, the focus is by and large the ruling classes: poems are either authored by elite poets or (in the case of folk attribution) represent the "people's" response to the ruler's activities.

For proponents of an idealized anthropological reading of the Odes, the Mao commentary creates, above all, a sexual repression: the large number of Odes that seem to deal with courtship and marriage must be reinterpreted, because in the imperial world (where marriage is controlled by the ritual institutions of kinship agreements, go-betweens, and segregation of the sexes), free association between men and women is inherently immoral. But this critique of hermeneutic change in turn encourages a simplistic history of women: that "Confucianization" of the text aims primarily to "put women in their place"—that the primary motive of the Mao historicist readings is to erase women from history. They indeed do that, to a large extent (or rather, they erase certain manifestations of women's social *presence*), but the reservations I have expressed about the courtship model of "Quince" suggest that the story is much more complicated. Rather, the redefinition and erasure of women is part of a much larger process, in which Confucianism is seeking to rewrite power relationships throughout society and locating history and literary traditions within that new power matrix.[8] An examination of how the Mao commentary rewrites "Quince" will suggest not merely a partial displacement of women from the imperial Confucian world but also new roles for the poem itself.

Mao's preface is brief and (characteristically) authoritative in tone:

> "Quince" praises Duke Huan 桓 of Qi 齊. When the state of Wei 衛 had been defeated by the Di 狄 barbarians and the ruler fled to Cao 漕, Duke Huan of Qi rescued him and enfeoffed him. The duke [also] presented him with chariots and horses, utensils and clothes. The people of Wei thought of the duke with gratitude and wished to repay him generously, and so they composed this poem.[9]

The events described here appear in greater detail in the *Zuo Commentary* 左傳 history (Duke Min 閔 2, 661 B.C.). When Wei was attacked by the non-Chinese Di, the government and a segment of the population fled. Shortly thereafter, the Duke of Qi sent his son Wukui 無虧 to suppress the enemy and succor the Wei court. Afterward, the duke was instrumental in persuading the other feudal lords to approve the enfeoffment of the exiled Wei ruler at Chuqiu 楚丘 and thus guarantee the perpetuation of his line.[10] Since "Quince" is the last poem in the three books that purport to relate the political history of the state of Wei, it takes on an elegaic though not entirely pessimistic note: it is rewritten as a text about faith between states, about gratitude, and about hope in times of adversity (all within the context of ruling class relations, of course).

Even the subcommentators working within the Mao tradition had some difficulty relating the preface's historical situation to the actual text of the poem; above all, they could not comprehend how the Duke of Qi's gift could be compared to mere fruit by a grateful population, nor how that impoverished population had the capacity to present gems in gratitude. Kong

Yingda 孔穎達, the seventh-century exegete, related event and poem ingeniously:

Because the people of Wei had obtained great merit from Huan of Qi, they longed to repay him generously, and yet they could not; and so they made use of a trivial example to speak of it: "If Qi were to throw us a quince, we would want to repay him—yet we cannot. Even if we could repay him with a girdle gem, we would still not dare to [consider] such a girdle gem as repayment for Qi's quince. We could only hope that Qi would take pleasure in it and so be tied to us in his affectionate thoughts."[11]

These convoluted conditionals (which can be deciphered by the classical reader mostly through context, since conditional conjunctions are often unrepresented) create a different poem in translation:

> If Duke Huan were to toss us a quince,
> We would like to repay with a girdle-gem;
> Yet such a girdle-gem would not be repayment enough—
> It only would guarantee forever our gratitude.

Although the Mao reading stabilizes the text within the context of Confucianized political history, it also exemplifies scholasticism at its most unconvincing—as Kong Yingda's elaborate defense indicates. Most interesting of all, the value of the gifts in the poem forces commentators to project a hypothetical exchange beyond the event the poem supposedly refers to: the people of Wei do not repay Duke Huan for his rescue of them from the hands of the Di. Ideally, such a refusal to repay is acknowledgment of the superiority of the duke: receiving anything from a superior represents such a free and unobligated act of grace on the superior's part that no gift from an inferior can ever equal it. Yet (as with the idealized world of our first translation), darker stories suggest themselves. The people project a new situation in which they are empowered to make the greater sacrifice, to upstage the duke (as our imaginary male poet of the earlier reading did to his lover) with a more impressive show of generosity. The dark alternative to Mao's act of textual control is the possibility of ingratitude, of resentment over favors, of depression over loss.[12] An "overreading," but a reading that is not prohibited by the limited and hypothetical context supplied by the Mao commentators.

As I have suggested above, the Mao reading, by denying a courtship reading, is not *explicitly* misogynist; it merely erases the possibility of female participation on either side of the equation in its implication that the most important moral lessons are found in the relations between states and their (male) rulers. Zhu Xi, however, relocated texts such as "Quince" within the courtship world by supposing that some Odes allowed for "immoral" readings; the introduction of the lascivious within a canonical text could be tolerated if one recognized the power of such poems to provide minatory examples.[13] "Probably also a poem about gifts exchanged between men and women, like 'Gentle Girl,'" he notes laconically, comparing it to an earlier poem (no. 42) involving a clandestine tryst beyond the bounds of ritual and formal betrothal.[14] Yet in spite of his distaste for unchaperoned courtship, Zhu publicly declares a meaning for "Quince" that fits the first reading I have discussed; since Mao and Zhu were the two most accepted commentators throughout imperial history, they generally provided the two standards against which other interpretations were working.

But were these two readings mutually exclusive? Normally we might think so; we would conceive the Mao reading as more or less accepted orthodoxy until it crumbled under Song revisionism, and then the Zhu reading held sway until the more adventurous Qing exegetes of the early modern era. Yet when we go outside the texts and their commentaries and observe the impact of the Odes on literature, we discover that the poems were viewed more flexibly. The modern commentator Zhu Shouliang observes that Zhu Xi was not the first to interpret "Quince" in courtship terms, citing two much earlier "love" poems that allude to it—one attributed to Qin Jia 秦嘉 (second c. A.D.), who is said to be addressing his wife by letter:

The poet, moved by a quince,	詩人感木瓜
Wished to answer with a bright greenstone.	乃欲答瑤瓊
Humbled I am that his gift passes mine—	愧彼贈我厚
Shamed at the worthlessness of my gift.	慚此往物輕

The second, by Lu Ji 陸機 (261–303), is written in the voice of the fruit-bearing wife:

How dare I forget the peach and plum so humble?	敢忘桃李陋
I think instead on your greenstone and pendant.[15]	側想瑤與瓊

Zhu Shouliang suggests that Qin and Lu possessed a prescience that saw through the artificial construct of the Mao commentary and anticipated Zhu Xi's own reading.[16] Yet such a conclusion is not inevitable. Rather, as Haun Saussy has so aptly shown, those educated in the early scholastic commentaries on the classics (that is, those who lived throughout the first millennium A.D.) may have seen no contradiction in reading an Ode as applicable to some general situation *as well as* having a specific political significance.[17] That is, it is quite likely that a reader of "Quince" would have recognized it as a poem involving an exchange between two "ordinary" people, quite possibly between a boy and girl linked by affection and desire; they could also recognize that the poem's significance within the realm of Confucian moral and ethical instruction was the expression of this poem (through composition or quotation) by the people of Wei in gratitude for Duke Huan's generosity. Readers may even have acknowledged that the poem's derivation from a situation describing personal relations could give emotional weight to it as a statement of gratitude expressed from a people to a duke.

In this respect, it is important to recall that before the rise of the Mao school of commentary, the Odes were often used as a source for timely quotations by educated scholars and diplomats. The *Zuo Commentary* gives numerous examples of statesmen quoting lines of the poems out of context to illustrate a situation or an emotional or moral stance. It has been argued that this quotational technique contributed to the Mao school's own interest in locating poems in specific events and reading them as allegories of those events; the difference, it is suggested, is that the Mao school thought the poems were *composed* specifically for those situations; they did not exist previously, they were not being "cited."[18] What the Mao scholars thought specifically of this issue will never be known, but the example of "Quince" suggests that the line between composition and quotation continued to be blurred in the minds of the educated. Surely (as the Mao school discussion of the Odes' significance argued) if poetry was a natural expression of the concerns of the politico-social self, then quoting an apt poem in a specific situation could be as much of a guide to and manifestation of one's own emotions and attitudes as a poem freshly composed.

My concern here is not to add to the already rich literature on Chinese poetic hermeneutics and aesthetics. Rather, I am suggesting that the early Chinese reader of poetry did not find it necessary to come down on one side or another when reading a poem—either to read it (particularly one that involves figurations of desire and eroticism) as a representation of "public" or "political" issues, *or* to accept it simply for what it supposedly said. Poems are rarely stable texts, and their meaning depends on their use—to whom they are sent, who reads them, who chooses to reply to them or to act upon their provocative gestures. Motives are important and a vital part of the poetic act for the Chinese reader; yet we must recognize that motives cannot be recovered from the text itself.[19] I will return to the fruitfulness of this line of thinking shortly after considering yet another reading of "Quince."

Third reading. "Quince" is the exchange of gifts as a demonstration of the ties of friendship: between man and man, woman and woman, or woman and man. Once stated, this hypothesis seems obvious for its simplicity. Why had not such a possibility emerged from the beginning? However, the consequences of this reading are even less easy to fathom than with the earlier interpretations. The courtship reading, whether in the idealist modern form or in Zhu Xi's moral critique, provides a certain institutional framework against which to see the act of exchange: even agrarian utopians assume some sort of marriage institution that will tend to regularize courtship behavior. The unspoken assumption of sexual consummation lies behind such readings as well. But how do we read "friendship" without any sort of institutional framework? Granted the subordinate position that women held, even in Zhou culture, can we conceive of a woman-woman relationship that would not be disapproved of and could be recorded in a song text? Or can we imagine a man-woman relationship free of the rituals of courtship and marriage? And what of male-male friendships? What sustained them? What shared ground of education or interests? Did class or age inequalities shape them?

The friendship reading became increasingly popular among Qing dynasty commentators, and some modern scholars take this position on "Quince" as well, in spite of the general acceptance of the courtship reading. Cui Shu 崔述 (1740–1816) even condemned earlier readers for complicating a poem whose meaning is so clearly self-evident.[20] Yet for these commentators (all of them, it may be noted, very late), friendship seems unproblematic. No one is interested in suggesting the role of such an institution in early Zhou society, because they assume that it cannot be different from what it is in their own times. If some citation to another text must be given to contextualize this exchange, the scholar could always turn to the *Li ji* 禮記, the early Han compendium of Confucian ritual practices:

> In the highest antiquity they prized (simply conferring) good; in the time next to this, giving and repaying was the thing attended to. And what the rules of propriety value is that reciprocity. If I give a gift and nothing comes in return, that is contrary to propriety; if the thing comes to me, and I give nothing in return, that also is contrary to propriety.[21]

Qian Zhongshu, however, offers a daringly cynical variant of the friendship reading. He notes the inequality of the gifts and wonders whether an act of self-aggrandizing commerce is being enacted here; then, citing Marcel Mauss on gift exchange as the fundamental cementing relationship in economic and social systems,[22] he argues that the giver of the gems politely undervalues his gift while at the same time forcing the giver of the more modest gift to reciprocate yet again with a more generous gift in order not to lose face.[23] The poem, in Qian's view, is not about an oath of friendship—but rather a contract for economic exchange. Goodwill conceals the motives of aggressive mercantile self-interest. He quotes the proverb of a Warring States merchant: "I take what others discard, and I give away what others take."[24]

This is not the only Maussian reading possible, of course; gifts are given not just to perpetuate an economic system but also to acquire power and prestige. Although Qian mentions the Pacific Northwest potlatch as an example of self-interested bestowal, he does not discuss the aggressive, destructive aspects of this economy of exchange. Within the series of intertribal games, festivals, and banquets, power struggles, almost as violent as any war, are re-enacted:

> Yet what is noteworthy about these tribes is the principle of rivalry and hostility that prevails in all these practices. They go as far as to fight and kill chiefs and nobles. Moreover, they even go as far as the purely sumptuary destruction of wealth that has been accumulated in order to outdo the rival chief as well as his associate (normally a grandfather, father-in-law, or son-in-law.). There is total service in the sense that it is indeed the whole clan that contracts on behalf of all, for all that it possesses and for all that it does, through the person of its chief. But this act of "service" on the part of the chief takes on an extremely marked agonistic character. It is essentially usurious and sumptuary. It is a struggle between nobles to establish a hierarchy amongst themselves from which their clan will benefit at a later date.[25]

It would seem that the dark side to an oath of friendship may very well be this struggle for hierarchical placement. And perhaps we can read Mao's account of the events behind the poem in this light as well: Duke Huan's bestowal of gifts on the Wei ruler becomes a form of intimidation, an exploitation of political opportunity in order to incur obligation.

"Quince," then, remains profoundly undecipherable, yet at the same time eminently readable. It is a site for others' meanings and thus demonstrates not so much the beginning of a tradition but rather the stuff out of which various traditions have been fashioned throughout the past two thousand years. And yet, even taking this into account, "Quince" often undermines attempts to stabilize its meaning. It not only potentially maintains other readings, thus implicitly questioning the "meaning of the hour," but also rhetorically questions any attempt to render it an instructive or pedagogical text, an illustration of "moral exchange." That this should be so is perhaps inevitable and would be true of any piece of writing, but it is a useful principle to keep in mind when observing the history of a poem's reading, the empirical account of how different interpretations emerge.

Although the Odes are infamous for their elusive qualities, I wish to suggest that they are merely exemplary specimens of a process that renders the reading of gender in Chinese literature similarly opaque—or rather, perceptible, but rendered ever more complex by layers of superimposed meanings. "Quince" in this case is particularly exemplary because it simplifies our many readings into three broad possibilities: male/female (heterosexual); sociopolitical ("public" homosocial); and friendship/alliance ("private" homosocial). How these three categories relate to each other and how they may be employed in a schema that discusses gendered texts of desire is the subject of the next section.

Positioning Gender

In any consideration of the relation between women's position in society and the expression of desire, we must inevitably begin with the subordinate position of women in most sex/gender systems, including that of traditional China. The maintenance of kinship structures, the political and social importance of marriage for social stability, and the role of women as an exchange medium that cements kinship relations are essential and irrefutable elements in traditional Chinese society. As a result, the social maintenance of differences between "male" and "female" is of the greatest importance and is in fact the most obvious aspect in the treatment of women—that they are to be separated from men and are obviously different from them. Although this perspective likely existed in some form from the earlier periods of the civilization, it later developed a more complex theoretical justification metaphysically in Chinese elite discourse through a theory of interactive polarities (*yin* and *yang*), and socially through distinguishing "outer" from "inner" (*wai* 外 and *nei* 内)—men function in the world of politics and larger social structures; women remain at home and manage domestic tasks.[26] Ideally, it was argued from the Han and after that men and women should not be allowed to intermingle before marriage. Even after marriage, their realms must be kept separate, and care must be applied not to allow gender transgression in assigned labor. Needless to say, these prohibitions were meant to apply mostly to women; as Patricia Ebrey notes, "Men . . . were rarely if ever told not to get involved with what their wives were doing; rather, their attention was directed to taking precautions to ensure that women did not intrude into the men's sphere."[27] In this ideological world, marriage is the essential defining act that clarifies the difference between men and women and supposedly defines the circumference of female roles: generally speaking, the rites of marriage garner more attention in ritual texts than any other rituals save funerals and mourning.[28]

Tani Barlow has pointed out one serious consequence of the kinship structure and the institution of marriage in traditional China: the absence of any clear, understandable term to denote an entity "woman" (as opposed to daughter, wife, mother).[29] However, the concept of "man" is equally vague as well, and men are also defined in terms of their social functions. Those functions far exceed those of son, husband, and father and include scholar, minister, nobleman, magistrate, and even recluse. The point here is not necessarily that traditional Chinese women (as opposed to men) were defined only relationally; rather, men's relational self-definitions reflected their greater freedom and their participation in the public world. As a result, women remain forever that which is *nei*, "within." Certainly not the least significant effect of this is their loss of self-articulation, since the use of classical Chinese is largely reserved during earlier periods for the expression of the public figure participating in society.

All of this is generally accepted as the *theorized* position of women in traditional Chinese society. But where does desire enter into this? Again, we may accept on the most basic level that the marriage/kinship system also attempts to prescribe a clearly articulated form of heterosexual desire as most socially accepted and that the subject's deployment of gender behavior and sexuality is inevitably affected by her/his experience of these societal "facts."[30] This does not mean, of course, that marriage as a social institution *creates* desire, although in fact texts that praise marriage in traditional China often carefully encourage the view that desire is a natural outgrowth of the *social* necessity of the marriage/kinship system, as if they must always confront the danger of dissidence. However, generally speaking, one consequence of a would-be compulsory heterosexuality is to make it seem inevitable or natural and to make gender behavior seem a secondary but understandable outgrowth of it. "The heterosexualization of desire requires and institutes the production of discrete and asymmetrical oppositions between 'feminine' and 'masculine,' where these are understood as expressive attributes of 'male' and 'female.'"[31] The establishment of traditional kinship systems works toward stabilizing sexual identity, whereby "socially constructive" desire must inevitably be directed to the "opposite sex":

> The internal coherence or unity of either gender, man or woman, thereby requires both a stable and oppositional heterosexuality. That institutional heterosexuality both requires and produces the univocity of each of the gendered terms that constitute the limit of gendered possibilities within an oppositional, binary gender system. This conception of gender presupposes not only a causal relation among sex, gender, and desire, but suggests as well that desire reflects or expresses gender and that gender reflects or expresses desire. . . . Whether as a naturalistic paradigm which establishes a causal continuity among sex, gender, and desire, or as an authentic-expressive paradigm in which some true self is said to be revealed simultaneously or successively in sex, gender, and desire, here "the old dream of symmetry," as Irigaray has called it, is presupposed, reified, and rationalized.[32]

Certainly in the Chinese instance, the polarities of *yin* and *yang* succeed in naturalizing inequalities, despite the lip service paid to the cosmologically mutual dependence of male and female; "the old dream of symmetry" conceals a practical asymmetry. However, what is most at stake in defining or interpreting *yin* and *yang* is not the equality or inequality of the sexes, but the idea that sexual behavior must be seen inevitably in terms of a male/female polarity at all: the act of power lies initially not in enforcing inequalities but in enforcing heterosexualization. Western scholars have not always helped through their continuing emphasis on popular *yin/yang* discourses on sexual hygiene, which should still be judged as products of sexual construction and not as faithful reflections of *all* Chinese forms of sexuality.[33] If marriage and kinship systems attempt to dictate not merely the subordination of women but also heterosexual behavior modeled on marriage (which they consider normative), then what happens to desire (of any kind) that falls outside this normative sphere? One consequence, perhaps, is that a desire of any kind *that can be spoken of* in any comprehensible way tends to use the marriage model (and its concomitant variants—courtship, seduction, even adultery) to articulate its own existence; if it does not, it remains unspoken or is represented in some "non-desiring" way.

Such issues engage us inevitably with certain modern Western theories of gender. Although I believe we cannot embrace these without reservation, it is possible to test certain models that have been proposed by modern scholars of premodern Western culture: for here the change of societal attitudes toward "non-normative" desire has been analyzed in its relation to broader historical and cultural changes. After all, one can talk not of "Western" or "Chinese" attitudes (if such dichotomies are ever appropriate) but rather only of specific social and historical formations that are varied and complex within the European and Chinese traditions themselves.

Foucault, in attempting to specify what constitutes modern (that is, Western bourgeois) sexuality, has in consequence clarified to some extent what preceded it. Defining the interests that protect the marriage/kinship system, he speaks of a "deployment of alliance," whereby the mechanisms that perpetuate the system are elevated, and other elements (including "non-normative" sexuality) are excluded, prohibited, or not spoken of. This he juxtaposes against the modern "deployment of sexuality," whereby a more complex alliance of discourses enforces more detailed and articulated concepts; one product of this is the "medical" classification of "non-standard" sexual behavior into such categories as "homosexual," "masochist," and so forth.[34] Although Foucault's detailed explication of these modern, complex discourses does not concern us here, his recognition of a completely different way to view sexuality is vital for understanding any non-modern or non-Western construction of such issues.

In this respect, we may consider Foucault's distinction in relation to the concepts of "sodomy" and "homosexuality" in the West and consider what such a discussion may yield in an analysis of early Chinese discourses on desire. Foucault notes in an oft-quoted passage that "homosexuality" as conceived in the West (and, one might add, as conceived in places recently influenced by its discourses) did not exist before the nineteenth century:

> The extreme discretion of the texts dealing with sodomy—that utterly confused category—and the nearly universal reticence in talking about it made possible a twofold operation: on the one hand, there was an extreme severity (punishment by fire was meted out well into the eighteenth century, without there being any substantial protest expressed before the middle of the century), and on the other hand, a tolerance that must have been widespread (which one can deduce indirectly from

the infrequency of judicial sentences, and which one glimpses more directly through certain statements concerning societies of men that were thought to exist in the army or in the courts).[35]

This is confirmed in concrete detail by Alan Bray, who notes that in Renaissance England the criminal charge of sodomy (and this could include bestiality, rape of minors, and anal sex of a heterosexual nature) tended to be leveled against those already perceived as dangerous outsiders: papists, heretics, sorcerers, traitors. What is of greater significance, Bray suggests, is that a wide variety of social practices that would now be considered "homosexual" were carried out in the ordinary course of interpersonal relations and that those who engaged in them did not choose to interpret them in the light of sodomy. Moreover, close contact among males over long periods of time (in servants' quarters, for example) might encourage a discreet form of sexual contact.[36] What is important for us is that such practices would not come to light in any sort of explicit form—not necessarily because they would be seen as wrong, and hence must be concealed, but because they would be manifested within another kind of discourse. In such a situation, the modern question "Were they just friends?" becomes problematic, because a clear division between sexual and non-sexual acts cannot be articulated in a way evident to the modern observer. The line into sodomy is crossed when the act becomes in some sense "outrageous"—when it seriously violates the important underpinning of the sex/gender system. For example, a nobleman who fathered several children but kept serving boys for his own amusement might not cause scandal unless he elevated those activities over his roles as householder, husband, and father.[37] It is only the connection in our own times of homosexuality with the person rather than the act that has resulted in the condemnation of individuals by society for what they *are*, rather than what they *do*.

I introduce this discussion not because I see a strict analogy between Renaissance England and traditional China; for one thing, no early canonical texts in China (no Leviticus, no Saint Paul) condemned "sodomitical practices." Rather, I believe that the Foucault-Bray conception of "sodomy" is useful because it illustrates how a *kind* of desire could be constituted and practiced in early China without generating a considerable *explicit* textual literature. There are, of course, references to male/male sexual liaisons in early Chinese texts, but they are rare in comparison to the European Renaissance and especially to ancient Greece and Rome. They are also rare in comparison with texts that mention or describe heterosexual desire. The reason, in Foucauldian terms, may simply be because such a sexuality stands *outside* an accepted code of representation, there was no social reason for writing about it. Even though Chinese canonical texts did not explicitly condemn such behavior, it would not have been a subject for explicit discussion either.[38] However, if we are to locate these texts within the context of the sex/gender system of China, within the problem of desire and sexualities in general, and within the various realms of power and their expressions, we must ask about this silence and what it portends. Is it possible that explicitly heterosexual texts at times "speak for" other forms of desire? And is it possible that certain texts that we have not thought to equate with desire do in fact express it in ways we can no longer understand? To consider these questions, we find ourselves not only considering the consequences of the silence but also wondering whether the expression of heterosexuality might not be more problematic than it seems from a modern perspective.

Let me approach the problem from another angle, from that of ritual and the construction of hierarchies in Confucian texts. Earlier, I argued that marriage was an essential institution in defining the woman's place as "within," as well as inferior to that of man. Marriage can represent more than that, however, and it was so seen by traditional Confucian exegetes. In writing about the epithalamium "Guanju" 關雎, the first poem in the *Odes* and a supposed wedding song of the first Zhou ruler, the Mao commentators write: "The 'Guanju' poem is the virtue of the queen. It is the beginning of the 'airs' and of moral persuasion. It is the means by which the world is transformed and the relation of husband and wife rectified. So it was performed in the villages and among the feudal lords."[39] Here, the moral influence of all the Odes is initiated through the wedding song of the king; not only is the king's behavior to act as a moral persuasion (*feng* 風) upon the people, but also the paradigmatic act inaugurating that influence is a wedding that exemplifies the virtues of bride and bridegroom. Why should marriage in particular be perceived as such a central action within a ritual system?

Some suggestion may be given by the *Li ji*:

Confucius replied [in answer to a question from the duke of Lu 魯], "With the ancients in their practice of government the love of others was the great point; in their regulation of this love of others, the rules of ceremony was the great point; in their regulation of those rules, reverence was the great point. For of the extreme manifestation of reverence we find the greatest illustration in the great [rite of] marriage. Yes, in the great [rite of] marriage there is the extreme manifestation of respect; and when one took place, the bridegroom in his square-topped cap went in person to meet the bride;—thus showing his affection for her. It was his doing this himself that was the demonstration of his affection. Thus it is that the superior man commences with respect as the basis of love. To neglect respect is to leave affection unprovided for. Without loving there can be no [real] union; and without respect the love will not be correct. Yes, love and respect lie at the foundation of government."[40]

Commentators identify the "great rite" mentioned here as the marriage of the ruler of the state—the marriage rite supposedly described in the "Guanju." A process of analogy is at work here: as the ruler extends love and respect to someone below him (his wife), he will also extend those virtues toward the treatment of his subjects. The centrality of this rite transfers "femininity" to his subjects and invites their own self-identification as "wife."[41] We have already noted that the political, public reading of desire in poetry was an early commentarial tradition; comparison with texts on rites suggests that such readings reconstruct desire as an essential element of the marriage relation and that unrequited desire is an expected emotional response when the socially higher member of the relation—the ruler/husband—fails to fulfil his "conjugal" duties.

How did literary interpretation move in this direction? For an answer, we might turn to the *Songs of Chu* 楚辭, the anthology of late Zhou and early Han texts written in the southern style of the Yangtze valley (home of the kingdom of Chu). The most famous of these verses, of course, is *Encountering Sorrow*, which we have noticed earlier. The language of desire here is explicit—particularly in the tone of melancholy the rejected statesman/suitor uses in addressing his ruler. Desire occurs in two notable places, although the gender of the speaker is still difficult to determine. First:

Though cast aside I made orchids my sash;	既替余以蕙纕兮
Added to it some angelica I had plucked.	又申之以蘭茝
Indeed this is what I love in my heart;	亦余心之所善兮
Though I'd die nine times I would not regret it.	雖九死其猶未悔
I resent the Fair One's wayward moods;	怨靈修之浩蕩兮
The Fair One never sees the heart of others.	終不察夫民心
All the women are jealous of my lovely brows;	衆女嫉余之蛾眉兮
In their gossip they call me lustful.[42]	謠諑謂余以善淫

The self-adornment here (as elsewhere in the poem, where it signifies an inner moral worth) is *not* gender-coded, at least as far as Chu culture was concerned. However, as we noticed before, later Chinese readers tended to see explicit self-ornamentation as predominantly a female attribute, and this passage might echo in later ears as a feminization. Gender roles do become clear in the last lines, when the poet very obviously places himself in the position of the maligned harem lady. The other ladies are, most likely, other ministers of the king. This passage is notable for balancing an *inclination* to identify gender against a refusal to clarify its precise claim for one sexual identity or another. The author portrays the act of desiring independent of clear gender categories or explicit heterosexualization.

The second passage occurs much later in the poem, when the persona resumes a male role as he searches for a more worthy object of his ministerial affections. Various rulers seem to be figured as maidens to whom he pays court; fluctuations in gender and desire have been replaced with a sanctioned form of heterosexual behavior, and political hierarchies have been inverted into a hierarchy of courtship.[43] However, as each prospective mate proves unsatisfactory, Qu resigns himself to despair and withdraws from society—either by suicide or by cultivating magical arts. However, in spite of a greater stability in gender representation in this second section, there is the continuing evocation of a world in which gender is not so stable. The author also adopts the personae of the "Nine Songs" 九歌, a group of religious hymns in the *Songs of Chu* that relate shamans' and shamanesses' attempts to seek or seduce deities and to have sexual/spiritual congress with them. The sexual nature of these texts has often been a subject of comment, but what is somewhat less noticed is their lack of sexual specification: the nature of classical Chinese often makes it unclear whether shaman or deity (or shaman possessed by deity) is speaking; gender is often in question (in the case of "Xiang jun" 湘君, for instance, commentators split on whether the deity is male or female), as is the issue of who desires whom.[44] Although one might compose dramatic scenarios that would hypothesize the religious ceremonies behind such texts, later readers are left adrift and must rely on commentaries to steer them to a "correct" reading; otherwise, the poem becomes a strange cooperative production of shaman/shamaness and deity, a fragmented utterance half-governed by a strange possession. But who possesses whom? Who is the ventriloquist? And how can one stabilize the desire expressed in such a poem, when only its existence and not its subject or object can be clarified?

Han and later readers read the "Nine Songs" as revisions by Qu Yuan of earlier hymns; he figures himself as the lonely seeker, the gods and goddesses as capricious rulers. Within the religious rites of Chu was concealed the drama of minister and ruler yet again. As with "Quince" (a poem of exchange that supposedly illustrated relations between Wei and the Duke of Qi), it was possible for the reader to see both the religious ritual and the politically and historically contingent allegory at the same time. Interestingly enough, as in *Encountering Sorrow*, the sexual ambiguities make gender roles difficult to specify. There is often no clear determination whether the ruler is female or male or whether the minister seeks or summons.

Perhaps the major difference in the movement from sexuality in the *Songs of Chu* to the interpretive tradition that reads politics into the sexual relation is the stabilizing of gender roles—the more or less determined representation of the lower member of the relationship as a woman, and the higher as a man. Here we may recall the *Li ji*'s formulation of marriage as the most important of all rites: although dealings between ruler and minister may be figured as sexual, the only *proper* way in Confucian terms of figuring these dealings is *hierarchically*. Private relations thus become a chief way of describing public ones. But why would any male, in a socially superior position as he is, willingly take on the attributes of the female, as Qu Yuan and many later poets did? To come back to the question we began to answer at the commencement of the introduction, what advantage lies in the "feminine"?

Again, a consideration of certain Western social structures may be helpful as a point of departure, for the gendering of social and political relations is not exclusively Chinese. Joan Kelly has suggested that Renaissance Petrarchism as described by Castiglione re-creates the courtier's relation to the ruler—not just because it figures one entity as subordinate to another, but also because the figure of the "female" stands in for a complex reversal of hierarchies that allows the courtier to be both subject and "user" of his ruler at the same time:

> The likeness of the lady to the prince . . . , her elevation to the pedestal of Neo-Platonic love, both masks and expresses the new dependency of the Renaissance noblewoman. In a structured hierarchy of superior and inferior, she seems to be served by the courtier. But this love theory really made her serve—and stand as a symbol of how the relation of domination may be reversed, so that the prince could be made to serve the interests of the courtier. The Renaissance lady is not desired, not loved for

herself. Rendered passive and chaste, she merely mediates the courtier's safe transcendence of an otherwise demeaning necessity.[45]

Of particular importance here is Kelly's perception of women as unspeaking and passive *mediators* of the courtier's role. If a figurative transvestitism marks the Chinese literatus's portrayal of heterosexual desire, then he is occupying the place of *nei* (inner chamber) to the ruler's *wai* (exterior world). As such, he is supplanting the woman's own position and potentially relegating her to yet another *nei*, a *nei* where she cannot be represented in the text. Her relative subordination and silence become to some extent necessary if the literatus is to have complete freedom to shift between sexual roles; he creates a sort of public interior space to get closer to his ruler and thus holds the power to represent himself as both poles: *yin* and *yang*, *nei* and *wai*;[46] this allows him to take on a specific "gender" in order to manipulate systems of power. Of course, he also has the power to reintroduce the "feminine" as "other," as the object of his own male heterosexual desire: the harem competitor, the court lady, can then become an object of control, a point of contention between minister and minister or ruler and minister that cements their own "homosocial" relations. Jealousy and desire merge, as well as the desire both to have and to be the female. Thus, the projection of the self as woman can never remain fully stable in every reading.

This male move may help to explain why "female" (*yin*) power only confirms another form of male power and does not represent in any way the "real" position of women in society. If the performance of any gender role is open to the literatus, he may manifest the affective aspects, both "male" and "female," of any of the Confucian virtues. It should be remembered that no one Confucian virtue is exclusively equated with the male or the female.[47] Hence, the transvestitism of much literati poetry is a manifestation of the poet's ability to change gender roles under different political contingencies.

I wish to carry this argument somewhat further, however, and reconstitute the subjects and objects of desire itself. What I said above about "figuring" minister/ruler relations as heterosexual desire presents the situation as it is accepted in the contemporary view: that sexual desire *substitutes for* political desire. But in a culture in which sexualities are not clearly defined and differentiated independent of other forms of social activity, can this distinction be made? Is it possible that the desire for the ruler itself remains one of the most important forms of desire in such texts? If such is the case, then the *figural* transvestitism of the literatus poet represents a performance of a sexuality that must be taken to some extent at face value. Here we can return to the idea of sexualities as constructed by culture, not inherent or unchanging, and we can recognize the presence of an environment free of Foucauldian scientific and medical discourses, in which sexual activity is perceived more in terms of actions rather than of actors. This is in keeping with Judith Butler's argument that gender is a form of performance that is culturally encoded and constrained, that such performances are intrinsically independent of any predetermined "sex," and that polarities of "male" and "female" are not metaphysically determined categories.[48] Of course, such performances are still often prescribed by social institutions. The restrictions a society places on sexual self-definition give rise not only to prohibitions but also to the degree to which sexual boundaries can be tested and different gendered activities "performed":

> The "performative" dimension of construction is precisely the forced reiteration of norms. In this sense, then, it is not only that there are constraints to performativity; rather, constraint calls to be rethought as the very condition of performativity. Performativity is neither free play nor theatrical self-presentation; nor can it be simply equated with performance. Moreover, constraint is not necessarily that which sets a limit to performativity; constraint is, rather, that which impels and sustains performativity.[49]

Butler is here suggesting that the subject comes into being through performance—this constitutes sexual identity. But if a society is constructed on the presupposition that various sexual identities are imposed on the body contingent to its relative position to others (through its position within a series of hierarchies and not merely through its allotted share of supposed unchanging male/female dichotomy), then the subject (if *he* is not at the very bottom of all hierarchies) has a prescribed *field* of gender activities that he can participate in (often more than one at the same time), depending on the situation.

Such a view would suggest that what Butler would call the body's metaphysical "materialization"—that is, the assumption that the body's "physicality" is the unchanging substrate of all signifying practices that involve it—is only partially present in early China. Such has indeed been suggested by some scholars, who see the body as constructed more by cosmological phenomena and social praxis than by physicality.[50] Roger Ames has suggested that the "fluidity" of the body in Taoism allows for a "monoandrogynous" ideal in political thought—a subject who can respond to contingencies with appropriate gendered behavior:

> It must be borne in mind that the Taoist conception of person is relatively lacking in the connotations of individuality, separateness and discontinuity that are implicit in its Western counterpart. The organismic definition of person would focus its interpretation of person and personal realization within a matrix of relationships rather than presupposing some notion of discrete and autonomous self. Although this interpretation of person would seem to militate against the necessity of any one "individual" developing the full complement of masculine and feminine gender traits within the parameters of one person, I still think that the Taoist ideal is a reconciliation and integration of these opposites. . . . Rather than the freedom to choose what combination of traits one prefers that characterizes the "polyandrogynist" theory, the Taoist ideal would seem to concern itself with the greater freedom inherent in the ability to meet all situations with imperturbable poise and assurance.[51]

As I have suggested, however, one of the principles directing gendered behavior in the literatus was social *hierarchy*, an idea that is sometimes thought to be missing from Taoist texts.[52] In Confucianism, however, the force that imposes behavior on the body is ritual and its hierarchical relations.[53] Here we can finally detect the system that enables male literati to maneuver between forms of behavior encoded as inferior or superior, male or female. It is the ritual system of Confucianism that enables the literatus/minister to perceive the ruler as one site among many of his own desire. This desire is not necessarily a mimicry of, or analogy to, the mandated heterosexual desire between husband and wife within the sex/gender system of traditional China; rather, the desire may be the *same* as that desire, because what dictates desire in these cases is not the adoption of male or female roles per se (although the performance of such roles clarifies such relations); it is the adoption of superior or inferior positions, as defined by ritual conduct.

In light of these discussions, it would be unavailing to ask the sort of question that might come to the modern Western mind: "Were the ministers sleeping with the ruler?" The analysis of the discourses of sodomy in Renaissance England tells us that the distinctions between genital and nongenital activity are somewhat unreadable to us, as far as representations of desire are concerned.[54] Moreover, such a question hints that sexual relations figured in a text must have some sort of enacted counterpart in the real world. However, the objects of desire in literature are largely phantasmatic;

the desired woman of the text is no more "possessable" than the ruler she figures. Of course, poetry may *rehearse* or *repeat* "real" enactments of its own desire. Such fulfillments are usually beyond our ken. We are more interested here in the figuring itself.

We can now see two readings of "Quince" in a larger context: the heterosexual "courtship" reading, and the public political reading involving the relations of ruler and subject. But what of the third potential relationship, that between equals—whether friend and friend, enemy and enemy, competitor and competitor? Ritual texts say little about friendship, even though they admit it within the basic group that defines human relations.[55] Part of the problem is that relations are seen as requiring reciprocal actions on two unequal sides: Confucian texts are fond of dividing virtues between those possessed by the inferior side of a relationship and those possessed by the superior.[56] It is difficult to include friendship in a code of ritually prescribed behavior, unless some other factor (age differential, teacher/pupil relations, pupil/pupil relations) can be superimposed on it. This may suggest that friendship, although admitted to be important for the literatus, was a potentially disruptive force that could escape the matrix of ritual power.[57] I will have more to say on this topic in examining individual texts. For now suffice it to say that the interrelations between two males of roughly equal social status may express not only friendship but also enmity: whether among the Chinese "knight-errants" of Warring States culture, who competed to gain honor and fame,[58] or among Six Dynasties aristocrats, who engaged in complex philosophical discussions that are indistinguishable from rhetorical debate, or among lower-level late Tang gentry, who saw the examination system as one key to increased prestige and power.

How did such patterns of amity/enmity emerge in texts that ostensibly tell of heterosexual desire or of the desire of the minister for his ruler? Partly through a continual awareness on the part of the reader that such texts were produced in a social world in which they circulated between men, who participated in complex games of criticism, judgment, and response. Here we are in the world of "homosocial" relations, whereby subjects of desire represented in writing serve to mediate potential desire or enmity among the male producers of that writing. And here, once again, one must undauntedly situate oneself in a world in which the modern distinctions between "friends" and "lovers" were simply not made. It is my interest to determine, rather, how males interact (in any way) in texts of desire.

CHAPTER TWO

The Traffic in Goddesses

The light on Mount Wu grows late,	巫山光欲晚
While on a sunlit terrace the light lingers.	陽臺色依依
That beauty, dwelling in a curve of the cliff—	彼美巖之曲
How can I know if her heart is true?	寧知心是非
Dawn clouds strike the stones where they rise,	朝雲觸石起
Dusk rain soaks my silken robes.	暮雨潤羅衣
I long to loosen her thousand-gold sash	願解千金珮
And bring her back to the great king.	請逐大王歸

—Fei Chang 費昶 (fl. 510): "Mount Wu Is High" 巫山高

The origins and development of the *fu* 賦 or "rhapsody" during the Han dynasty are issues of literary history that continue to exercise scholars. Is the genre a development of Warring States oratory (as Zhang Xuecheng 章學誠 first argued)?[1] Or does it derive from the southern traditions of lament found in the *Songs of Chu*? Or does it reflect Han society's increasing interest in centralization and cultural homogenization—poem as cultural encyclopedia, as "word hoard" for the educated? The answer must lie in a dynamic interaction of all these factors as they manifest themselves in the world of Han culture and society during the second century B.C. Far from attempting a rationalizing history of the genre, however, I examine here some of the consequences of rhapsody rhetoric in the representation of the female, as well as the participation of that rhetoric in the demarcation of power relationships: male/female, emperor/courtier.[2]

As has often been noted, the early rhapsody tradition tends to follow two styles: relatively brief laments focusing on the supposed misfortunes of the poet, and long, descriptive set pieces that often serve as panegyrics of the ruler. The laments are often seen as deriving from Chu poetry, and the panegyrics, from Warring States oratory. Wai-yee Li has mapped this difference in style along an axis of "lyrical" versus "rhetorical."[3] Such a distinction is difficult to prove without more information on the social context of poetic composition and performance. But one point is self-evident: many of the more public rhapsodies are explicitly concerned with persuasion and rhetorical debate, in that they actually dramatize persuasion and debate in the text itself. Most famously, in Sima Xiangru's 司馬相如 (179–117 B.C.) *Shanglin Park* (*Shanglin fu* 上林賦), a number of courtiers come together to decide which of their masters has the most impressive hunting preserve; in Ban Gu's 班固 (A.D. 32–92) *Rhapsody on the Two Capitals* (*Liangdu fu* 兩都賦), two men debate the superiority of Chang'an 長安 and Luoyang 洛陽. In other rhapsodies, the persuader addresses the king or emperor himself and attempts to bring him around to his own way of thinking. This suggests the continuing power of oratorical traditions in the Han, as well as the public role of rhapsodist as courtier-poet; and it also foregrounds a more problematic aspect of rhapsodies, their power to persuade.

Persuasion, of course, is not wrong in itself, but it is potentially subject to amoral manipulation. If the rhapsody indeed derived from the speeches of wandering persuaders, then its genealogy was particularly suspect. The surviving text that best preserves this oratorical tradition is the *Zhanguo ce* 戰國策 (Intrigues of the Warring States), a work often viewed disapprovingly since the Eastern Han. Interested primarily in the powers of persuasion and the skill of sophistical reasoning, the text presents an unusually cynical view of oratory and its power to move the ruler to reasonable or unreasonable decisions. This perspective would inevitably be rejected by promoters of a Confucian tradition, as the statesman Qin Fu 秦宓 (fl. A.D. 230) made clear:

> If the sea receives any pollution, within a year it cleanses and clears itself; and if a gentleman is of broad knowledge, he will refuse to consider any conduct not in keeping with the Rites. *Intrigues of the Warring States* repeatedly brings up the art of the rhetoricians Zhang Yi 張儀 and Su Qin 蘇秦, in which one kills others to preserve oneself. Surely such actions were condemned in the Classics![4]

Important here is the claim that the debates of Warring States orators were not "in keeping with the Rites." Such (non-ritual?) use of language for less than upright purposes could be even more suspect under an imperial system, where the conditions that helped produce oratory in the first place—the intense and violent competition among states and advisors to discover the best policy for conquering the rest—no longer held. Of course, the rhapsody poets of the Western Han did offer a moral defense for their own imperial paeans and court panegyrics: they meant to persuade the ruler *away from* indulgence in imperial wealth and glory through exhausting those pleasures in words alone, thus satiating royal desires and focusing the ruler's attention on the concerns of a just government. But as Wai-yee Li has suggested:

> From Yang Hsiung [Yang Xiong] to the critics of this century, there is a deep suspicion of the dense and elaborate verbal surface of *fu*. The standard denigration of *fu* is centered around two partially overlapping issues: stylistic excess and dissimulation, that is, a content incommensurate with the resplendent verbal resources being marshaled, and a discrepancy between the avowed moral purpose and the insidious pandering to the pleasure of the reader-listener.[5]

As Li mentions, the attacks began most clearly with the great rhapsody poet Yang Xiong 揚雄 (53 B.C.–A.D. 18), who rejected the genre in his later years and turned instead to philosophy. Yang believed the language of the rhapsody was suspect in its use of "ornament" and its supposedly indirect criticism of imperial conduct; in the end, indirection could only condone imperial vice.[6]

We cannot know how influential Yang's critique was on other rhapsody poets and on the prestige of the genre in general. However, an examination of Eastern Han responses to the flamboyance of Western Han rhapsody style suggests that literati were groping toward a rhetorical anti-rhetoricism, a way of persuading that ostentatiously rejected the earlier traditions of persuasion. One might see a classic statement inscribed within a rhapsody itself, Ban Gu's *Two Capitals*. The first in a tradition of rhapsodies describing capital cities, Ban's poem presents a fictional frame of two interlocutors, one from Chang'an (the Western Han capital, symbol of the past) and one from Luoyang (the Eastern Han capital, symbol of the new, "ethical" present). The former boasts of the glories of the earlier site and bemoans its abandonment by the court after the usurpation of Wang Mang 王莽 (45 B.C.–A.D. 23). The latter defends Luoyang and proposes in defense a view of the imperial court that emphasizes harmony and good government instead of ostentation and display. Of special note is the different way the two imaginary speakers present their chosen worlds: Ban Gu (who is obviously supporting the Luoyang ideal) carefully structures the Chang'an passage so that it displays what he considers to be the excesses of the Western Han court rhapsody, whereas the Luoyang section seeks to reformulate the genre as a source of intellectual and moral guidance for the educated readers of his own day.

One point is of particular significance. As the Chang'an supporter displays a knowledge of the topography of the capital area and praises its location, he gradually introduces the numinous influence of the imperial presence: first through the construction of the city, then through the construction of imperial palaces, then finally through the detailed enumeration of imperial activities within the scope of those palaces. As David Hawkes has remarked of another rhapsody:

> We are made to feel that the purpose and function of the enormously elaborate account of palaces, gardens, parks, lodges, and so forth is merely to provide a setting in which the Great Man, the emperor, who is the heroic protagonist of this little cosmos, may be revealed in power and splendor. Essentially this kind of rhapsody is not the description of a place but the epiphany of a person.[7]

When the enumeration of Chang'an's wealth approaches the "rear palaces," or habitations of the emperor's harem, a description of architectural and ornamental marvels suddenly blends into a description of the women themselves:

In red gauze and swaying sleeves,	紅羅颯纚
In painted silk sashes in profusion,	綺組繽紛
Purest brilliance glowing brightly,	精曜華燭
They soar and drop like divine spirits.	俯仰如神
The ranks of the Rear Halls	後宮之號
Were fourteen in all.	十有四位
Secluded and shy, but blooming in abundance,	窈窕繁華
Flourishing and honored in turns.	更盛迭貴
Those who found rank among them	處乎斯列者
No doubt numbered in the hundreds.[8]	蓋以百數

These women are engaged in a fine choreography centered around the royal person: those dangling sleeves and bodies that "soar and drop like divine spirits" are emblems of restlessness and change. The eighth line in the passage quoted above articulates their movement within a hierarchy—the rise and decline of individual careers, the fickleness of imperial favor.

Immediately, without break, the speaker changes focus to other, more public spaces:

To left and right in the courtyards,	左右庭中
In the audience halls, the posts of the hundred officials,	朝堂百寮之位
Xiao, Cao, Wei, Bing,[9]	蕭曹魏邴
Made their plans at the head.	謀謨乎其上
Aiding the Mandate and passing on the Rule,	佐命則垂統
They assist so that moral transformation is perfected.	輔翼則成化
They make flow forth the comfort of Great Han,	流大漢之愷悌
Shake out the poisonous vermin of fallen Qin.	盪亡秦之毒螫
....	
There also were	又有
Tianlu and Shiqu,	天祿石渠
Storehouses for all writing.	典籍之府
He commanded those	命夫
Old gentlemen, hardworking teachers,	惇誨故老
Renowned scholars, instructors, tutors;	名儒師傅
They lectured and expounded on the Six Classics,	講論乎六藝
Examined and collated the textual variants.	稽合乎同異
There were also	又有
Received Light Court and Bronze Horse Court:	承明金馬
Places for writing and composition.	著作之庭
The dignified, the deeply learned	大雅宏達
Formed a crowd here.	於茲爲群
They got to the root of things,	元元本本
Saw to the utmost, heard all,	殫見洽聞
Revealed the truth of all books and chapters,	啓發篇章
Collated and put in order the library's texts.	校理秘文
....	
They brought together the Master of Rites' first-class pupils,	總禮官之甲科
Assembled the pure and filial of a hundred districts.[10]	群百郡之廉孝

In the eyes of Ban Gu's fictitious speaker, the collection and maintenance of scholars are other ways of displaying imperial prestige, not unlike the display of palace women with which this passage is juxtaposed. The scholars elucidate the Classics through debate and study, and nearby is the court where assemble the newly recommended men from the provinces, the "first-class pupils," pure and filial, who will themselves rise to positions of favor with the ruler (again, not unlike the palace ladies).

It is important to remember here that this is not a description of the "way things were" in the Western Han, but rather the way Ban Gu would like us to see them. The portrayal is all the subtler for being (by and large) a positive one, representing as it does the point of view of a man advocating the "good old days." However, when the other fictitious speaker, the man of Luoyang, takes his turn, he abruptly attacks the first for excessive attention to territory and buildings and for insufficiently stressing the Han founder's virtuous accomplishments:

"You boast of your lodges and chambers, and make rivers and mountains your fortress and border. You are truly familiar with King Zhaoxiang 昭襄 and the Qin Emperor [both rulers of the authoritarian Qin state that preceded the Han]. But how could you have looked into the words and deeds of the Great Han? When the Great Han was established, a commoner soared upwards and took the emperor's throne. After several years he had established an immortal empire."[11]

The Luoyang advocate cleverly avoids the tropes of location and enumeration that characterize his opponent's description: to continue to describe in such a way would be merely to rearrange these aspects of the ruler's power while including them within a similar system of evaluation. Instead, he emphasizes the *emperor's actions* and locates them within a detailed explication of ritual behavior that ensures the correct governance of the realm. For example:

Doubly brilliant, perpetually gracious, 重熙而累洽
He set forth the Great Ritual in the Three Harmony halls.[12] 盛三雍之上儀
He cultivated the ceremonial robes with coiling dragons, 修袞龍之法服
Displayed great refinement, 鋪鴻藻
And relied on his bright luster. 信景鑠
He set up the ancestral temple, 揚世廟
Put right the ceremonial music. 正雅樂
The harmony of man and spirit was full, 人神之和允洽
The ranks of all the ministers were properly ordered. 群臣之序既肅
Then he set forth in his royal carriage, 乃動大輅
Pursued the imperial avenues, 遵皇衢
To examine customs on a royal tour. 省方巡狩
He personally examined what all states had and lacked, 躬覽萬國之有無
Investigated where his famed teachings reached, 考聲教之所被
Dispersing his royal brilliance to enlighten the benighted.[13] 散皇明燭幽

Here the ruler does not own the empire; rather, he moves about *within* it and performs correctly at each stage. Banquet and hunt are no longer opportunities for display but ceremonies that guarantee order, in keeping with preimperial models. In such a world, the minister and state-employed scholar seem to disappear (for to describe these groups collectively would be to imply, by the precedents of earlier rhapsodies, that they are merely representations of imperial prestige); rather, they are sublimated into the ritual grid, to emerge here and there as an intersection of ritual obligations: a master of ceremonies at the banquet, a spear bearer at the hunt, a keeper of the protocol.[14] It is as if the scholars of the Six Classics, whom the ruler previously collected in Chang'an, have now entered into their own books, from which they manifest themselves at the correct moment. The description of Luoyang itself becomes a ritual text; it no longer evokes the efforts of those scholars endeavoring to discover the way things ought to be (their topic for research) and instead articulates the way things *are*, where study is no longer needed because the ideal has been obtained. The speaker then steps outside this world and perpetrates a violence on the genre of rhapsody itself:

> Before the host was finished his speech, the guest from the Western Capital grew startled. He withdrew back down the stairs, fearful and at a loss. Holding up his hands respectfully, he started to take his leave. "Return to your place!" said his host. "I'm going to instruct you with five poems." When the guest had finished learning them, he praised them: "How beautiful are these verses! Their principles are more correct than those of Yang Xiong, and their content is more substantial than Xima Xiangru's. My host not only takes pleasure in study but also has found a successful place in this age. I, but a mere child, am wild and careless and didn't know how to comport myself. But after hearing of the proper way, I beg leave to chant these poems until the end of my life."[15]

The poems follow, written in the archaic temple-ode style of the *Classic of Odes*. They replace the rhapsody now, offering alternative genres superior to the work of the great Western Han poets, Sima Xiangru and Yang Xiong. Ban Gu brushes the genre aside and opens the doors of literature to new, more morally inspired approaches.

"My host not only takes pleasure in study but also has found a successful place in this age." Ban means us to see the two factors as mutually dependent. The first century B.C. saw a gradual increase in the ideological coherence of the *ru* 儒 or "Confucian" position in scholarship, as well as the position's bid to monopolize cultural and literary discourse. Although widely different perspectives on public, literary, and intellectual roles continued to surface in Emperor Wu's reign (r. 140–87 B.C.) and those that followed, the increasing imperial patronage of Confucian scholarship soon guaranteed the difficulty of engaging in cultural work without to some degree considering oneself a *ru*.

This is different from the supposedly "Confucian" position that is said to arise in the works of, for example, Sima Xiangru (the poet) or Sima Qian 司馬遷 (the historian; ca. 145–ca. 85 B.C.). For them, public-minded participation in court and in politics still belonged to a generalized role shared by educated men.[16] By Ban Gu's generation, however, a *ru*ist position increasingly implied totalizing claims on meaningful public life and discourse. For many educated men of the later Han, the right to participate in public affairs was accompanied by a belief that mastery of the Confucian tradition gave them the moral and technical knowledge to administer the state.[17] Devotion to Confucian scholarship comes to some extent to replace the oratorical and literary training of the rhapsody poet; it is no surprise to see scholars engaged in prestigious court competitions flaunting their erudition as flamboyantly as had the rhapsody writers.[18]

The later critiques of the rhapsody probably derive from a perspective that posits an inevitable conquest by morally thinking, self-conscious "intellectuals" who forged a state Confucianism. This development has generally been accepted by literary historians of every ideological stripe: those who approve of the Eastern Han developments see Western Han court poets as decadent aesthetes dependent on an imperial court system and Eastern Han *ru*ists as progressive humanists; those who disapprove view the Western Han poets as literary romantics squelched by the excessive moralism of a stultifying Confucianism. But it is more useful to see the early rhapsodies produced at the court as—simply—court poetry. Traditional literary history's obsession with the poets' intentions detracts from the significance of the text as a representation of imperial power. Consider, for example, the supposed moment of "moral intention" in one of Sima Xiangru's greatest poems, *Shanglin Park*: a fictional nobleman is describing an elaborate and violent imperial hunt followed by a luxurious banquet at which the beauty of the court ladies (the culminating stimulus to his soon-to-be-satiated desire) drives the ruler into the calming and soothing arms of moral policy:

Then, while in his cups, amid music and drunkenness, the emperor was suddenly lost in brooding, as though he had lost something. "Alas! This is all a great luxury! In the time left over from my imperial duties, I spend my days away in leisure. I follow the prescriptions of Heaven in my hunting [i.e., he kills in the right season and in the ritually correct way] and sometimes take my rest here. But I fear that later generations will be dissolute and will pursue this same course without returning [to the Way]. This is not the way to establish a legacy and pass on good governance for later generations!" Thereupon he dismissed the banquet and hunting party and commanded the responsible officials:

"May the land [i.e., the hunting park] be put under cultivation	地可墾闢
And all become farming land,	悉爲農郊
To supply the needs of my humble subjects.	以贍萌隸
Tear down the fences, fill in the moats,	隤墻塡塹
And let the inhabitants of hills and swamps come here.	使山澤之人得至焉
Stock the pools and do not restrict them;	實陂池而勿禁
Empty my lodges and halls and do not staff them.	虛宮館而勿仞
Open the granaries to aid the poor and desperate.	發倉廩以救貧窮
Supply their lack!	補不足
Pity the widowed!	恤鰥寡
Preserve the orphaned and childless;	存孤獨
Issue virtuous commands,	出德號
Examine the system of punishments and fines;	省刑罰
Reform our institutions;	改制度
Change our ceremonial garments,	易服色
Reform the calendar,	革正朔
Turn over a new leaf with all the world!"[19]	與天下爲更始

It is the seeming ambivalence of such a stance—the evoking of pleasure through the power of language, on the one hand, and the call to a return to imperial duties, on the other—that seems to have annoyed Yang Xiong and many more recent readers. But such a reading deliberately ignores some of the implications of the compositional context. Considering Sima's popularity as a court poet and Emperor Wu's own commitment to the patronage of *ru*ist scholars *as well as* large statist projects, it would be unlikely that the poet's ending here is *primarily* an attempt to persuade the ruler to do as his fictional counterpart does. Rather, Sima gestures toward his own loyalty and his celebration of the ruler by figuring in the text certain generally accepted actions that are expected of virtuous rulers. He covers all the bases of imperial virtue: Emperor Wu can have his hunt, his banquet, and his virtuous renunciation afterward. The imperial prestige can only be enhanced by the magnitude of his unselfish gesture. If, of course, Emperor Wu decided on hearing this passage to "go and do likewise," all the better, but we do not have to assume that the whole poem's raison d'être is to slip in a critique of his rule. Rather, persuasion here has become another entertainment, another way to dazzle the ruler and to compliment him on his own ability to select court poets who can so dazzle him.

Cuckolding the King

Court poetry is public poetry; and if its ambivalences and paradoxes do not lie in the failure of moral persuasion to overcome a rhetorical excess, it is nonetheless true that ambiguity and a language that engages with the arbitrary power of the court may be detected within it. I would like to turn my attention here to a number of poems that represent a more playful (or at least self-referential) consideration of the imperial poet and his position: namely, *Rhapsody on a Beautiful Woman* (*Meiren fu* 美人賦; attributed to Sima Xiangru) and *The Wind* (*Feng fu* 風賦) and *Master Dengtu the Lecher* (*Dengtu zi haose fu* 登徒子好色賦; attributed to Song Yu 宋玉).[20] These poems attract the reader interested in rhapsodic language mainly because they make such language (the attempt to persuade the ruler) their actual subject. Without a doubt they show a thorough familiarity with the rhapsody tradition of the Western Han; although the latter are attributed to Song Yu (trad. dates: 290–223 B.C.), a semi-mythical disciple of Qu Yuan who is said to have been active in the Chu court shortly before its fall to the state of Qin, they are most probably products of the Western Han at least, and quite possibly later.[21] But unlike the explicitly moral criticism of the rhapsody poets Yang Xiong and Ban Gu, these poems do not castigate; rather, they parody the poet's power to manipulate and move the ruler. In each, the poet himself ascends the stage as a fictional character, taking it upon himself to lecture the emperor or to elaborate on his own preposterous behavior. In this sense, the poet becomes not just a jester, but a "trickster" figure, no longer turning the ruler toward moral conduct but rather converting him to a radically irrational position through the use of dubious persuasion and argument.

The simplest of the Song Yu rhapsodies is *The Wind*, which begins as follows:

> King Xiang 襄 of Chu 楚 was relaxing in his palace at Orchid Terrace, accompanied by Song Yu and Jing Zuo 景差. A gust of wind came in, and the king faced it, opening his gown to it. "What a delightful breeze! This is something the common people and I can share!" Song Yu replied, "Only you possess the wind. How could the common people share it with you?"[22]

The poet then proceeds in elaborate poetic description to differentiate the wind of the king from the wind of the common people: the former brings pleasure, whereas the latter is only dirty and unpleasant. The baldly flattering and counter-intuitive message implied by the poem has, of course, aroused comment. Burton Watson suggests that a type of debate is suggested in the poem, modeled on Warring States–era persuasions, and such would be in keeping with the "rhetorical" origins of the genre.[23] This might also make the poem a particularly obnoxious example of the praise rhapsody, wherein Song Yu is driven to making an outlandish claim for the king's inherent superiority. There are, however, more likely readings. The poem displays a comic and satiric verve aimed not only at the king's love of luxury but also at the hypocritical sanctimony of his claim to "share the wind" with his people; the poet then may even be parodying the sort of logic found in a number of passages in *Mencius*, in which the philosopher is willing to indulge princes in their love of luxury and displays a seemingly "un-Confucian" ability to manipulate language for his own devices:

King Xuan 宣 of Qi 齊 granted Mencius an audience at the Snow Palace. The king said, "Do worthy rulers also have pleasures like this?" Mencius replied, "If the people should not be allowed to share in pleasure, then they will denigrate their superiors. Of course it is wrong to denigrate one's superiors when one does not have his share in pleasure. But to be in a position over the people and yet not share your pleasures with them is also wrong. If you take pleasure in the people's pleasure, then they will take pleasure in yours; if you worry about their worries, then they will worry about yours. To make your pleasures and worries shared throughout the world—there has never yet been one who did thus and yet did not rule as a true king."[24]

One thing is clear: the fictional frame of *The Wind*, the outrageousness of its "moral," and the implicit criticism not only of luxury but of the disingenuous defense of it place this poem neither quite in the Western Han rhapsody tradition nor in the tradition of later Confucian critique. It is instead a parody of sorts, possibly provoked by the very constraints poets may have had to face in their ordinary public compositions.

Witty sophistical persuasion combines with pornographic stimulation in *Rhapsody on a Beautiful Woman*, attributed to Sima Xiangru:

Sima Xiangru was handsome and elegant, and he traveled to see the Prince of Liang 梁.[25] The Prince of Liang delighted in him, but Zou Yang[26] 鄒陽 maligned him to the king. "It's true that Xiangru's good-looking, but he makes a seductive display of his beauty; he is charming but disloyal. Perhaps he will wish to use seductive speech to take his pleasure and dally in Your Majesty's private residences. Will Your Majesty not investigate him?" The prince asked Xiangru, "Are you lecherous?" Xiangru replied, "I am not." The king said, "If you are not lecherous, how do you compare with Confucius and Mozi?" Xiangru said,

"Thus did the ancients avoid sex	古之避色
(Men like Confucius and Mozi):	孔墨之徒
When they heard that Qi bestowed singing girls they left for distant parts;	聞齊饋女而遐逝
When Zhaoge came into view they turned aside their carriage.[27]	望朝歌而回車
As if they kept off fire with water,	譬於防火水中
or fled from drowning in the crook of hills.	避溺山隅

But since they never saw what they might have desired, how could they make clear that they weren't lecherous?

But as for me—	若臣者
When young I grew up in the western realms,	少長西土
I dwelt a bachelor and lived alone.	鰥處獨居
My chambers were vast and empty,	室宇遼廓
I took my joy with no one.	莫與爲娛
My neighbor to the east	臣之東鄰
Had a certain daughter:	有一女子
With cloud-like hair and voluptuous charms,	雲髮豐艷
Mothlike brows and gleaming teeth.	娥眉皓齒
Her features full and beauty flourishing,	顏盛色茂
Her bright radiance emanating.	景曜光起
Ever lifting her head And gazing westwards,	恒翹翹而西顧
She wished to detain me so I might tarry with her.	欲留臣而共止

She climbed her wall to gaze on me—and behaved thus for three years. But I spurned her and would not assent.

Then I admired Your Majesty's Lofty Principles,	竊慕大王之高義
And I ordered my carriage to come east to you.	命駕東來
My path passed through Zheng and Wei,	途出鄭衛
My road came through the mulberry groves.	道由桑中
At dawn I issued from the streams of Zhen and Wei,	朝發溱洧
And at dusk dwelt in Shanggong.[28]	暮宿上宮
At Shanggong I lodged in leisure,	上宮閒館
Lonely in the cloudy void.	寂寞雲虛
In the daytime I shut my gate,	門閤晝掩
Dark it was, like a dwelling for a spirit.	曖若神居

Then I opened her door 臣排其戶
And I came into her hall, 而造其堂
Where fragrance wafted forth 芳香芬烈
And figured curtains hung from on high. 黼帳高張
And there was a woman dwelling alone, 有女獨處
So charming upon her bed. 婉然在床
Marvelous beauty, untrammeled grace, 奇葩逸麗
Pure flesh shedding seductive light. 淑質艷光
She looked on me, hesitated, 睹臣遷延
Smiled slightly and said, 微笑而言曰
'What land do you hail from, honored sir? 上客何國之公子
Haven't you come from distant parts?' 所從來無乃遠乎
Then she set out fine ale, 遂設旨酒
Brought in a sonorous harp. 進鳴琴
I then stroked the strings. 臣遂撫絃

I played 'Hidden Orchid' and 'White Snow.' 爲幽蘭白雪之曲
The lady sang. 女乃歌曰

'I dwell alone, alas,
desolate, no one to rely on. 獨處室兮廓無依
I long for a lovely man, alas,
my heart is sore afflicted. 思佳人兮情傷悲
There is a lovely man, alas,
how slow he is in coming! 有美人兮來何遲
The sun draws to dusk, alas,
and my beauty starts to fade. 日既暮兮華色衰
Dare I give my body, alas,
to my love forever?' 敢託身兮長自私

and her jade hairpins caught on my cap, 玉釵挂臣冠
and her gauze sleeves brushed my robes. 羅袖拂臣衣
At the time the sun drew to eve in the west, 時日西夕
And black shades grew dark and dim. 玄陰晦冥
Rushing winds were bitter and chill, 流風慘冽
The white snow blew in gusts. 素雪飄零
Her idle chamber was lone and silent, 閑房寂謐
No sound of human tongue. 不聞人聲

Then she 於是

Put out the bedclothes, 寢具既設
Garment and ornament precious and rare, 服玩珍奇
Golden burners with fragrant scent, 金鉔薰香

Figured curtains hanging low. 黼帳低垂
Cushions and coverlets piled one on another, 裀褥重陳
And spread out was her pillow before me. 角枕横施

She then 女乃

Undid her dress, 馳其上服
Displayed her lingerie. 表其褻衣
Gleaming body was exposed, 皓體呈露
Fragile frame and fuller flesh. 弱骨豐肌
Then she grew close to me, 時來親臣
Tender and smooth as rouge. 柔滑如脂

I then 臣乃

Controlled my pulse within,[29] 脈定於内
Set right my heart in my breast. 心正於懷
I pledged my faith sincerely, 信誓旦旦
I held to my will and did not turn. 秉志不回
I rose up gracefully, 翻然高舉
And parted with her forever."[30] 與彼長辭

This is possibly the earliest surviving specimen of the "stilling the passions" (*ding qing* 定情) theme. The poet illustrates his ability to resist the attractions of a seductive and willing woman. There may have been moral, social, and quasi-religious motives behind this theme: a man gains power through the exercise of self-control. Most of the surviving fragments we have of the genre suggest that this was the main "purpose" of such poems.[31] But even in this straightforward example, we can detect a playfulness, a pleasure in inverting logic. If the sages were so conscientious about avoiding the beautiful at all costs, says the poet, how do we know that they could avoid the emotions of lechery altogether? Virtues must be tested! The poet then introduces the western neighbor, whose advances threaten to undo his self-control. He then describes in much greater detail an encounter with a high-born lady: the luxuriousness of her boudoir and her physical charms are evoked in a slow striptease. It soon becomes evident that Sima's performance in turn tests the Prince of Liang and the reader; Sima has the self-control to evoke these images and dares the listener to hear them without losing that same self-control. This dynamic will occur over and over again in later erotic verse, as we shall see.

A more sophisticated transformation of the *ding qing* theme occurs in *Master Dengtu the Lecher*, perhaps the most outrageous distortion of the rhapsodist's power to persuade.[32] It falls into two brief sections: the first is Song Yu's defense against the charge of lechery:

Once when the High Minister Master Dengtu was attending on the King of Chu, he maligned Song Yu: "If I had to judge Song's conduct, I would say that his form and features are possessed of a refined beauty and that his speech is mostly subtle phrases. Moreover, he is lecherous in nature. I suggest that Your Majesty not allow him to frequent the Rear Palaces."

The king asked Song Yu about Master Dengtu's accusations. Song Yu replied, "Heaven granted me the refined beauty of my features. The subtle phrases of my speech were learned from my teachers. But lechery is one thing I do *not* possess." The king then said, "If you are not lecherous, then you certainly should be able to persuade me of that fact. Do so, and that will be the end of it. But if you cannot persuade me, then you shall be dismissed."

Song Yu then said,

"Of all the dazzling women of the world,	天下之佳人
None compare with those of Chu.	莫若楚國
Of all the lovely women of Chu,	楚國之麗者
None compare with those from my village.	莫若臣里
Of all the beauties from my village,	臣里之美者
None compare with the daughter of my neighbor to the east.	莫若臣東家之子
As for this daughter of my eastern neighbor:	東家之子

Add an inch and she'd be too tall,	增之一分則太長
Subtract an inch and she'd be too short.	減之一分則太短
Powder her face and she'd be too white,	著粉則太白
Rouge her cheeks and she'd be too red.	施朱則太赤
Brows like kingfisher feathers,	眉如翠羽
Flesh like gleaming snow,	肌如白雪
Waist like a bleached silk sash,	腰如束素
Teeth like white cowry shells.	齒如含貝
With a single captivating smile	嫣然一笑
She brings confusion to the city of Yang,	惑陽城
Leads the state of Xiacai astray.	迷下蔡

Now for three years this girl has climbed the wall around her house to peek at me; yet up to the present I have never granted her desires. But Master Dengtu is nothing like me. His wife has

Tousled hair and crooked ears,	蓬頭攣耳
A cleft palate, crooked teeth,	齞脣歷齒
She sidles sideways when she walks, a real hunchback,	旁行踽僂

She's got a dose of scabies and piles as well. 又疥且痔

Yet Master Dengtu delights in her and has fathered five children on her. Now, Your Majesty, consider carefully. Which of us is more lustful?"

As the poem begins, Master Dengtu, a minister who speaks for the social proprieties, warns the Chu king of the deceit of surfaces as manifested in his other minister, Song Yu: physical attractiveness combined with an ability to manipulate language (thus repeating the accusation of Zou Yang in *Beautiful Woman*). It goes unsaid, but these qualities may also be identified with the "feminine": as we saw earlier in such texts as Mencius, although women themselves must ideally remain silent, the power to persuade through trivial effect rather than through reasoned argument is thought to be "feminine," the "way of the concubine." Song Yu is thus placed in a position analogous to that of the ruler's sexual favorite. As such, he may indeed "seduce" the ruler through language—an ability Dengtu most likely does not approve of, but which at least may be acknowledged by the court. However (and here is the important point, claims Dengtu), Song Yu is lecherous; literally, he "likes surface/sex." The word for "sex" here, *se* 色, obtains its meaning not through reference to a biological function but through the sexual arousal caused by visual stimulus: *se* means "color," "surface," "visual attractiveness." Song Yu is thus at once the producer of an attractive surface (in appearance and words) and the one who covets such a surface. He may belong, as it were, to the imperial harem and has access to it, but he also becomes an androgyne who has the power to take on any gendered characteristic. He emasculates himself so that he may "enter and leave" as he pleases, but unlike the historian Sima Qian, who served in the harem after his castration at the order of Emperor Wu, Song Yu's castration is figurative and temporary, a ruse to gain access to *other* women. In this guise, he can use his feminine qualities in turn to seduce the ruler's females and to threaten the legitimacy of the royal line.[33] Of course, the primary witty irony of the poem is that Song Yu must defend himself from these charges through the use of language: the ruler gives Song Yu a choice of either talking himself out of his dilemma or of facing the consequences, and the very quality that makes him dangerous ensures his safety. This ultimately is not meant to be taken very seriously: Song Yu's threat to the king's honor may exist only in Dengtu's mind, and the king himself may well be in on the game. But the poem *is* an excuse to let Song Yu exhibit his marvelous rhetoric.

The poem employs the same trope as the *Beautiful Woman* rhapsody: an eastern neighbor who attempts to seduce the narrator. When the poet creates these spaces—his existence within his own garden and his own house—he shifts the seducer's role from himself (as Master Dengtu would have it) to the woman. This allows him to take on the role of the female, but not in the negative sense that Master Dengtu suggests when he attributes beauty and seductive language to the poet. Rather, he becomes an imperial concubine who remains loyal to the ruler. On the contrary, the neighbor girl, like Song Yu, has "form and features possessed of a refined beauty"; no doubt she, too, would use subtle phrases to attract him—although such an attempt might be outdone by the poet's own superior control of language. *She* is the female turned male, breaking into the ministerial harem, threatening the symbolic sexual loyalty the poet owes to his ruler.[34]

The poet now inverts another distinction: having rooted much of his language in the persuasiveness of surface attraction, he breaks into a strong invocation of the ugly as he outlines the surface qualities of Master Dengtu's wife. The peculiar meaning of *se* here creates a resonance that cannot be translated: the English suggests that Dengtu's sexual drive overcomes any possible revulsion he might feel for his wife's unattractiveness, but the original implies that Master Dengtu has foolishly found this "surface" attractive and cannot resist it. Master Dengtu cannot read surface correctly and falls victim to it; Song Yu understands it and masters it. Surface is language as well; Song Yu is master of that too, and Dengtu cannot hope to match him.

The poem is not over, however; Song Yu now discovers an ally who will complete this deconstruction of moral reason as embodied in the person of the unfortunate chief minister:

At that time, Minister Zhanghua of Qin 秦章華, who had been standing to one side, took the opportunity to step forward and speak in approval: "Just now Song Yu commended his neighbor at great length, considering her to be a beauty. I, an obtuse, ignorant, and wicked person, used to consider myself capable of guarding my virtue. But now I say that I cannot compare myself to him.

"Now the maids from the back alleys of the towns of Southern Chu are hardly worthy of being described in Your Majesty's presence; and someone as base as myself dares not speak of those I have seen with my own eyes."

"Nonetheless, you must attempt to describe them to us," said the king.

"As you wish," replied the minister.

"When I was young, I traveled far:	臣少曾遠遊
I roamed to my content in all the lands,	周覽九土
Gazing upon all the great cities.	足歷五都
One day I'd set out from Xianyang,	出咸陽
The next I'd delight in Handan,	熙邯鄲
I'd take my leisure in Zheng and Wei,	從容鄭衛
By the rivers of Wei and Zhen.	溱洧之間
One time,	是時
When the end of spring	向春之末
Met the rising force of summer,	迎夏之陽
And when orioles twittered and sang,	鶬鶊喈喈
All the maidens went picking mulberry leaves.	群女出桑
The girls of that place	此郊之姝
Were blooming, filled with a brilliance,	華色含光
Their bodies lovely, their features, charming;	體美容冶
No need had they for adornment.	不待飾裝
I gazed on the loveliest	臣觀其麗者
and addressed her in a poem:	因稱詩曰
'I go along the broad road And take you by the sleeve—'	遵大路兮攬子袪
And I gave her a fragrant blossom, Speaking to her with fine-sounding words.	贈以芳華辭甚妙
Then	於是
That maid grew dim and distant,	處子怳
As though appearing far off but never coming;	若有望而不來
Grew vague,	忽
As though she had come but would not be seen.	若有來而不見
Her thoughts were hidden, her form remained apart.	意密體疏
Then of a sudden she altered her glance,	俯仰異觀
And full of joy, she smiled, slightly,	含喜微笑
Looked to me shyly, full of tenderness.	竊視流眄
She replied to me in a poem:	復稱詩曰
'Awake to spring wind, the new blossoms open.	寤春風兮發鮮榮
Proper and chaste, I await a few kindly words.	絜齋俟兮惠音聲
Yet to take presents from you in such a manner— better rather to never have been born!'	贈我如此兮不如無生

Thereupon she moved away reluctantly and withdrew herself from my sight. With only subtle phrases we had moved each other to affection, and our own spirits had joined together; our eyes had expressed a desire for each other's features, yet our hearts still minded the higher principles. While we uttered our poems, we preserved our propriety, and never did we fall into wrongdoing. This is worthy of relating."

The King of Chu then complimented them both; and he did not dismiss Song Yu.

Zhanghua now tells his own erotic tale, suggesting a more reasonable method of sexual "self-restraint" than that practiced by the virtuous Song Yu. In doing so, he moves the poem into a different literary context: that of the *Classic of Odes*. As the minister tells of his travels throughout Warring States China, he makes the same move as the *Beautiful Woman* rhapsody, evoking legendary lands of desire and illicit sex—the lands of Zheng and Wei. But Zhanghua makes much more of these associations. Zheng and Wei provide him with a different intertextual tradition, a different discourse with which to seduce.

The women he meets are picking mulberry leaves, which are to be fed to the silkworms before they spin their cocoons; this harvest was one of the essential steps in the cycle of sericulture, and it was a moment in the agrarian calendar when unmarried women were exposed to the gaze of male admirers, as many early texts attest. Unlike Song Yu's eastern neighbor, who trumpets her availability through her own initiative (she after all is part of the dynamics of inside/outside) and violates proprieties in doing so, the peasant maids are (potentially) easy pickings for the powerful gentry. Here we are in the world of the *pastourelle*, where fast-talking aristocratic poets and girls of the lower orders fight verbal battles of seduction and resistance.[35]

Yet Zhanghua does not rely on his own words; instead he quotes a poem from the *Classic of Odes* (no. 81) to the maid of his fancy. Why he should use this avenue of approach is open to interpretation. Does our anonymous poet recognize some original purpose of many *Classic of Odes* poems that identifies them with courtship and seduction? Or is he using the text as a rhetorical enhancement to his effectiveness, as Warring States diplomats did?

Regardless, Zhanghua's maid rejects his suit and remains distant and cold; yet she melts and replies seductively herself when she realizes that she can substitute a counter-text for her body. She replies with a poem of her own, written in *Odes* style (the last couplet is taken from no. 233). Zhanghua then asserts that this union of minds brought on by a union of texts replaces sexual pleasure and keeps both of them within the bounds of propriety. After this act of oratorical lovemaking, the two of them go their separate ways, seemingly satisfied.

What is Zhanghua's point here? "With only subtle phrases we had moved each other to affection." Dengtu had used the term "subtle phrases" (*wei ci* 微辭) to describe the dangerous seductive hypocrisy of Song Yu; now it is being used to describe this sublimated sexual encounter between two pastoral lovers. Surely, suggests Zhanghua, Your Majesty has nothing to fear from Song Yu; even if he does not resist your ladies as he did his eastern neighbor, he will merely exchange "subtle phrases" with them, and all will be content with that. The ruler himself is the last to be seduced: he is compelled by his pleasure in the rhetoric or simply by his sense of humor to grant Song Yu and his partner the victory.

Dream-Panders and Impotent Lovers

The two "goddess" rhapsodies attributed to Song Yu, *Gaotang* (*Gaotang fu* 高唐賦) and *The Goddess* (*Shennü fu* 神女賦), cannot be divorced from *The Wind* and *Master Dengtu*. Although there is no way of knowing whether they were written by the same author, all four focus on the dramatic figure of Song Yu as court poet and trickster. In fact, an awareness of *Gaotang* and *The Goddess* clarifies the way in which sex is figured in *The Wind* and *Master Dengtu*; some of the same power relations between ruler and minister are refigured in the erotic realm.

As has often been noted, *Gaotang* is not really an erotic poem, although the opening became the most famous erotic scene in Chinese literature:

Once upon a time King Xiang of Chu 楚襄王 went roaming with Song Yu near the terrace at Cloud-Dream Marsh. He gazed out toward the shrine built at Gaotang, and there rose steeply above it a lonely cloud of mist. Suddenly it changed its appearance, transforming endlessly within a single moment.

The king asked Song Yu, "What is this mist?" Song Yu replied, "This is what is known as 'the cloud of morning.'" The king asked, "What does that mean?" Song Yu then explained, "Once upon a time, a former king roamed here. Growing tired, he took rest during the day and dreamed he saw a woman, who said, 'I am the woman of Wu 巫 Mountain, and I roam here at Gaotang. I heard that my lord was enjoying himself here, so I wish to offer you pillow and mat.' The king then took his pleasure with her. When she departed, she said, 'I dwell on the south side of Wu Mountain, at the steep side of a high hill. In the dawn I become a cloud of morning, in the evening, a passing shower. Dawn after dawn, dusk after dusk—so it is beneath the sunlit terrace.'

"At dawn the king looked out, and it was as she had said; so he established a shrine there in her honor and called it 'cloud of morning.'"[36]

The sexual implications in the meeting of mortals and deities are, of course, not new; as we have seen, they are suggested in the shaman songs collected in the *Songs of Chu*. However, there the discourse oscillates between a language of courtship and a language of worship; shamanic possession may indeed be a sexual ecstasy, but the encounter is somehow more, encompassing an experience of non-human worlds and incommunicable pleasures. In *Gaotang*, the encounter is rendered baldly, unequivocally—"the king took his pleasure with her." The relation between sexuality and male power seems embarrassingly obvious here: after Song Yu's description of the dream, he engages in a description not of the goddess and her charms but of the land itself, whose attractions are under the ruler's sway. The poem peaks not with sexual climax but with the appearance of the royal court at the hunt and "represents the king's mastery over the landscape."[37] A goddess is never fully controlled, but her body, reinterpreted as her surroundings, becomes the lord's demesne; it may be that the area itself ("Cloud-Dream" Marsh) has received its name (and hence its role in the relationships of possession and control) from this sexual act. One is reminded of John Hay's suggestion that the boundaries of the body are permeable, always threatening to transpire into some form of *qi* (mist, clouds, breath).[38] Song Yu is offering King Xiang a public and political substitute for the inevitable postcoital depression that results from the ephemerality of the private sexual encounter.

The reader (and the king, no doubt) at first expects an erotic poem, and in fact the erotic opening is the part most people remember (one may note Xiao Tong's 蕭統 [501–31] controversial decision to place the text in the chapter on "passions" [*qing* 情] in his anthology, the *Literary Anthology* [*Wenxuan* 文選]). Yet if the poem mostly seems to be an act reinforcing the power of the king, one must keep in mind that (as usual) Song Yu is the master of words here. This was not King Xiang's sexual encounter; it was a "former king's." The poet is deliberately moving away from the subject of sex, in order to frustrate a lubricious ruler. Is this, like Sima Xiangru's evocation of virtue at the end of his rhapsodies, an attempt at persuasion? Or is it something darker?

The sequel to *Gaotang, The Goddess,* follows directly in the *Literary Anthology,* the oldest extant source for these poems.[39] Here the poet's control of the erotic situation is more obvious and more complex:

King Xiang of Chu was roaming with Song Yu at the edge of Cloud-Dream Marsh when he had Song Yu compose a rhapsody on the affair at Gaotang. That night, as Song Yu [or the king] slept, he dreamed he had an encounter with the goddess. Her features were extraordinarily lovely. Song Yu [or the king] marveled at this and told the king [or Song Yu] about it the following day. The king [or Song Yu] said, "What was the dream like?" Song Yu [or the king] replied,

"Toward evening,	晡夕之後
My spirit felt in a daze,	精神怳忽
As if there were some reason for joy.	若有所喜
In a flurry, all aroused—	紛紛擾擾
But I didn't know the meaning.	未知何意
At first my eyes could just make out colors,	目色仿佛
But suddenly I could tell what it was.	乍若有記
I saw a woman before me,	見一婦人
Wonderful was her form.	狀甚奇異
As I slept, I dreamt her—	寐而夢之
But when I awoke I knew her no more.	寤不自識
Worried then, and joyless,	罔兮不樂
I grew sad and disheartened.	悵然失志
Then I stroked by breast, settled my spirit,	於是撫心定氣
And could see before me my dream once more."	復見所夢

What does this opening tell us when compared to the opening of *Gaotang*? First of all, it suggests that someone (either the poet of *Gaotang,* or some later reader) felt that the erotic possibilities of *Gaotang* had been left untapped and decided to expand those possibilities into an independent poem. In *The Goddess,* the poet will be the interpreter not of the landscape but of the female body. His role will be just as disingenuous as it was in *Gaotang,* but in a different way—here he will tantalize the reader, tease him with drawn-out descriptions, start and stop at interesting points to maintain his arousal. The mock virtue of the first poem has been dispensed with.

Yet another essential question suddenly confronts us at the beginning of the text. Who is doing the dreaming? The original text in the *Literary Anthology* has the king as the dreamer, but since at least as early as the Song dynasty, many commentators have been convinced that this reading is in error and that the position between king and minister should be reversed. The reason for confusion is that the character for Yu, the poet's personal name (and the name the text calls him by, as well as the name he uses in Chinese fashion to refer to himself) is the character for jade (玉); the character for "king" is identical to it, except for the absence of a small stroke in the lower right corner (王). Many later readers, assuming that the king would not need a poet to describe his own dream to him, believed that a copying error is responsible; the two characters were often used for each other in very early texts. However, the Chinese reading community is still more or less split on this issue.[40] One could find plausible reasons to support either reading: for example, *Gaotang* suggests that the goddess prefers royal lovers; moreover, Song Yu's position here and in other poems is that of court flatterer and entertainer. Perhaps the idea of the king needing Song Yu to tell him his own dream is not so strange either, granted the ruler's impotence in language and the dependence he shows on his poet in the rest of the Song Yu corpus. However, there are ultimately much better reasons to make Song Yu the dreamer, beyond the fact that it makes it easier for him to be the enumerator of the goddess's charms. Even at the beginning, there is an identification between the dream of the ancient Chu king's in *Gaotang* and this later one, although the first visit of the goddess was one unbidden by the dreamer; here Song Yu might seem to have evoked the goddess through the power of his words—his description of her in *Gaotang* has led to his own vision of her here, thus suggesting some secret advantage he possesses as master of language to summon others.[41] Perhaps it is best to imagine a reading community in which both options were held open as possibilities, since we have no observations on the problem of interpretation for over five hundred years after the compilation of the *Literary Anthology*. What the textual ambiguity creates is the king and poet as doubles—so easily confused that only a small mark possessed by one of them makes the difference.

Either way, and as in the other Song Yu poems, poetry is displayed as a speech on the part of a character/poet in the poem—not as a direct statement from an implied author. A fictional character *within the poem* must relate events; we can know nothing beyond what this fictional point of view believes and chooses to tell. Thus, "what happens" as the plot of a rhapsody must always be mediated by a public voice addressing a listener: rhapsodies were defined in part by this aesthetic of performance, whereby the assumption of a public speaker determines how one "reads" what he says. This means that experience can never be private. Not only that; it is exposed to outside scrutiny twice: first, to the royal audience within the frame of the poem and second, to us, the readers, as the unaddressed (seemingly uninvited) audience. This will have important consequences in the depiction of sex.

After an initial description of the dream itself, Song Yu pauses, waiting for the king's response. The latter is eager for details:

The king said, "What was her appearance?" Song Yu said:

"Flourishing! Lovely!	茂矣美矣
Possessed of all felicities;	諸好備矣
Shapely and so gracious in form	盛矣麗矣
I could not plumb the depths of her beauty.	難測究矣
Such as she have never been before,	上古既無
Nor have such been seen in the present world.	世所未見
Jasper grace, bejeweled gestures,	瑰姿瑋態
A source for endless praise.	不可勝贊
When she first came,	其始來也
She glowed like the white sun first striking roof and rafter.	耀乎若白日初出照屋梁
When she advanced slightly,	其少進也
She gleamed like the bright moon spreading its light.	皎若明月舒其光
Within a brief moment,	須臾之間
Her lovely visage arose suddenly before me,	美貌橫生
Bright as a flower in blossom,	曄兮如華
And welcome as the glow of jade.	溫乎如瑩
All colors shone forth from her,	五色並馳
Yet I cannot capture her form in words.	不可殫形
When I looked closely at her,	詳而視之
She took away the very spark from my eyes.	奪人目精
As for her plentiful ornaments:	其盛飾也
Gauze silks, painted satins,	則羅紈綺
Profuse with fine patterns.	繢盛文章
Finest clothes marvelous,	極服妙采
Glowing in all directions,	照萬方
Flourishing embroidered robe,	振繡衣
Clothed in vest and outer skirt,	被袿裳
Generous fabric, never cut too short,	穠不短
Fine cloth, not cut too long.	纖不長
Her stride graceful, she glowed in palace and hall.	步裔裔兮曜殿堂
Then she suddenly changed expression again,	忽兮改容

Magnificent, like a sporting dragon winging its way through the hanging clouds."	婉若遊龍乘雲翔

From this point, confusion between Song and the king ceases; the former is in control. Either the king is eager to hear of the dream of his minister, or he cannot comprehend his own dream unless interpreted by his poet; either way, his enjoyment of sexual experience must be mediated by text.

The machinery of rhapsody description begins to hum as Song Yu gears up for his elaborate account. In English, we are forced back on the same handful of adjectives to reflect a large variety of Chinese words meant to evoke feminine beauty: "charming," "lovely," "magnificent," and so forth. Yet the passion and excitement underlying Song Yu's account becomes believable, not because he makes us see what he saw, but precisely *because* they have unleashed this torrent of language. Song Yu himself hyperbolically gestures toward the uncapturability of the experience: "I could not plumb the depths of her beauty," "I cannot capture her form in words." The first phrase here is particularly curious: he uses the language of surface/depth and unconsciously admits the inadequacy of *se*, "surface/sexual allurement," to substitute for a potentially more "profound" experience.

"Splendid in her apparel,	嫷被服
Charming, her light and filmy robes,	侻薄裝
Orchid hair oil,	沐蘭澤
full of pollia scent.	含若芳
Her nature all amiability,	性和適
Suitable to serve at my side;	宜侍旁
Compliant in her manner,	順序卑
In harmony with my heart."	調心腸

In spite of the excitement of the poet and his frustration over the gap between experience and language, he does not forget that his poetry is the only way to carry out his central task: entertainment of his ruler, who is reliving his own ancestor's encounter vicariously through the medium of his poet. Song Yu pauses a moment to stress the availability of the goddess—an aspect that will redound to his own ultimate benefit. But we see even more clearly that descriptive elaboration may have other motives than the conveyance of an exciting experience; Song Yu begins to write not as a man but as the king's servant and sees the goddess, too, from the perspective of his own servility.

The king said, "If she was so lovely, then try to compose a rhapsody for me about her." Song Yu said, "As you command.

"How lovely and bewitching is the goddess!	夫何神女之姣麗兮
Clothed in ornament of light and darkness.	含陰陽之渥飾
Adorned by her charming blossoms,	被華藻之可好兮
Like the flashing wings of the emerald kingfisher.	若翡翠之奮翼
Her form, without peer;	其象無雙
Her beauty, without end.	其美無極
Mao Qiang hides behind her sleeve,	毛嬙障袂
Incapable of matching her.	不足程式
Xi Shi covers her face,	西施掩面
In contest with her, quite plain.	比之無色
As I near her, how seductive!	近之既妖
But even from far away, I would continue to look.	遠之有望
Even her frame is marvelous,	骨法多奇
Fit complement to a prince's features.	應君之相
To gaze upon her fills the eyes,	視之盈目
Who is worthy of surpassing her?	孰者克尚
Secretly in my heart I cherished her alone;	私心獨悅
Took pleasure in her without measure.	樂之無量
Such meetings are rare, her grace seldom granted—	交希恩疏
It cannot be fully related.	不可盡暢
Let others not look upon her—	他人莫覩
But Your Majesty may behold her form."	王覽其狀

As Song Yu hints at the goddess's availability, the king decides that such apparent willingness promises a potential for future trysts; he presses Song Yu for details. We realize now that the preliminary poetic description was merely an introduction and that a more elaborate evocation is to come. Yet we also notice that this is the third time Song Yu has begun the account of his dream, and the overall effect is to prolong the suspense and lust of the king. Song Yu now plays upon the fact that the king cannot gain sexual satisfaction without him and so now combines the roles of divine lover, of procurer, and of substitute for the goddess herself—he enables the king's sexual congress, who is incapable of a masturbatory fantasy without his poet. A similar relationship arises as well between the poem's author and his audience.[42] Thus, when Song Yu compares her to two earlier royal lovers (Mao Qiang 毛嬙 and Xi Shi 西施) and suggests his own role as conduit of experience ("Let others not look upon her— / But Your Majesty may behold her form"), the poem's author speaks to the audience as well—momentarily raising them as well to the level of voyeuristic but impotent royalty.

The same irony surrounding the indescribable continues, however, as well as a comparable stress on acts of looking, gazing, of luminosity, blindness, visual infatuation. The language of rhapsody continues to capture the surface excitement of the experience, but not the inner qualities of it.

"Her form towers above me in her splendor,	其狀峨峨
How can it be described completely?	何可極言
Features full and graceful, makeup lovely—	貌豐盈以莊姝兮
A jade-like face, warm and glowing.	苞溫潤之玉顏
Pupils flashing forth her spirit,	眸子炯其精朗兮
Glances beauteous, worthy of your gaze.	瞭多美而可觀
Eyebrows gracefully arched in a moth-like curve,	眉聯娟以蛾揚兮
Crimson lips shining like cinnabar.	朱脣的其若丹
Silk-white body generous and full,	素質幹之醲實兮
Her will was easy, and her body at rest.	志解泰而體閑
Graceful amid her quiet reclusion,	既姽嫿於幽靜兮
Sinuous in step now as she walks among men.	又婆娑乎人間
A high palace is fitting for her generous thoughts	宜高殿以廣意兮
That went winging everywhere in freedom and space.	翼放縱而綽寬
She shook mist-silk in her slow progression,	動霧縠以徐步兮
Brushes stairs with the sound of the swish of her robe.	拂墀聲之珊珊
She looked to my bed curtains, craned neck to see;	望余帷而延視兮
Her gaze flowing ripples turning to waves.	若流波之將瀾
She would shake her sleeves, straighten robe's folds,	奮長袖以正衽兮
Stood in hesitation, ill at her ease.	立躑躅而不安
Then, clear and still, kindly and gracious,	澹清靜其愔嫕兮
Her nature grew serene and undisturbed.	性沈詳而不煩
For the moment, amiable, as she moved slightly,	時容與以微動兮
But as for her intentions—I could not reach their source.	志未可乎得原
She seemed wishing to approach, yet remained distant,	意似近而既遠兮
As if about to come—yet again she turned.	若將來而復旋
She lifted my curtains, begged to do service;	褰余幬而請御兮
Wished to show her heart's deepest feelings to their depths.	願盡心之惓惓
Cherishing the pure clarity of light and purity,	懷貞亮之絜清兮
At last, Alas! she found fault with me.	卒與我兮相難
In reply I laid forth my finest phrases,	陳嘉辭而云對兮
Emitted fragrance of pollia and orchid.	吐芬芳其若蘭
Our essences joined as they came and went,	精交接以來往兮

Our hearts rejoiced in pleasure and joy.	心凱康以樂歡
But my spirit persevered alone and it never connected;	神獨亨而未結兮
Soul was left companionless, no end to its seeking.	魂耿耿以無端
She once wished to assent, but now I found it impossible,	含然諾其不分兮
I emitted a sigh, grievously lamented.	喟揚音而哀歎
She turned red in anger, took control of herself:	頩薄怒以自持兮
Always would she stay inviolate."	曾不可乎犯干

Much of the poem up to this point has been confused in terms of sequence of events; Song Yu has been too overwhelmed by her appearance overall to relate the actual plot of his dream. Yet here he begins to sketch events for us. Though confined to his bed of dreams, the poet has the ability common among dreamers to see her from afar—first giving us an increasingly obsessive dwelling on her eyes, eyebrows, lips and body, then noticing her arrival as she "walks among men." Closer and closer she draws, proceeding through hall and palace, up stairs, and into the sleeping poet's bedroom. Here she hesitates, so that the poet may emphasize her modesty on the one hand and prolong the sexual tension on the other. Finally, however, she changes her mind and "finds fault" with the poet.

Why does the sexual encounter fail at this point? Primarily, one might say (thinking of *Master Dengtu*), because a mock respect for the superior position of the ruler actually undermines the ability of a poem in bringing the ruler pleasure. Song Yu realizes that it is the king's right to have relations with a goddess and that as a loyal servant he should not bring the relationship to culmination; on the other hand, he destroys the ruler's satisfaction, since the ruler is vicariously experiencing the sexual encounter through Song Yu. Of course, this potentially applies to the motives of the "real" author as well, and his extended audience.

However, this motive also introduces factors that turn back upon the poet and attack him. The goddess's refusal here arises not as an example of ambiguous, quixotic, or "divine" behavior (that is, a repetition of the enigmatic actions of the *Nine Songs* divinities) but rather from Song Yu's singular *awareness* that an orgasm within the text is imminent—an orgasm that now becomes a form of public performance. We may well note the figurative representation of the goddess' invitation: "[She] Wished to show her heart's deepest feelings to their depths. / Cherishing the pure clarity of light and purity." Such a statement suggests that the goddess is concerned with the exchange of "profound experience" as well as intent on preserving her purity.

Under these conditions, it is not difficult to see why Song Yu is figured here as guilty of bad faith. The purpose of his description cannot have escaped him—he can only relate his dream in language, and his use of language is predicated on the production of verse in the service of his ruler. He cannot keep his sexual experience private, because the public rhapsody compels him to cater to his patron and thus to compromise him in the eyes of his divine lover. He can do nothing more than pander. The result is a curious failure of masculine power, exposed as it is to the gaze of a greater male authority. As his language fails to penetrate the surface, his own physical encounter fails as well; we may not read too much into things if we see the goddess's anger is further increased by her realization of the impotence of both the poet and the king who stands leering behind him.

At this point, it should not surprise us that Song Yu now relies on verbal persuasion within the dream itself, much as he has employed the poetry to attract his ruler. Yet the attempt, as might be expected, fails. His words, a vain attempt at sexual persuasion, "emitted fragrance of pollia and orchid"—an unintentionally ironic reinterpretation of these symbols for the inner beauty of the righteous man.[43] At first it has some brief effect; as with Zhanghua and his mulberry maiden, the poet and the goddess achieve a kind of poetic/spiritual unity. But the poet is left by himself in the end.

Yet there are other interpretations. The absence of pronouns gives us new possibilities. It is the poet, not the goddess, who "lifts the curtains, begs to do service, wishes to show heart's deepest feelings to their depths." And it is the goddess, not the poet, who excuses herself by "laying forth finest phrases."[44] Her refusal mimics that of the mulberry maid, who replaces sexual congress with conversation. Does the goddess refuse the poet because he is not the ruler? Or is this alternative yet another deflation of the ruler's/audience's expectations—another refusal to commit oneself to private actions when others are looking?

Regardless of how we read it, the goddess soon departs in a hurry:

"Then she shook her waist's pendants,	於是搖珮飾
Let her jade bells ring;	鳴玉鸞
Straightened her garments,	整衣服
Composed her features;	斂容顏
Looked back to her matron,	顧女師
Summoned her tutor.	命太傅
Never had our pleasure joined,	歡情未接
She took her leave and departed.	將辭而去

I stretched out to her, betook myself,	遷延引身
But could not draw close again.	不可親附
She seemed to leave, yet did not depart—	似逝未行
In mid-path she turned back to me.	中若相首
Her eyes again flashed the slightest gleam;	目略微眄
Essence's glow passed once more to me.	精彩相授
Intention and features spread forth—	志態橫出
I cannot record it fully.	不可勝記
Before our loving thoughts were parted for good,	意離未覺
My soul and spirit sunk in fear.	神心怖覆
For courtesies there was no leisure,	禮不遑訖
Nor were words of farewell completed.	辭不及究
I wished to snatch a final moment,	願假須臾
But the goddess claimed she needs must hurry.	神女稱遽
My bowels twisted within, my inner breath injured—	徊腸傷氣
I was overturned without support.	顛倒失據
Things grew darker in the gloom,	闇然而暝
At once I knew not where I was.	忽不知處
My love I kept private in my heart,	情獨私懷
For who is there who could tell of it?	誰者可語
I grew sorrowful, I shed my tears,	惆悵垂涕
Searching until the dawn came."	求之至曙

At this point, the meter in the Chinese changes to abrupt four-character lines, to simulate the rapidity of the goddess's departure and the poet's frenetic attempts to halt her.

As with the other Song Yu poems, *The Goddess* dramatizes the court poet's involvement with (and entrapment by) political and power relations. There is great unease in this poem; as the poet finds himself compelled by the ruler into the position of composer of textual masturbations, he finds himself continually vulnerable as his powerless mistress, subject to the same rejections from his employer as from the goddess—whose autonomy and power of sexual choice masculinizes *her*. Will his descriptions ultimately be found unsatisfying, and his persuasions fail with a hearer who can see through them to a supposed depth that does not exist? Will he be seen as shallow?

There is a further lesson to be learned, however. Although it is true that rhapsodic language was condemned for being "feminine," the condemnation merely concealed the fact that language and its manipulation was an exclusively male prerogative, derived as it was from male moral, political, and social discourse from the time of the Zhou dynasty; it was women in fact who seemed to stand entirely outside literary language and its role in the dynamics of power. Dwelling in the realm of the inarticulate, women became the object of male imagination. Perhaps on the one hand "inarticulate" meant "ineffable": women might ultimately be admired because they would have a special relation to the Tao, the reality that could not be spoken of. On the other hand, they could represent a special threat, a form of being not subject to male discourse; to negate language was to negate male power. Infinitely attractive and frightening, they easily became seemingly capricious goddesses trapped within the text.

Reading Rivalry

There are many goddesses and many sexual encounters in later poetry. Many of them are derived from *The Goddess* and show the power of its influence.

Following *The Goddess* in the *Literary Anthology* is the most famous of them all, Cao Zhi's 曹植 (192–232) *Luo Goddess* (*Luo shen fu* 洛神賦). Cao claims to be writing an imitation of Song Yu, and so (although he speaks in the first person) he projects himself as a fictional character within his own poem. Is this a mark of taboo, a disingenuous disclaimer of a "lived" experience? Or an acknowledgment of a necessary social-poetic mediation for a sexual encounter, even if that very mediation guarantees an ultimate impotence?

In the third year of Huangchu 黃初 [A.D. 222], I attended court in the capital then returned home, crossing the Luo River. Of old, men have said that the spirit of this river is named Fufei 宓妃. Moved by the matter of the goddess that Song Yu told of to the King of Chu, I then wrote this rhapsody. It runs as follows:

I took my way to the capital land,	余從京域
Returning then to my eastern fief.	言歸東藩
My back to the Yi Guardtower,	背伊闕
I crossed Winding Peak,	越轘轅
Crossed Tong Valley,	經通谷
Passed over Jing Mountain.	陵景山
When the sun had sunk to the west,	日既西傾
My carriage slipped and my horses tired.	車殆馬煩
And so I unharnessed my carriage at Duheng Marsh,	爾迺稅駕乎蘅皋

Fed my team at the Mushroom Fields, 秣駟乎芝田
Took my rest at Willow Wood 容與乎陽林
Cast my gaze toward the Luo Stream. 流眄乎洛川
Thereupon my soul shook, my spirit startled, 於是精移神駭
All at once my thoughts scattered; 忽焉思散
I looked down—still nothing clear; 俯則未察
I looked up, saw something strange: 仰以殊觀
I beheld a lovely woman 睹一麗人
On the river's bank. 于巖之畔

I then turned to my driver and asked, "Do you see her over there—that woman? How lovely she is!" My driver answered, "I have heard that the deity of the Luo is named Fufei. So wouldn't she be the one My Lord sees? What does she look like? I wish to hear." I told him.[45]

And so on. The text has been frequently translated and discussed, and I will not give a detailed reading here;[46] suffice it to say that Cao's tryst with the goddess fails as well, although they do exchange love gifts as a symbol of their continuing affection.

What interests me is the story that came to be associated with the poem. Cao Zhi was a younger son of Cao Cao 曹操 (155–220), warlord and empire-builder; Cao Cao's older son, Cao Pi 曹丕 (187–226), was to overthrow the Han and establish the Wei 魏 dynasty in 220, taking the name Emperor Wen 文. Cao Zhi had been known through his youth as a lad of considerable talent and had even aspired to the succession, but supposedly due to his excessive behavior, he earned the enmity of both father and brother. Repeatedly persecuted by his brother (or so the story goes), Cao Zhi saw one friend after another exiled or murdered.

Legend has it that this was rendered all the worse by his infatuation with one of his brother's consorts, Empress Zhen 甄. Our earliest surviving texts of the *Literary Anthology* have a note attached to the title of *Luo Goddess*; some attribute this note to the *Anthology* commentator, Li Shan 李善 (d. 689), although there is good reason to suspect it is an interpolation in the text possibly as late as the eleventh century. It may reflect earlier traditions, however.

An account says: Prince Dong'e 東阿 of Wei [i.e., Cao Zhi] around the end of the Han sought to marry the daughter of Zhen Yi 甄逸, but it didn't work out. [His father] Cao Cao instead gave her to [Cao Zhi's brother] Cao Pi. Zhi was extremely upset; he longed for her both day and night and could neither eat nor sleep. He came to court during the Huangchu reign [220–26]; the emperor [Cao Pi] showed him a pillow of Zhen's, embroidered with jade threads and gold embroidery. When

Zhi saw it, he broke into tears. (By then, Zhen had died from the slander of Empress Guo 郭.) The emperor suddenly realized the cause of his grief, and so he took the opportunity to invite Zhi to stay and banquet with him. He presented Cao with the pillow as a gift.

When Cao Zhi left for home and passed by Winding Peak, he stopped and rested for a time by the banks of the Luo River, where he lay thinking of Empress Zhen. Suddenly he saw a woman come to him. She said, "I first gave my heart to you, but our love came to nothing. I first gave this pillow to your brother before I married and was still living at home; now he's given it to you." She then offered herself to him, and they joined in their pleasure and joy. Ordinary words could not describe their feelings.

"Because when I died Empress Guo stuffed my mouth with chaff and my hair was loose, I'm ashamed to be seen by you." Then she disappeared, and he never saw her again. She sent someone with a gift of a pearl to the prince; he responded with a jade girdle pendant. He was powerfully moved, both sad and happy; he then wrote *Rhapsody on My Love for Zhen*. [Cao Pi's son] the Emperor Ming 明 saw it after and changed the name to *Goddess of the Luo*.[47]

Pretty much every scholar writing on the subject since the eighteenth century has agreed that the Empress Zhen infatuation is apocryphal—for one thing, Cao Zhi would have been a mere boy when the empress married her first husband.[48] But that makes little difference to the average reader when it is enshrined in the most respected commentary to the *Literary Anthology*; after all, scholars have also felt compelled to debunk the myth repeatedly every time they address the subject. There are reasons for its attraction.

Empress Zhen of Wei was clever and quite attractive. First she was the wife of Yuan Xi 袁熙, who treasured her greatly. When Cao Cao put the city of Ye 鄴 to the sword, he quickly sent someone to bring Lady Zhen, but his retainers reported, "The Commandant of the Five Offices [Cao Pi] has already taken her away." Cao replied, "I destroyed these rebels all for her sake!"[49]

So runs one earlier romantic anecdote on how Cao Pi acquired his empress. She was created in the literature as a woman everyone desired, including every member of the turbulent Cao triangle, Cao, Pi, and Zhi. Through this fable the story of "Luo Goddess" becomes a clandestine desire concealed amid the dangers of the court; the dream-woman becomes real, only to become a ghost or the projection of an unrequited longing. And most prominently of all, a woman becomes the focus of desire for two men, one in power, one not. The one in power can wound his impotent brother with the pillow on which once lay the receptive female body, but fantasy can restore that body to the pillow as a promise not just of sexual fulfillment but of real power. Even without believing in Empress Zhen, one can see that Cao Zhi is inscribing a dream of empowerment and potency, in which sex and power become one. His empress; his goddess; the female form dispersed into the territory of the empire. Yet in the end there is only another failed liaison, another failure of the text to substitute for power. As a late rhapsody, Cao's poem is already subject to the failure of language brought on by the triumph of Yang Xiong's critique—erotic language is surreptitious, guilty, ultimately blocked. The only escape is to euhemerize the divine body and turn it into a sentimental fable.

CHAPTER THREE

The Competitive Community

> 19.5. Lady Zhao 趙姬 [a lady-in-waiting to Sun Quan 孫權, the ruler of Wu 吳, r. 229–52] married off her daughter. When the latter was about to depart, her mother instructed her: "Be careful not to do good." The girl replied, "If I don't do any good, then may I do evil?" Her mother said, "If it is not permitted to do good, how much the less is it permitted to do evil!"
>
> —*Worldly Tales, New Series*

Let us return to Cao Pi, Cao Zhi, and their tumultuous relationship:

4.66. Emperor Wen [Cao Pi] once commanded the Prince of Dong'e [Cao Zhi] to compose a verse before seven paces could be taken. If he should not complete it in time, he would suffer the supreme penalty. The prince immediately responded:

"You boil the beans to make a gruel;	煮豆持作羹
You strain the pulse to make a broth.	漉菽以爲汁
Bean stalks burn beneath the pot	萁在釜下燃
While the beans shed tears within.	豆在釜中泣
Stalks shared their roots with the beans themselves—	本自同根生
Yet how eagerly they boil them now!"	相煎何太急

The emperor appeared deeply ashamed. (134)

This apocryphal account, from the early fifth-century collection of anecdotes, the *Shishuo xinyu* 世說新語 (Worldly tales, new series),[1] demonstrates that the legend surrounding Cao Zhi was in place well before the Tang dynasty. As we saw in the preceding chapter, the legendary enmity between the poet and his brother could take many forms and involve different kinds of jealousy: Zhi's jealousy over Empress Zhen, Pi's over his younger brother's greater talent. Here, Pi targets talent in a cruel attempt to trip up his rival. Of course, his order is a monstrous variation of a common social phenomenon: the ruler requests a poem from one or more of his attendants, to be composed orally in his presence, usually at a banquet or imperial outing. Today such verse is generally thought to be merely ceremonial and superficial; limited in subject matter, overly concerned with decorum and proper vocabulary and with "empty" descriptions, it supposedly tells more of the sycophantic relation of poet to patron than of the socially engaged "lyric self" so praised by later critics.[2] Yet such compositional contexts were very important indeed. A poet's future in society, his self-respect, and his reputation with his companions often rested on avoiding gaffes, on the gracious performance of his cultural duties.

To this degree, the *Worldly Tales* anecdote is a virtuosic rewriting of the society poet's plight, forced as he is to compose within spatial and temporal limitations. Zhi's poem is inscribed within his pacing, a small, shrinking, interior space, the prison cell of his own talent. He is "in the soup"; trapped within the iron chamber of the cooking kettle, he weeps boiling tears while his brother rages outside. One thinks of John Donne's pun on *stanza* as a "little room" or Keats's comment that the sonnet is a dance in chains, but the cozy pleasure the Western poets derive from working under constraint does not apply here. There is too much at stake. The anecdotist exposes his own sympathy as the underdog thwarts the assassin's plot; the emperor is made to reveal his shame.

The "Seven Step Poem" suggests a sensitivity in literati culture to the interrelations of cultural production, danger, and power. Although these factors do not always involve the representation of gender, they do affect such representations, most evidently in the sixth-century anthology of erotic verse, the *Yutai xinyong*, which is the subject of the next chapter. In this chapter, I make a detour from gender and discuss the new factors in the cultural construction of the literati in the third and fourth centuries, their representation in texts, and the evolution of the competitive mentality that produced the tensions of social poetry composition.[3] I am interested in ideologies: the unreflective conditions for daily life and activity among elites, as well as the ground for more explicit notions of meaning and value expressed through literature and what we think of as "philosophy."[4] We will explore not only the evolving resources of the "lyric" (*shi* 詩) genre but also the literati's cultural universe—most prominently, as it is reflected in the *Worldly Tales* collection itself.

Poetry and Literati Self-fashioning

How did lyric poetry become a dominant genre in Chinese literature? To answer, we must return to the ancient *Classic of Odes* and its hermeneutics (at the risk of taxing those already familiar with the rich scholarship on the subject).[5] The Han dynasty reading traditions associated with the Mao 毛 school established a method of scriptural interpretation and laid the groundwork for a "secular" aesthetics: a poetics for current practice, not merely a method of reading the sanctioned texts of the past. This emerges most clearly in the section of Mao school commentary appended to the first poem in the anthology, the *Guanju* 關 雎, and traditionally termed the "Great Preface." Its first part is usually considered the most important:

> The Ode is where one's aspirations go. In the mind it is aspiration. When it comes forth in words, it is a poem.
>
> Emotions move within and take shape in words. When words are insufficient, then one sighs it. When sighing is insufficient, then one sings it. When singing is insufficient, then without even realizing it one's feet and hands dance it.
>
> Emotions come forth in sounds. When sounds form a pattern, we call them "tones." The tones of a governed age are peaceful and joyful; its administration is harmonious. The tones of an age in chaos are resentful and angry; its administration is perverse. The tones of a doomed state are grieving and brooding. Its people are in trouble.
>
> Therefore to put right gain and loss, to move heaven and earth, to move ghosts and spirits, nothing is more appropriate than poetry. The former kings used it to regulate husbands and wives, to perfect filiality and respect, to make human relations firm, to make moral education and transformation attractive, and to alter local customs.[6]

Until recently, Western students of East Asian poetry have been influenced by James J. Y. Liu's categorizations of literary purpose, which were inspired by M. R. Abrams's *The Mirror and the Lamp*. Liu considers the Great Preface's argument to represent the poetic act as essentially "didactic" (in that the text stresses the educative functions of the poems) and as "expressive" in a primitive way (in that it claims that poetry is "unmediated by culture," so to speak, and that a poet possesses no conscious "art").[7] However, over the past several years a number of scholars have read the Great Preface in other ways and stressed not so much its "didactic" nature as its fusion of the emotive/aesthetic and the moral. Writing about the Great Preface's mention of the "aim" or "will" (*zhi* 志) as well as "emotion" (*qing* 情), Steven Van Zoeren suggests the text is arguing that a "sincere" or "authentic" *zhi* seeks to direct the emotional powers of the individual and that the Confucian "project" is involved in the training of its students in developing just such a sincere aim. In this project the affective power of the Odes is paramount:

> If this hypothesis about the relation of the aim and the emotional nature is correct, we can see something of the significance of the Odes for the Confucian project of moral self-transformation as it was conceived in the late Warring States and Han. The Odes inscribed—preserved and made available for the student—the aims or *zhi* of their authors, and these aims are in every case paradigmatically normative ones. By "studying" the Odes—that is, by attending to their discursive exposition, by memorizing, reciting, and internalizing them—the student is able to "take on" those aims, at least temporarily.[8]

Earlier, Van Zoeren notes that Confucian ethics asks not "'What is the good,' but rather 'How can one *be* good?'"[9] As a result, Confucianism turned to the problem of training the individual so that he could respond to the world spontaneously in a moral way.[10] This emphasis on training followed by spontaneity is a trademark of much of Chinese ethical thinking in general—notably in the work of Zhuangzi 莊子 (who stresses that harmony with the natural must be internalized) and Chan 禪 (Zen) Buddhism. Ethics is not a matter of rational choice and decision-making, but an internalization of correct modes of behavior.[11] The *Odes* (ideally) offer a direct aid to such training; they bypass intellectualization altogether and speak to the emotions and to the "aim."

The early commentators on the Classics were well aware that such an ideal was not realizable. With the decay of the early Zhou system, they argued, the human capacity to internalize norms declined as well. The loss of the music that accompanied the Odes may also have contributed to their practical failure as effective morality. As Stephen Owen has noted,

> Poetry occupied a very important place in the Confucian cultural program, but its instruction is not supposed to be coercive. Instead, according to the K'ung Ying-ta [Kong Yingda] commentary, when combined with their music, the poems of the Book of Songs [i.e., the *Classic of Odes*] were supposed to influence people to good behavior unconsciously: listeners apprehended and thus came to share a virtuous state of mind, and the motions of their own affections would be shaped by that experience. But this Edenic power was possible only in those days when the poems still had their music; in this later age, with the original music and only the naked

texts remaining, commentary is required to show the virtue that was then immanent in all the poems.[12]

However, to understand how the interpretive tradition dealt with the "loss" of poetry's original power, one must turn to the "Little Prefaces" 小序 of the Mao school, the contextualizing headnotes that introduce each poem. I have already discussed these in part in my explication of "Quince"; now we need to see how they contribute to a praxis of composition for the late Han literati.

As noted above, the Mao reading tradition insisted on a precise historical situation "behind" each poem; the commentators located the text's significance in an awareness of the moral consequences of history. Thus, the "morally paradigmatic" attitudes that are to be internalized can be understood only if the motives of the composer at a single historical moment are comprehended. Pauline Yu has made the intriguing suggestion that the understanding of a sequence of such contexts creates a narrative, perhaps one that served an "epic" purpose.[13] My interest, however, lies in the social consequences of the contextualization.

Although perhaps originating in the Western Han, the Mao interpretations did not join Confucian mainstream thought until the Eastern Han; they were then elevated to canonical status with the additional commentaries of Zheng Xuan 鄭玄 (127–200). It is perhaps not surprising that this happens simultaneously with the development of a class of literati who defined themselves partly as the possessors of specialized textual knowledge.[14] With the Han's fall and the rise of a literati culture no longer completely dependent on imperial institutions, the identification of Confucian knowledge as a mark of superior class status did not disappear; rather, aristocratic groups became "specialists" in the Confucian classics and pointed with pride to long lines of teacher-student transmission, often within families.[15] In such an atmosphere, the poetics of the Mao commentary, with its emphasis on an emotional response to a specific current event, could have resonated fully only for a person educated in the Classics and in the history of elite culture. Although the Great Preface suggests an interchange between "those above and those below," the attributions of the Little Prefaces lean either toward elite authorship or toward a folk authorship whose attention is always directed toward the actions of the elite. Given the way the Mao reading tradition ties poems to history, it necessarily precludes a history of or for the "underclasses."

I am not here making the vulgar Marxist claim that the Mao commentary was a self-conscious move on the part of the "feudal rulers" to repress the contributions of the "people," nor do I believe that the Odes are necessarily folk poems or have their origins in some idealized folk consciousness. But clearly, the Great Preface creates an ideology of solidarity for the literati, as well as a blueprint for their poetic composition. Most cultural elites (particularly before commercial printing) undertake to make their own forms of literary production exclusive and dependent on elite knowledge available only through rigorous and expensive training—their own form of "cultural capital." This strategy may operate through erudite allusion or rhetorically elaborate compositional structures; at this point in the Chinese case, historicization and contextualization of poetry play much the same role.[16]

As lyric poetry began its rise to ascendancy at the end of the second century A.D., it seems to have derived its style and themes from formulaic traditions of composition as well as urban popular song traditions (anonymous "old poems," *gushi* 古詩).[17] Early literati poets composed poetry orally, in a social context, portraying themselves as participants in popular cultural activities (drinking parties, flirtations with women, *carpe diem* songs on aging and death).[18] However, the Mao commentary's inherent bias toward elite historical contextualization facilitated the eventual creation of a poetry community freed from the formulaic qualities of the earlier poems. This is most obvious in the creation of a literary language that makes use of allusions to the Classics and to history, but it is also manifest in the contexts of literati poetry production, the epistolary and social exchanges that soon became characteristic.

Many of the poems composed by the so-called Jian'an 建安 poets (that is, by the warlord-poet Cao Cao, his sons Cao Pi and Cao Zhi, and the literati circle patronized by them during the Jian'an reign period [196–220]) are characterized both by their use of generic "conventionalized" imagery and by the poets' need to mark their works off as "elite" productions.[19] They are operating as poems within the process of history, commenting upon it, and aware of their own past. Popular poetry did not need to speak for the specific, identifiable moment, the historical event: it spoke supposedly of "universal" themes, or at least themes that had a conveniently transferable use-value, ones that could be adopted by any performer to his own needs:

I drove my carriage out Upper East Gate,	驅車上東門
Gazed afar at the tombs north of the walls.	遙望郭北墓

How desolate seem the white poplars! 白楊何蕭蕭
And pines and cypress line the broad cemetery lanes. 松柏夾廣路
Below lie those who have died long ago, 下有陳死人
Vague and dim now, in their eternal dusk. 杳杳即長暮
In sunken sleep they lie in Yellow Springs, 潛寐黃泉下
Never to wake for a thousand years. 千載永不寤
Vast the cycles of the seasons— 浩浩陰陽移
While our years are but as morning dew. 年命如朝露
Human life is a brief sojourn, 人生忽如寄
Our span not as hard as metal or stone. 壽無金石固
Ten thousand years follow, one by one, 萬歲更相送
And not a sage or a worthy can outlast them. 賢聖莫能度
Some take drugs and seek the Immortals, 服食求神仙
But most are betrayed by these poisons. 多爲藥所誤
Better to drink the finest wine, 不如飲美酒
And dress oneself in satins and silks![20] 被服紈與素

This, the thirteenth of the anonymous "Nineteen Old Poems" 古詩十九首, most likely composed during the second century A.D., elaborates a series of common sentiments that arose on going out the city gate and "gazing afar." The recurrence of the gesture in other poems suggests we have a trope common to popular poetry. The images are concrete, but of general significance: the pessimism, the cynical reference to alchemy, and the injunction to "enjoy life" are generally perceived as characteristic literary stances of the period.[21]

As literati poets manipulated this shared language, they had several ways of making it their own. Many restate standard popular situations, with their pessimism and fatalism, in the context of the poet's personal and historical experiences. By the Jian'an period, poets could not simply "represent" their experience in any sort of direct way (if such representation is ever possible, unmediated by culture and previous texts); they must instead recognize the validity of standard tropes within individualized experience. Perhaps the most outstanding work to do so in this period was Wang Can's 王粲 (177–217) first "Seven Sorrows" 七哀 poem:

The Western City thrown into unprecedented chaos: 西京亂無象
Jackals and tigers come to plot its ruin. 豺虎方遘患
So I abandon the capital and depart, 復棄中國去
Consign myself to the barbarians of the South. 遠身適荊蠻
My close kin all face me in sorrow, 親戚對我悲

Dearest friends follow and cling to my robes.	朋友相追攀
As I leave the city gates I see nothing before me	出門無所見
Save white bones covering the level fields.	白骨蔽平原
On the road, a starving woman	路有飢婦人
Abandons the child of her breast in the weeds:	抱子棄草間
She turns to hear his screams and howls	顧聞號泣聲
Then wipes her tears and goes on alone—	揮涕獨不還
Saying "I hardly know where *I* shall die!	未知身死處
How can I look after both of us?"	何能兩相完
I drive on my horse, abandon her and depart,	驅馬棄之去
For I cannot bear to heed her words.	不忍聽此言
Southward I climb the Tomb of Ba,	南登霸陵岸
Turn my head, look to the city.	迴首望長安
I know now how the poet of "Falling Stream"	悟彼下泉人
Moaned and felt his heart break within![22]	喟然傷心肝

Commentary identifies rather precisely the event behind this composition: the fall of the warlord Dong Zhuo 董卓 in the year 193, which created a struggle for power around the capital city of Chang'an.[23] Wang Can works within certain parameters of popular poetry that were likely to have already existed: the "parting poem," which usually expresses the grief of wife and husband or friend and friend, as one of the two departs for distant and dangerous territory; the ballad in which straitened circumstances compel the speaker to leave his home and take up a desperate occupation; and, most strikingly, the poem in which one emerges from the city gates and notices a burial ground. Perhaps for the modern reader familiar with Western conceptions of the art of lyric poetry, Wang Can's narrative could be taken as the experiences of a crafted fictional "persona" who moves through certain actions that seem a commentary (potentially ironic) on the tradition within which he works.[24] But an early literati reader would more likely see the poem as a relation of the author's "real" experiences, although mediated through the language of a literary tradition that speaks for the universality and repeatability of those experiences—and that defined them as fit occasion for poetry when one encountered them in one's own life. To look out from the city gate means to see death; if Wang sees unburied corpses, rather than the graveyards of the earlier "Old Poem" persona, he more than likely undertakes a hybridization of tropes: perhaps here the other source is the lament over soldiers left unburied on the battlefield in antiwar *yuefu* poems. Yet on these stock themes (parting, contemplation of death, war), he composes variations that locate him within the historical literati community and also grant him the specificity to speak as Wang Can, the literatus with these particular experiences. Certainly the emotional response theory of the Great Preface allows here for an intersection of event and response that we might be inclined to call "psychological": the abandonment of family and country creates a guilt that allows him to notice and hence represent the woman abandoning her child. It also perpetuates itself powerfully through the constant reiteration of the verb *qi* 棄, "to abandon."

This language of guilt and abandonment is addressed specifically to Wang Can's role as a "responsible literatus" and clearly identifies him as a member of his class, with ideological commitments to governing, to awareness of the needs of the "other." The capital stands here for centrality, for the human community (it is named here *zhongguo* 中國, "the central state"). Threatened by warlords whom the poet represents as beasts (and here the poem draws on associations of the Confucian virtue *ren* 仁, "humanity," with its homonym, *ren* 人, "human"—those who act in a "non-humane" manner are no better than animals), the community cannot protect Wang, nor can he protect it. He is cast to the outskirts, to live with the barbarians (moving, perhaps, to the lesser of two "non-human" evils). If his first "abandoning" is a public one, it is soon refigured as personal, as friends and relatives try to restrain him from departure. Here, too, there is an echo of literati discourse that reads the state as the family writ large. The vision of the unburied dead demands that Wang must take responsibility in part for the people's suffering and leads in turn to the most strikingly inhuman act of all: the mother abandons her child, thus contributing the little body to the bones that surround it. There is an extraordinary moment of identification here, when Wang figures her as both Self and Other: a peasant who should be helped (a child who needs the assistance of the father/literatus, who must look after her) and a nurturing parent who abandons its own (already the term "mother and father of the people" had become a common term for the official or ruler). Yet it is his identification with the mother, the adoption of her gender, that brings about his own "impotent" gestures at this point: at the moment that he figures himself as a mother, as nurturer, he similarly figures himself as helpless, without institutional support: "I hardly know where *I* shall die! How can I look after both of us?" The words are his own as well as the woman's. The gesture of abandoning her continues his guilt while it seeks excuses.

Yet this text continues its creation of the literatus author by allowing him the reflexive gesture of climbing a hill and looking back—a meditative, self-conscious maneuver that privileges his cultural sensitivity. As Ban Gu once said (enigmatically), "He who can climb high and chant a poem may become a high official";[25] here, Wang performs the act as a recognition of the trope and its supposed literati origins. Moreover, he becomes further enmeshed in history at this point; the hill he climbs is Emperor Wen's burial mound, and Wen oversaw the most peaceful era of the Han empire (180–156 B.C.). As Wang looks, he recalls the lines of a poem from the *Odes* (no. 153), "Falling Spring":

How cold the falling spring 洌彼下泉
That soaks the lush wolf-tail grass. 浸彼苞稂
In grief I sob and cannot sleep— 愾我寤嘆
I recall the city of Zhou.[26] 念彼周京

Interpreted as a man's nostalgic longing for the golden days of the early Zhou empire, "Falling Spring" enables Wang to find a canonical situation behind his own narrative and to reinforce his own connections with history. His quotation not only betrays his possession of a cultural capital not fully available to folk or popular poets but also validates his own existence as a literatus and reads the "Falling Spring" poet as a literatus like himself.[27] It also holds open the possibility of future recognition and alliance with literati of the future: as future cities suffer their own catastrophes, future poets will look back to his own verses.

I have gone into this degree of detail to illustrate how much the Jian'an poems (generally praised for "spontaneity," "realism," and "manly sorrow") nonetheless express the specific social and ideological concerns of their authors: stylistically indebted to the limited subject matter and tropes of folk and popular verse, they manipulate the seemingly "universal" nature of these images into the service of "self-confession." This self-confession does not, however, form an "individual" (in the modern sense of the term) so much as a literatus, a man defined by particular social and cultural concerns and with certain privileges within his own society.

This formation of the literatus "self" in poetry helps explain a further dichotomy in Jian'an composition: Jian'an poems can broadly be divided between ballads and poems of "direct experience," on the one hand, and highly ceremonial and eulogistic banquet poems, on the other. Although many modern critics reject the latter out of hand as merely representing the desire to win position with warlord masters and thus of little value compared to the "realist" verse, the two categories are really alike at a higher level in that the literatus uses both sorts to define his own social status.[28] It must be remembered that for several centuries after the Jian'an period, it was the social event—especially the banquet—that came to be associated with Jian'an poets; critics such as Liu Xie 劉勰 (ca. 465–522) recognized that the social community created by the salon enabled literary talent to meet, exchange verse, and interact. Poetry could be born only within such a community. Xiao Tong recognized the Jian'an writers as a literary salon by including a significant number of their formal banquet poems in the *Literary Anthology*.[29] As the imperial court broke up, and talented literati associated themselves first with warlords and then with the more "personal" courts of the south, the ceremonial eulogy became another form of self-fashioning, of rendering the literatus self comprehensible in the text. And, of course, eulogy poets could, by quoting from numerous Odes that served similar functions, equate their own court with the Zhou court and their own persons to the "poet-historians" of the golden age.

Thus, elite composition in a social context by no means implies mere "socializing"; on the contrary, the elite poet saw a bond between himself and the "contextually located" poets of the Odes: both could use verse as a means to protest or celebrate imperial policy or to demonstrate social ills. Verse is a badge of privilege *and* a moral/critical tool. An illusory historical continuity is drawn between morally upright citizens of a "golden age" and the present-day literatus. A knowledge of the past through mastery of texts grants the elite student the equipment he needs to practice virtue and verse.

It is clear, then, that the birth of a new lyric poetry in the third century grew not so much from the need to find a distinctive poetic voice (as in post-romantic poetics) as from a need to find new techniques to differentiate literati status. As we have seen, the important point is not that every literati poet is inevitably "different" from his peers in every way but, on the contrary, that good literati poets often respond in the same way to the same sort of situation. Poetics thus participated in the new interest in typologies and evaluations, the compulsive need to "rank" or "categorize" individuals, their moral or creative capacities, or their intellectual abilities (as Christopher Connery has noted). Such evaluations often contain within themselves ideologies of social distinction and class status: after all, they arose simultaneously with the "Nine Rank" 九品 system, a method for judging suitability for office that evolved into a system of inheritable quasi-aristocratic titles.[30] Connections between evaluation and literary production came early—with Cao Pi's "Discourse on Literature" 典論論文, in which the author-emperor attempted to characterize the writers of the Cao family salon:

> Ying Chang 應瑒 [d. 217] is pleasant but not robust. Liu Zhen 劉楨 [d. 217] is robust but is not subtle. Kong Rong's 孔融 [153–208] style and force are lofty and marvelous, and he has qualities that surpass others, but he cannot make an argument, and his reasoning powers do not match his diction. He even goes so far as to include gibes and jests in his work. But when he is at his best, he is a match for Yang Xiong and Ban Gu.[31]

The argument moves to literary genres quite easily:

> Literature shares the same root but is different in its branches. Memorials and deliberations should be elegant and dignified; letters and discourses should be logical and thought out; epitaphs and eulogies should respect accuracy; poems and rhapsodies want to be lovely. These four categories are not the same, and so those who have skill tend to have their preferences for certain genres.[32]

It is implied in such texts that since emotions and qualities are categorizable, they are not infinite in number; in other words, it would be absurd to presuppose that every text is unique simply because its author is an "individual." It is also suggested, to some extent, that a writer tends to write in the genres whose modal qualities are most in harmony with his own personality. Thus, one can perceive literary genres as arising out of emotions or personality traits that are exhibited by various writers at various moments in time. The genre they write in depends on themselves, their situation, and the social and emotional validity of the particular genre for the situation.[33]

However, as genre theory develops in the fourth and fifth centuries, culminating in Xiao Tong's prose and poetry categories in the *Literary Anthology*, it becomes clear that the "appropriateness" of a genre is mostly a social phenomenon. Moreover, the increasing popularity of poetry grants it the capacity to subdivide into further authorized "subgenres" (for example, the *Literary Anthology* specifies "exchange" poems, "travel" poems, "laments," etc.). In these later centuries, these various subgenres become the social contexts that demand poetry's composition. But granted the development of an earlier

Odes poetics and the organization and categorization of modal qualities, it is wrong to suppose that sixth-century authors were "formalists" or concerned only with composing set pieces based on certain rules for certain genres—any more than Jian'an poets were "spontaneous" or "realists." Rather, literature could be seen as a form of cultural capital that cemented social ties while expressing through exclusive language the modal emotions appropriate for the social occasion. If poems repeated the sentiments (or even the diction) of earlier texts, they necessarily validated their own "authenticity." Thus, repetition and variation of compositional models could be seen as a mark of social distinction *and* as a form of self-representation. As poetic language becomes increasingly refined after the Jian'an period, this aspect of repetition becomes increasingly ritualized; the "occasions" for composition become more clearly defined, and expectations of what the texts will say become set as well. This was likely a manifestation of the growing importance of poetry as a bond within the literati community: it was a form of social interaction, of competition, for a closely knit elite society.

Competition and Violence

I have anticipated myself somewhat, however. It was not a foregone conclusion that lyric poetry would become the significant cultural phenomenon that it did by the fifth and sixth centuries. Although literary histories have tended to paint a smooth, organic development from the Jian'an period, these same histories also acknowledge failures and gaps—most significantly, during the century following the fall of the Western Jin 西晉 dynasty in 317, when non-Chinese peoples occupied the Yellow River plain and much of literati culture fled to the southeast. Rather than read this gap in poetry as the product of literary-historical vagaries, we might seek the reasons for it in the cultural concerns of literati themselves throughout the period. If poetry seems to have declined somewhat in its function as a form of social exchange during the Eastern Jin 東晉 (317–420), did any compensatory pursuits take its place? And what accounted for poetry's recovery and triumph in the subsequent centuries?

To address these broader issues, we can turn again to the *Worldly Tales* collection.[34] As has frequently been pointed out, the relative scarcity of materials for this period and the disruption of traditional historiography until the Tang have made historians dependent on this text; indeed, the seventh-century compilers of the official Jin history relied on it. But rather than trust the text as a guide to biography, history, or even to social and intellectual developments, I prefer to see it as a guide to literati ideology: to their sense of self-definition and value, as well as to their anxieties and fears. Of course, we must acknowledge the peculiar circumstances of its compilation. Its authors were a group of literati active in the circle of Liu Yiqing 劉義慶 (403–44), a prince of the Liu Song 劉宋 dynasty, during the 430s.[35] Since it is thought that this group idealized the literati culture of the fourth century to a great extent, one must be cautious not to see the work as an authentic account of elite cultural attitudes throughout the period it purports to describe. It would be equally pointless, however, to claim that the anecdotes reflect the attitudes solely of Liu's circle and have no deeper roots. Rather, we can assume with some justification that *Worldly Tales* is a text that does interpret the past, but that its interpretation has nonetheless evolved from earlier attitudes.

Even a casual reading of *Worldy Tales* reveals its extraordinary interest in defining approved or admired behavior in members of the literati: its subjects are constantly judging themselves and others or measuring certain codes of conduct against competing ones. This may not come as a surprise, granted the culture of "evaluation," and one may tentatively oversimplify by suggesting that this insecurity came about in part from the need to find a replacement for the collapse of the imperially sanctioned bureaucratic order toward the end of the Han.[36] Regardless of the source of the obsession, its working out in literati culture is transparent throughout *Worldly Tales*. The interactions of literati (their treatment of each other, their speech, their evaluations of their activities) give a strong impression of the importance placed on the male social community. Validation of one's activities, whether these are "conventional" or "unconventional," must come from others, and the strain or even violence of competition is always near the surface. Moreover, this cultural interaction is rehearsed against a violent political environment that elevates its importance all the more: competitive cultural activities threaten to escalate, to turn into feuds, to implicate the struggles of political cliques. Not a few competitions end in violent death. As a result, considerable emphasis is placed not merely on cultural and social self-definition but on control and self-restraint: concealment becomes a prized characteristic, as well as a form of cunning and grace under pressure. A selection of *Worldly Tales* anecdotes will help clarify literati discourse along these lines.

VIOLENCE ON THE POLITICAL SCENE

The instability of dynasties between the Han and the Tang is obvious to any observer, as is the impact of such violence on the literati, who were compelled either to walk the dangerous tightrope of factional politics or to attempt to withdraw from political life altogether. Such dangers often issued from the warlords and their families, who struggled to replace weak emperors and establish their own dynasties: the Caos and Simas 司馬 in the third century, Wang Dun 王敦 (266–324) and the Huans 桓 in the fourth, and Liu Yu 劉裕 in the fifth. Literati often threw in their lot with such figures in the hope of finding power or safety; other literati remained loyal to the ruling dynasty, speculating that warlord schemes would ultimately be defeated. *Worldly Tales* itself emphasizes the difficulty of avoiding political turmoil; it suggests that the ideology of "judging and employing talent" that arose in the Cao circle during the Jian'an period so permeated politics that any literatus reputed to have any abilities whatsoever might be courted or compelled to serve by the leaders of various factions. In such an environment, a refusal to serve might look like a gesture of allegiance to the rejected lord's opponents.[37] We cannot be sure how true this actually was, or how much *Worldly Tales* constructs a legendary world in which literati are always in demand, even if always in danger. Nonetheless, we see the somewhat novel notion that talent in itself can become dangerous and that skill in judgment can result in disaster and death:

14.1. Once when Emperor Wu of the Wei dynasty [Cao Cao] was about to receive an emissary from [the nomadic people] the Xiongnu 匈奴, he felt that his own features were too plebeian, making him unsuitable for the task of overawing this remote land. He then had [his subordinate] Cui Yan 崔琰 take his place, while he himself stood guard by the dais with sword in hand. When the audience had ended, he sent a spy to ask the emissary, "What did you think of the Prince of Wei?" The envoy replied, "The prince was exceptional in culture and refinement. But the man who stood guard with sword by the dais—he was a true hero!" When Cao heard of this, he sent someone to overtake the emissary and kill him. (333)

This anecdote is part of a whole series of legends that surround Cao Cao. Most of them emphasize his ruthlessness and wiliness in equal measure. The story is equally important, however, in demonstrating that a perspicacious judge of character may need to keep his skills and judgments to himself, even if he is a barbarian envoy: one can be too talented to live. Literati forced by powerful figures to make judgments would often hedge their comments to avoid the envoy's mistake:

1.15. Prince Wen 文 of the Jin 晉 [Sima Zhao 司馬昭; 211–65] praised Ruan Ji 阮籍 [210–63] for his extreme caution; for whenever he spoke with him, Ruan's words were mysterious and remote, and he never praised or denigrated personages. (10)

9.86. After [the warlord] Huan Xuan 桓玄 [369–404] served as grand mentor, he held a great assembly. When the court ministers had all gathered and taken their seats, he asked Wang Zhenzhi 王楨之, "How do I compare with your seventh uncle [Wang Xianzhi 王獻之; 344–88]?" Everyone present gasped in apprehension for Wang, but he calmly replied, "My late uncle set a standard for an age, but you, Sir, are a hero of a thousand generations." Everyone was happy. (299)

9.87. Huan Xuan asked Liu Jin 劉瑾, "How do I compare with the former grand mentor Xie An 謝安 [320–85]?" Liu replied, "You are lofty, while Xie was deep." "And how do I compare with your distinguished uncle, Wang Xianzhi?" "The hawthorn, the pear, the orange and the tangerine—each one possesses its own special quality." (299)

Those who judge character might in fact consider disingenuousness and evasiveness as exceptional virtues:

8.44. Contemporaries judged Yu Ai 庾敳 [262–311]: "He excelled at entrusting himself only with great matters, and was talented at self-concealment." (245)

8.4. Gongsun Du 公孫度 [d. 204] judged Bing Yuan 邴原 [d. 211]: "He's what you call a white crane in the clouds—not the sort you can snare in a net for swallows or sparrows." (228)

In this last case, one may notice how the diplomatic avoidance of danger is evoked through a Taoist image of transcendence, the white crane. When explaining the rise of "neo-Taoist" speculation during this period, one should not forget the usefulness such philosophy had in its taste for "otherworldliness" or even (more cynically) for rhetorical obfuscation.

Political violence and the danger that it presented to literati could easily betray family loyalties and affections; in the following anecdote about the great calligrapher Wang Xizhi 王羲之 (309–65?) in childhood, precocious brilliance (a quality greatly admired in *Worldly Tales*) is demonstrated in a number of striking ways. Wang wisely refuses to trust his kin and displays an uncanny willingness to undergo personal humiliation to save his own life:

27.7. When Wang Xizhi was scarcely nine years old, the Generalissimo Wang Dun was particularly fond of him and often had the boy spend the night within his bed curtains. One morning the general arose first. Before Xizhi could get up, Qian Feng 錢鳳 happened to arrive, and the two men, having dismissed their attendants, began to discuss important affairs. Forgetting the presence of his young relative, Dun turned the conversation to his plans for rebellion. When Xizhi woke and overheard the discussion, he knew that they would not let him escape. He then forced himself to vomit, soiling his face and bedclothes, and then feigned unconsciousness. Suddenly in the midst of [discussing] his plans, Dun realized that Xizhi had yet to rise. In consternation he and Qian Feng resolved that he must be done away with. However, when he pulled back the bedcurtains and saw the vomit everywhere, he believed that the boy had been unconscious the whole time. Xizhi thus managed to escape, and all the people of the time admired his resourcefulness. (456–57)

CULTURAL STRUCTURES OF STABILIZATION: FAMILY PRIDE

However, the clever evasion so admired in *Worldly Tales* represents merely one form of protective coloration. The text also shows literati devising other mechanisms that could either mask the dangers they faced (making them more tolerable) or counteract those dangers by allowing a space in which defenses might be constructed and in which the forces that threatened them could be intimidated into submission or alliance.

The most superficially obvious of these are family pride and aristocratic snobbery; these are responsible in part for later literary clichés about "great families" like the Wangs and Xies. However, *Worldly Tales* prefers to demonstrate that "aristocratic" status was difficult to maintain in fourth-century cultural circles. In fact, the period saw several waves of "aristocrats": from the remnants of Han families who looked askance at parvenus to new, powerful families that appeared after the loss of the North.[38] The ideological construction of the "great family" was just as important as the possession of ancestors, power, or territories that any one family could claim. As Jean-Pierre Diény has pointed out, the Xies, from the modern viewpoint the quintessential "aristocratic" family, were initially insulted and ignored by many members of older lineages.[39]

24.9. Once in the company of his older brother [Xie An], Xie Wan 謝萬 [ca. 321–61] rose to look for a chamber pot. Ruan Yu 阮裕 [ca. 300–360] was present at the time. He later remarked, "These parvenus may be sincere, but they have no sense of the proprieties." (414)

24.12. Once Xie An went to the capital in the company of his brother Xie Wan. When they passed through Wu 吳 District, Wan wanted to go and visit [the governor] Wang Tian 王恬. An said, "I fear he won't receive you. You're just not significant enough to warrant his attention." Wan insisted, however, and since An refused to turn back, he went to Wang's house alone. After he had sat with Wang a short time, Wang rose and withdrew into his private chambers. Wan was overjoyed, thinking that Wang was about to offer him more informal and intimate courtesies. However, after some time had passed, Wang emerged with his hair washed and unbound. He did not return to sit with his guest but instead perched on a Tartar stool in the courtyard, where he proceeded to dry his hair in the sun. His manner was haughty and distant, and he would make no reply to Wan's polite small talk. Wan then left and went off to rejoin his brother. Before he reached An's boat, he hailed him from afar. An replied, "It didn't work out with Wang, did it?" (415)

The issues here are not whether there is an "aristocracy" and who belongs to it; rather, there seems to be a continual need in many families to *create* prominence for their own genealogy as a way of delineating status. For example, families attempt to exclude "non-elite" individuals from social intercourse (as with Wang Tian's treatment of Xie Wan).

There is also the assumption that heredity will show in one's elevated behavior—often manifesting itself after a "period of decline" and thus demonstrating that blood will out, in spite of evil fortunes:

12.4. Minister of Works Gu He 顧和 [288–351] loved to engage in philosophical debate [lit. "pure conversation"; see below] with the worthies of his time. Once when his grandsons Zhang Xuanzhi 張玄之 and Gu Fu 顧敷 were six years old, they were playing beside his couch. When they were listening to the conversation, it seemed as if they were not concentrating on it, but when dusk fell, they sat under the lamp and repeated to themselves the arguments of both sides, not leaving a single detail out. Gu came over to their seat, and leaning forward, he said into their ears, "Who would have thought that our declining house should produce such treasures?" (324)

Most evident of all in *Worldly Tales* is the caution with which "great families" treat marriage alliances.[40] Parvenus or military men were for the most part unwelcome:

5.25. Zhuge Hui's 諸葛恢 [284–345] eldest daughter married the son of Defender-in-Chief Yu Liang 庾亮 [289–340]; his next daughter married the son of the Regional Inspector of Xuzhou 徐州, Yang Chen 羊忱 [ca. 255–311]. When Liang's son was killed by Su Jun 蘇峻, Hui remarried the daughter to Jiang Bin 江虨 [d. ca.

370]. His son married the daughter of Deng You 鄧攸 [d. 326]. At that time, Xie Pou 謝裒 sought the youngest in marriage for his son, but Hui replied, "The Yang and Deng families have been allied to us by marriage for generations. The Jiangs have esteem for us, and we have esteem for the Yu family. We simply can't contract a marriage with a son of Xie Pou's." However, after his death the marriage went through anyway. When Wang Xizhi went to the Xie house to inspect the new bride, he found that she preserved her father's principles. She was dignified and circumspect, neat and proper in her manner and dress. Wang sighed with admiration. "To think that she acts just as if her father were still alive!" (173–74)[41]

5.58. When Wang Tanzhi 王坦之 [330–75] was acting as senior clerk for Huan Wen 桓溫 [312–73], Huan sought Wang's daughter for his son. Wang promised to discuss it with his father, Wang Shu 王述 [ca. 303–68], and went home. Shu was quite fond of Tanzhi, and although he was full-grown, Shu would often dandle him on his knee. When Tanzhi took this opportunity to seek permission to marry his daughter to the Huans, Shu was furious. He pushed Tanzhi off his lap and said, "I regret to see you have become such a fool! Are you so frightened of Huan Wen and his army that you would give away your daughter?" Tanzhi returned to Huan and reported. "I'm afraid my family's already arranged for the girl's marriage." Huan said, "I understand. Your father's just not willing." Later a daughter of Huan's married Tanzhi's son. (189)

As is often the case, families of high status must agree among themselves to remain exclusionary, to consort and intermarry only with each other. However, there is always the assumption that a distinguished family will be recognized by conduct and not exclusively by blood. *Worldly Tales* remains ambivalent toward these claims. Often it confirms "superior breeding." It is just as likely, however, to satirize snobbishness by setting up a kind of alternative "anti-snobbery snobbism" and asserting that a true elite is recognized by certain patterns of "spontaneous" behavior. This holds particularly true when family status is vaunted through a display of wealth:

23.10. Ruan Xian 阮咸 (234–305) and Ruan Ji lived on the south side of the street, and the other Ruan families lived on the north. All the northern Ruans were prosperous, while the southern Ruans were poor. On the holiday of the Seventh Day of the Seventh Month, the northern Ruans set out all their family robes for an airing; all of them were of fine silk and gauze, brocade and painted satin. Ruan Xian in his turn hung a large loincloth on a bamboo pole in the middle of his courtyard. When someone commented about this, he replied, "I've done this because I just can't avoid following custom." (393)

Such anecdotes suggest that although the distinction between literatus and commoner was becoming clearer and more difficult to cross,[42] the distinctions between the most elevated gentry and the minor gentry were less clearly drawn and more fluid.[43] If we are looking for the techniques whereby the literati of the *Worldly Tales* defined their own self-worth and social distinction, it is by no means clear that the construction of aristocratic bloodlines was among the most successful.

CULTURAL STRUCTURES OF STABILIZATION: INTELLECTUAL ATTAINMENTS

The third and fourth centuries were a golden age of Chinese philosophical speculation, giving rise to the Chinese adaptation of Buddhism, to the metaphysical strain of thought often termed "neo-Taoism," and to new types of Confucian exegesis. The attraction of these pursuits during this period has led Charles Holcombe to suggest that a man who could live the "philosophic life" of retirement and self-cultivation could claim a certain respect from his contemporaries and dependents and thus contribute to social and political stability.[44] Such a man could be said to rule through the force of his *de* 德, a quality somewhat equivalent to the old Roman *virtus*. Could this, then, represent the most typical way the literati proclaimed their own importance, their greater authority in comparison with "military men" or ineffectual rulers?

To a certain extent, yes. But we must be careful not to elevate this form of intellectual prestige to a level where it attains authority in and of itself. A reading of *Worldly Tales* suggests we should see philosophical self-cultivation not as an idealized form of social control but as a material and practical one that often interacted with and mirrored political tensions and struggles. Many anecdotes describe *qingtan* 清談, or "pure conversation," a type of sophisticated philosophical debate that enabled literati with intellectual pretensions to hone their logical skills and often win followers and allies. In fact, *Worldly Tales* is the largest single source for reconstructing the thought of the time, since most of the writings have been lost.[45] Yet when we turn to these anecdotes, we see that philosophical conviction is not nearly as important as the powerful impression one makes as a debater or the grace with which one counters one's opponents. Quality of mind is admired more than content of argument. This holds true as well for early Buddhist debates influenced by "pure conversation."

4.40. Zhi Dun 支遁 [314–66], Xu Xun 許詢, and others were staying at the retreat of the Prince of Guiji 會稽. Zhi Dun was serving as "instructor in the dharma," and

Xu acted as commentator and discussant. Whenever Zhi explained a principle, all the guests were rapt with attention; whenever Xu suggested a difficulty, they all danced with glee. But there was not one who did not sigh in admiration over their excellence, regardless of what their actual positions were. (123–24)

4.11. During the Western Jin, a certain enthusiast of Taoist thought sought out Wang Yan 王衍 [256–311] so that the latter could explain some points he had doubts about. It so happened that Wang had debated much the previous day and was a little tired. Unwilling to engage in an interview, he said to his visitor, "I'm a little unwell today. Pei Wei 裴頠 [267–300] lives close by. Why don't you go ask him?" [Pei usually disagreed with Wang's own philosophical positions.] (108)

In at least one case, the debater shows his skill not only in defeating his opponent but also in taking up both sides of the argument.

4.6. When He Yan 何晏 (ca. 190–249) was serving as the minister for personnel, he had attained high position and was admired by all. At the time conversationalists would crowd the seats of his house. When Wang Bi 王弼 (226–49) was still in his teens, he went to visit him. Yan had heard of his reputation and so he selected superior arguments from previous conversations and said to him, "I consider these arguments to be the best. I wonder if you can find any objections to them?" Bi then made his objections, until everyone there decided that Yan had been defeated. Bi then took up several further rounds of conversation, this time taking both sides at once. No one there could surpass him. (106)

We are dealing here not with transcendent self-cultivators but with social competitors; these men impress and overawe with an elite cultural skill, like chess playing or fencing. Yet this is no mere sport. Other anecdotes demonstrate the seriousness of the debates, the violence of the competition, and the humiliation, shame, and loss of prestige that accompany defeat.

4.38. When Xu Xun was young, people used to compare him to Wang Xiu 王脩 [ca. 335–58]. Xu was discontented with this. Once several gentlemen and the monk Zhi Dun went to a lecture on the dharma at a temple west of Kuaiji. Wang was among them. Xu, who was greatly annoyed, followed them so that he could debate Wang on philosophical principles and finally settle who was superior. After they had ground away at each other with considerable effort, Wang was decisively defeated. Xu then took up Wang's position, and Wang took Xu's. Again they went at it, and again Wang was defeated. Xu said to Zhi Dun, "What did you think of my arguments just now?" Zhi Dun calmly replied, "It's true your arguments were good. But why should both of you grow so bitter? You certainly weren't seeking out the real truth of things, were you?" (122)

Zhi Dun makes a claim for competition as a form of solidarity, as a way of resolving aggression and seeking the truth in common—something akin to our sense of "fair play" and "sportsmanship." However, the anecdotes that attracted the *Worldly Tales* compilers largely portray competition as merely a reworking of violence on a smaller scale.

One interesting if apocryphal anecdote suggests how the Southern Dynasties fascination with competition, danger, and aggression refashioned the biographies of early Confucian scholars retroactively. Zheng Xuan and Ma Rong 馬融 (79–166) were second-century exegetes; yet here they play the roles of hero and villain in an apocryphal, folk tale setting.

4.1. When Zheng Xuan was a disciple of Ma Rong's, he did not succeed in obtaining an audience with him for three years. Only the highest-ranking disciples would receive instruction. One day Ma could not get some calculations he was making on his celestial sphere to come out right. Someone said that Zheng Xuan could manage it. Rong then summoned him and ordered him to work out the calculations. Zheng solved the problem with a single turn. All acknowledged his superiority with astonishment. When Xuan had completed his studies and took his leave, Rong sighed and said, "The Rites and Music are going eastward with you." He envied Zheng, fearing that he would supplant him in reputation. Zheng guessed this, and fearing pursuit, he went and sat under a bridge over the water, standing on a pair of wooden clogs. Rong had a session with a divining board to discover where he was hiding. He told his diviners, "This says Zheng is below earth and above water, resting on wood. He must be dead." He gave up the search, and Zheng escaped. (103–4)

Why should these skills seem so important to their practitioners? Perhaps "pure conversation" is not the visible manifestation of *virtus* so much as the display of expertise in a specialized cultural field. Literati society had originally defined itself in the Han through its literacy and its knowledge of Confucian texts; these qualities were acknowledged, at least theoretically, by their employment within the bureaucratic hierarchy of the Han imperial order. In the chaos of the following centuries, this cultural expertise refigured itself into a form of closed competitive exchange *independent of* the ostensible political system. Although literati continued to hold office, officeholding contributed much less to their sense of worth, their sense of *identity*. A means of valuing literary and cultural knowledge free of imperial sanction and bureaucratic structure had been devised. "Pure conversation" as portrayed in *Worldly Tales* may seem a game, but it is a game that partially determines value.[46] These skills are exclusionary devices that define elites within the society: only those trained at an early age in the rules of the competition have a good chance at excelling. The elite then perpetuates itself without necessarily creating an explicit ideology of heredity or aristocracy. Pure conversation potentially serves as a more powerful force for defining the literati elite than kinship. However, the game does not merely create solidarity; as we have seen, it also accentuates competition and antagonism, creating alliances and cliques. The very factors that make the literati class superior to parvenus and military men also threaten to tear it apart, to drag it into the realm of social violence and instability from which it hopes to escape.

FRIENDSHIP, COMPETITION, AND VIOLENCE

If much of literati cultural production as portrayed in *Worldly Tales* contributes to the creation of elite self-awareness and demonstrates its desire to constitute itself away from the contentious violence of politics, it nonetheless forms a male community in which violence constantly re-enters and intersects with that political realm. It can hardly escape the reader that most of the *Worldly Tales* anecdotes quoted so far incorporate elements of denigration and humiliation—as if one's own superiority can be acknowledged only through the demonstration of others' inferiority. Yet those others must always be validated as members of the community so that they may in turn recognize that superiority: recognition of talent requires cultural competence. The literati community is thus united by bonds of amity and enmity, friendship and hatred. Controlled by such tensions, the interaction turns sour, resulting in recriminations and violence; the rite of guest and host deteriorates into feud.

Worldly Tales protagonists are striking for the passion of their likes as well as the harshness of their dislikes. Some idealize their heroes; some are brought to physical illness through their passion and enthusiasm.

8.9. When Yang Hu 羊祜 [221–78] was returning to Luoyang, Guo Yi 郭奕 [d. 287] was serving as magistrate of Yewang 野王. When Yang reached the borders of his district during his travels, Guo sent someone to invite him for a visit and then went out himself to greet him. After he had seen Yang, he sighed in admiration and said, "One could hardly say that he is inferior to me." Later he went to the place Yang was staying, and after a short while he left. Again he sighed. "Yang surpasses others by far." When Yang left, Guo accompanied him on his way for several days, going several hundred *li* before he turned back. Because he had absented himself from the district, he was removed from his post. Again he sighed. "Yang is hardly inferior to [Confucius' greatest disciple] Yan Hui 顏回!" (230–31)

4.20. When Wei Jie 衛玠 first came south, he went to visit Wang Dun and sat with him deep into the night. Wang Dun took the opportunity to have Xie Kun 謝鯤 [280–322] join them. Jie was delighted with Xie and took no further notice of Wang. He discussed abstruse topics with him until dawn, and Wang could not make a single comment. Jie was always rather sickly and had been frequently forbidden stressful activities by his mother. Due to the excesses of this evening, however, he fell seriously ill and never recovered. (113)

Reinforcing this passionate admiration is a language of male show and desirability. Although women are praised for their beauty, *Worldly Tales* frequently creates an aura of splendor and attraction around its male protagonists, most typically by comparing their physical presence to glittering jewels:

14.12. Pei Kai 裴楷 [237–91] had a superior appearance and manner. Even when he doffed his cap of office, appearing in rough clothing and disheveled hair, he was still fine. The people of the time called him the "man of jade." One who saw him said, "Looking at Pei Kai is like strolling atop a jade mountain—that's how the light from him shines on you." (336)

14.14. The cavalry general Wang Ji 王濟 was the uncle of Wei Jie. Valiant and straightforward in character, he was possessed of grace and bearing. But whenever he saw Jie, he would sigh and say, "When such pearls and jades are by my side, I feel my own body unclean." (337)

Another anecdote explicitly relates Wei Jie's attractiveness to the feminine:

14.16. When Wang Dao saw Wei Jie, he said, "You evidently do have a frail body. Although you can remain open and friendly all day, it's as if you 'can't bear the weight of gauze and silks.'" (337)

Wang Dao here jokingly quotes a line from Zhang Heng's 張衡 (78–139) *Rhapsody on the Western Capital* describing the seductive movements of female dancers. Wei, the young literatus so taken with Xie Kun's conversation that he fell ill and never recovered, is thus emblematic of qualities generally associated with certain beautiful women: fragility, ill health, delicate grace.

The most famous example of male beauty was the poet Pan Yue 潘岳 (247–300), who (*Worldly Tales* claims) attracted crowds of women when he entered the street. Liu Jun's 劉峻 (462–521) sixth-century commentary to the text quotes an alternative version of this legend, one that amusingly echoes the *Odes* poem "Quince": women reward with fruit the jade-like beauty of the man himself:

Pan Yue was very beautiful; every time he went out, old women would throw fruit at him until it filled his carriage. Zhang Zai 張載 (d. ca. 304) was very ugly; every time he went out, little boys would fill his carriage by throwing tiles and rocks. (335)

However, Pan Yue's jewel-like beauty was best shown off through his friendship with another beauty:

14.9. Both Pan Yue and Xiahou Zhan 夏侯湛 [243–91] had beautiful features. They enjoyed going out together. Those of the time called them the "linked jade disks." (336)

Readers now may interpret this as analogous to a modern "homosexual" relationship, but the intimate connections that link talent, desirability, attraction, and friendship are much more significant: they further cement the male social community characteristic of the *Worldly Tales* perspective.

Worldly Tales goes further in its descriptions of friendship by suggesting that it can have a positive effect on political and military affairs. Ideally, friendship can be bestowed only on men of comparable virtue or artistic/intellectual attainments. In this ideal literati world, even the powerful may be rejected, and sometimes the powerful humble themselves before their "moral superiors."

5.2. Zong Cheng 宗承 [d. ca. 230] of Nanyang 南陽 was a contemporary of Cao Cao's but thought little of the latter's conduct and would not associate with him. When Cao became director of works and took control of the court's policies, he amicably asked Zong, "Will you have dealings with me now?" Zong replied, "My pine and cypress nature is still with me." Because Zong had such disdain for Cao, Cao generally overlooked him, and his position never matched his virtue and ability. However, whenever Cao's sons Pi and Zhi would visit him, they would each do obeisance before his couch. Such was his regard for the rites. (153)

Yet these anecdotes give only the "positive" side of friendship; other male-male relations can turn easily to violence when codes of friendly conduct are strained or violated. Social harmony among elites is always threatened by the need for prestige, and just as one may idealize one's better nature in another, one may also see the other as the force that blocks his own achievements. Vendettas arise from old wrongs, and not even an emperor can bring the parties to peace.

31.7. Both Wang Chen 王忱 [d. 392] and Wang Gong 王恭 [d. 398] were once at a banquet of He Cheng's 何澄 [d. ca. 404]. At the time Gong was governor of Danyang 丹陽; Chen had just been appointed to Jingzhou 荆州. Just when the banquet was breaking up, Chen asked Gong to drink a toast. Gong wouldn't drink. Chen continued to urge him, and things turned bitter. Finally each belligerently wrapped a sash around his fist. Gong ordered his staff into the room—around a thousand men altogether. Chen's followers were fewer, but he ordered them forward also, intending to fight it out to the death. Cheng was at wit's end. He rose and moved his seat to a space between them, and only then would the two sides disperse. This is why "the ancients disdained friendships based on power or profit." (475)

36.3. When Generalissimo Wang Dun had captured Prince Min 愍, he sent Wang Yi 王廙 (276–322) to take the prince out in a carriage at night and kill him. At the time no one knew all of this story—not even all the members of the prince's household. His sons, Wuji 無忌 and Wuji's brother, were still very young. Later, [Wang Yi's son] Wang Huzhi 王胡之 became intimate with Wuji after they had grown. One day when the two were together, Wuji went in to tell his mother that Huzhi had arrived and to ask her to fix something for them. The mother replied, weeping, "Wang Dun had your father cruelly killed, and he had Wang Yi carry it out. The reason why I never told you is because the Wang house is powerful, and you children were still young. I didn't want it to get around in case there was trouble." Wuji cried out in horror. He drew a sword and rushed out, but Huzhi had taken off long before. (494–95)

5.16. When Xiang Xiong 向雄 [d. 283] was chief clerk for the grand warden of Henei 河内, a certain incident occurred. Although Xiong was not implicated, the Grand Warden Liu Zhun 劉准 was furious with him. He had him beaten and sent him away. Xiong afterward became a palace attendant, while Liu had become a personal attendant to the emperor. From the first they would not talk to each other. Emperor Wu 武 [236–90; r. 265–90] heard of this, and he ordered Xiang to renew the good relations that should exist between minister and subordinate. Xiang had no choice but to go and visit Liu. Bowing to him several times, he said, "I come because I was ordered to. But since the relations between minister and subordinate have already been severed, nothing can be done about it." He then left. Emperor Wu heard that they were still enemies. He angrily said to Xiang, "I told you to restore relations with him. Why are they still broken?" Xiang said, "The gentlemen of old approached others with propriety and rejected others with propriety. The gentlemen of today approach others as if they were about to set them down in their laps and reject others as though they were going to cast them into an abyss. Liu should consider himself lucky that I haven't taken up arms against him. How could I restore good relations?" The emperor finally desisted. (166)

In this last example, one participant in a vendetta actually recognizes the change of custom that has occurred since the time of the ancients; although he suggests the Golden Age was one of tolerance, he cannot but hold to the passions and sensitive honor of the "gentleman of today."

WOMEN IN THE HOMOSOCIAL COMMUNITY

Although the code of male beauty in *Worldly Tales* has its greatest impact on male-male relations, we have seen the role women can play as judges and fruit-throwers. In fact, women constantly emerge throughout the text as reinforcers of the cultural codes of literati society and of the homosocial community.[47] One might compare this active role to that of Western theories of homosociality, in which women become objects of contention over which men work out their own competitions, jealousies, and alliances. Such a role is not completely absent in the Chinese context:

23.37. Yuan Dan 袁耽 [ca. 315–39] had two younger sisters. One married Yin Hao 殷浩 [306–56], and the other one married Xie Shang 謝尚 [308–57]. He said to Huan Wen: "I'm sorry I don't have another sister to marry to you."

However, it is much more typical for women to make their opinions known. For *Worldly Tales* culture, the most distinguished woman is the lady who is aware of her role in cementing family relations and of her own worth as a social commodity. Thus, the text admires the cleverness of women who enforce judgment through (often harsh and satiric) comments on their husbands or relations.

19.9. Wang Guang 王廣 (ca. 210–51) married the daughter of Zhuge Dan 諸葛誕 (d. 258). When he visited her house and talked with her for the first time, he said, "You're rather plebeian, aren't you? Not at all like your father." She replied, "You can't even imitate your own father, but you want your wife to take after a great and distinguished man!" (367–68)

19.25. Lady Chi 郗, the wife of Wang Xizhi, said to her two younger brothers, Chi Yin 郗愔 and Chi Tan 郗曇, "When the Wang family sees Xie An and Xie Wan, they are so eager to greet them they overturn baskets and put their sandals on backwards in their haste. But when they see you coming, they're quite self-contained. You might as well not bother going there again." (377)

19.26. After Xie Daoyun 謝道蘊, the wife of Wang Ningzhi 王凝之 (d. 399), went to live with the Wangs, she felt the greatest of contempt for her husband. When she returned to visit her parents, she gave off a discontented air. Xie An consoled her.

"Master Wang's the son of Wang Xizhi, and he's not so bad on his own account. Why are you so resentful?" She replied, "When I lived in this household, I had Ada 阿大 and the Central General for uncles; and I had Feng 封, Hu 胡, E 遏, and Mo 末 for cousins.[48] [With kin like them,] I never thought about any 'Master Wang' living in the world!" (377)

Xie Daoyun, herself a poet of talent, here contributes to the literati discourse on cultural distinction based on something more than family status; her comment is spoken against the understood background that the Wangs were a more prestigious family of the time than the Xies.

Women thus become conscious facilitators of the homosocial community—creating yet another space in which men can meet. Women are "inside outsiders"—nonparticipants in male competitive strategies but familiar with the rules. They are thus often chosen as evaluators, since their position grants a certain distance:

25.8. Wang Hun 王渾 [223–97] and his wife Lady Zhong 鍾 were sitting together one day. They saw [their son] Wang Ji pass through the courtyard. Wang Hun, pleased, said to his wife, "To give birth to a son like this is certainly a cause for satisfaction!" His wife laughed. "If I had had the opportunity to marry [your brother] Wang Lun 王淪, my sons would have been more than this!" (424)

9.21. Song Yi 宋褘 was once the concubine of General Wang Dun. Later she belonged to Xie Shang. Xie Shang asked her, "How do I compare with Wang?" She replied, "Wang is to you as the dwellers in the fields are to noblemen." This was because Xie was attractive and handsome. (281)

Here, the judgment is somewhat ambiguous; is the anecdotist condemning the concubine for being influenced by "superficial" qualities? Probably not, since qualities of male attractiveness were highly esteemed, and the military overlord Wang Dun is generally not treated positively throughout the text.

In one telling anecdote, a wife is invited to spy on an intimate friendship so that she may recognize the superior judgment and character of a husband who would have such friends—a fascinating homosocial inversion of the Candaules-Gyges motif:

19.11. Once Shan Tao 山濤 (205–83) had met Xi Kang 嵇康 (223–62) and Ruan Ji, the three of them formed a friendship as firm as metal and as fragrant as orchids. Shan's wife, née Han 韓, felt that their alliance was quite different from that of ordinary men, and asked him about it. Shan replied, "These are the only two men I can consider my friends these days." The wife said, "Once upon a time Fuji's 負羈 wife got to observe his companions Hu Yan 狐偃 and Zhao Cui 趙衰 personally.

I'd also like to get a peek." The following day the two arrived. Madame Han urged her husband to detain them for the night; she then put out wine and dishes for them and dug a hole in the wall so she could watch. She was so intrigued she stayed there all night. Her husband went into her and said, "So what did you think of them, anyway?" She said, "You fall far short of them in talent. You're a friend of theirs only because you're such a good judge of character." Shan replied, "They, too, respect my ability in that." (369)

This anecdote is a good example of how modern interpreters are eager to engage in an anachronistic division between "sexual" and "non-sexual" realms of activity. Some assume that the men are engaged in some sort of "active" homosexual activity (though what that might be is not speculated on).[49] But the more significant point is that the text is simply not interested in the form of the male interaction. Rather, Shan allows his wife to take a position as a voyeuristic judge of his friends and to confirm their worth. This active role of female judgment will become much more important in the Tang era, as we shall see in later chapters. Significant here as well is Shan's honest agreement on his own inferiority: it is likely that the text introduces this final comment of his to illustrate his genial nature, his ability to make friends, and his refusal to let jealousy and envy control his actions. Although Shan is occasionally criticized in other places, he is often praised as a judge of talent as well.

The female consciousness of male competition and social niceties also allows the female community to "reflect" or "echo" the concerns of the male. One of the most telling anecdotes (which serves as the epigraph to this chapter) suggests a code of living for Chinese literati in general, where passivity can be an asset, a strategy for survival:

Lady Zhao married off her daughter. When the latter was about to depart, her mother instructed her: "Be careful not to do good." The girl replied, "If I don't do any good, than may I do evil?" Her mother said, "If it is not permitted to do good, how much the less is it permitted to do evil!"

"CULTIVATED TOLERANCE"

Granted the ever-present threat to elite society of competition, contempt, and violence, it is not surprising that self-control achieves a high value as well—a quality suggested by Lady Zhao's advice to her daughter. Whether it is the superiority of his hereditary background, his philosophical and conversational abilities, or the power of his *de* that a literatus is asserting, he is likely to do so in a skillful, light, epigrammatic way: manipulating language for the playful putdown, demonstrating his indifference to the desires of the vulgar, and maintaining a cool self-control in the face of adversity. The *Worldly Tales* editors termed this quality *yaliang* 雅量, translated by Mather as "cultivated tolerance." This is a fine rendering for something untranslatable into English—although one should emphasize that *yaliang* also implies the maintenance of one's own calm in the presence of threatened violence and that the display of *yaliang*, rather than exhibiting "tolerance" for the other, is often a display of one's own superiority over his vulgar emotionalism.[50] *Yaliang* arises out of the need to control the passions that can destabilize oneself and one's companions and escalate into more threatening situations. One cannot afford to care too much.

6.9. Once when Pei Xia 裴遐 [fl. 300] was staying at Zhou Fu's 周馥 [d. 311] house, Zhou held a party. It so happened that while Pei was in the middle of a game of Go, an adjutant of Zhou's served him wine. Since Pei was in the middle of the game, he had no time to drink. The man grew angry, and he dragged Pei over onto the ground. Pei resumed his seat, showing no change in manner or expression. He resumed play. Wang Yan asked him how he had managed to keep his temper. He answered, "He was just interested in getting a private feud going, that's all." (197)

Yaliang can manifest itself in the face of natural disaster; but it also allows one to keep calm amid political danger—and sometimes to escape it.

6.3. Xiahou Xuan 夏侯玄 [209–254] was once leaning on a pillar practicing calligraphy. A storm came up and a bolt of lightning split the pillar. His clothes were scorched and yet there was no change in expression as he resumed his writing. All his guests and retainers were deeply shaken and would not stay. (195)

6.7. When Pei Kai was arrested, there was no change in his expression, and his manner remained calm. He asked for a brush and paper and composed a number of letters. After the letters were finished and sent out, many came to his aid, and so he was rescued. Later he was given a ceremonial rank equal to the three Grand Dukes. (196)

6.29. Huan Wen held a banquet with armed men lying in ambush; he invited many ministers of the court, hoping to do away with Xie An and Wang Tanzhi. Wang was terrified and asked Xie, "What are we going to do?" Xie showed no change of expression. "The survival of Jin rule rests upon our conduct now." The two of them then went ahead. Wang's fear was increasingly evident in his face, while Xie's features became even more calm. Going up the stairs and proceeding to his seat, he began to chant a poem in the Luoyang manner: "How mighty the flooding waters . . ." Huan was intimidated by his calm and distant manner and hastily dismissed the soldiers. Formerly Xie and Wang had been of equal reputation. From this time, people could judge their relative merits. (206)

Xie An was in fact the embodiment of *yaliang* in literati society; the most famous manifestation of it occurred at the moment of his famous victory over Northern invaders at the battle of the Fei 肥 River.

6.35. Xie An was playing Go with someone when he suddenly received a letter from [his nephew] Xie Xuan, encamped with the army by the Huai 淮 River. After he had finished reading, he silently and calmly returned to the game. A guest asked how the army at the Huai had fared. He replied, "My boys have decisively defeated the bandits." His expression and manner did not alter at all from the ordinary. (209)

Beyond its abilities to modulate hostility and to aid the literatus in the face of personal disaster, *yaliang* also allows him to project indifference to "less important" matters and thus appear superior or transcendent. Sometimes it shows itself as an indifference to the politics of marriage and clan alliance; at other times it shows itself as an utter disregard of bureaucratic procedures and responsibilities.

6.19. When Chi Jian 郗鑒 [269–339] was in the capital, he sent a retainer to Wang Dao 王導 [276–339] with a letter, asking whether one of his nephews could be married to his daughter. After Wang had read the letter, he led the man to an eastern veranda and told him to select the lad he desired. When the messenger returned, he informed Chi, "All the boys in the Wang family were exceptional. When they heard that I had come to look for a son-in-law, all came out and held themselves with great dignity and self-control. Only one boy continued to lie with his belly showing, on a couch by the east side of the room, as if he had heard nothing." Chi said, "I want that one." He went to inquire further, and discovered that it was Wang Xizhi. He then married his daughter to him. (201–2)

24.11. Wang Huizhi 王徽之 [d. 388] was appointed a lieutenant in cavalry to Huan Chong's 桓沖 [328–384] army. Huan asked him, "So what office did they assign you to?" Wang replied, "I have no idea. At the time I saw somebody bringing in some horses, so I suppose it's a horse stable." "And how many horses are you in charge of?" "I didn't ask about the horses, so how should I know?" "How many horses have died then?" "We don't even know about life, so why should we know anything about death?" (414–15)

Wang is being witty as well as disdainful. The *Analects* says of Confucius that when he heard about a fire in the stables, he asked if anybody had been hurt, but did not inquire about the horses (*Analects* 10.17). Another time, someone asked him why he never spoke of the afterlife. He responded, "We don't even know about life—how can we know anything of death?" (*Analects* 11.12)

23.31. Yin Xian 殷羨 [fl. 300] was assigned the prefecture of Yuzhang 豫章. When he was about to leave for his post, various people in the capital entrusted him with over a hundred letters to deliver. When he reached Shitou 石頭 [immediately outside the capital], he dumped them all in the water, chanting:

"If they sink, let them sink.
If they float, let them float.
I, Yin Xian, am nobody's mailman!" (400)

It is this quality that has most struck later students of Chinese literature and culture; when one turns to *Worldly Tales*, one immediately thinks of the Seven Worthies of the Bamboo Grove and of Xie An, of their indifference in the face of danger and their unrestrained love of wine and unconventional behavior. Yet this is only a small part of the picture; insufficiently contextualized, this *yaliang* seems merely a philosophical rejection of "Confucian" or societal repression or a simple desire to retreat from political instability. Only when one places it against incidents of excess, pride, humiliation, and violence does its further significance become apparent.[51] As part of literati ideology as reflected in *Worldly Tales*, *yaliang* is a fantasy of empowerment—a belief that one's own personality and will can control one's environment.

WEALTH AND WASTE

When competition and friendship, enmity and amity, go hand in hand, then gifts become powerful foci of social significance.

1.32. When Ruan Yu was living in Shan 剡, he was the possessor of a fine carriage, which he would lend to anyone who asked for it. There was a man who lost his mother. He wished to borrow it [for the funeral] but didn't dare ask. Ruan heard of this afterward. Sighing, he said, "If I have a carriage and yet have others not dare to borrow it, what use is it?" And he burned it. (20)

Ruan may seem simply conventionally "altruistic" here until one realizes the waste of his final gesture: the true altruist would keep the carriage on hand until another opportunity for its use presented itself. Ruan, deprived of the initial gesture, must *immediately* compensate with this display of destruction. It is more important for Ruan to seem indifferent to his own possessions than for him to do good for others.

Ruan's action is minor and may still be seen as a manifestation of his inner virtue. But it leads the way to a form of gift giving that becomes competitive, obsessive, and destructive, as Marcel Mauss demonstrated with the potlatch. At its worst, it becomes no longer the graceful gesture of the well educated and culturally sophisticated but a brutal demonstration of the desire to compete, to dominate, and to control.[52] Surely the continuing fascination of *Worldly Tales* with the poles of amity and enmity, with companionship and competition, accounts for the attention it pays to the stories surrounding that monster of excess, Shi Chong 石崇 (249–300). Drawing on incalculable wealth, Shi's conspicuous consumption and display are meant to exhibit the excellence of his taste and his supposed indifference to possessions, as well as his inability to put limits to his desires and passions. In a well-known anecdote of his rivalry with Wang Kai 王愷, waste is allied with a gesture of aggression: the rival's goods are destroyed so that they may be replaced with even more wealth.

30.8. When Shi Chong and Wang Kai vied in display, they exhausted every elegance and beauty to ornament their carriages and clothes. Emperor Wu was Wang's nephew and would always help him out. Once he gave Wang a coral tree two feet high with spreading branches—truly a world treasure. Wang showed it to Shi. After Shi had looked it over, he took up an iron staff and smashed it to pieces. Wang was heartbroken and thought that Shi was jealous of his treasure. His expression grew fierce. Shi said, "This isn't worth getting in a huff over. I'll repay you now." Then he ordered all his retainers to bring in coral trees. Six or seven of them were three or four feet in height, with unsurpassed trunks and branches that shone with a brilliance that blinded the eye; there were many every bit as good as Wang's. Wang was stunned and at a total loss. (471–72)

Wang Kai's need to compete (abetted here, one may notice, by the emperor himself) leads later to espionage and further aggression—with the result that violence is turned from inanimate objects to animate possessions, the bondservants.

30.5. When Shi Chong had bean gruel made for a guest, it would be ready as soon as he called for it; and throughout the winter he had a supply of leek and duckweed pickles. Moreover, although the appearance and vigor of his carriage ox were inferior to Wang's, it ended by winning in races: Whenever the two of them would go out, they would set out late and then race each other back to the Luoyang city gates before they would shut for the night. Shi would start out several dozen paces behind, but he would always zoom ahead like a bird, and Wang's ox could never catch up, no matter how hard it tried. These three things made Wang crazy with frustration. Finally he secretly bribed Shi's steward and his charioteer. . . . Wang acted in accordance with their information and started to come out on top in their competitions. When Shi Chong heard of the reason, he had the informers put to death. (470)

However, flamboyant display is not contained merely within the dynamic of exchange: Shi Chong exploits the cultural expectations of his peers to create scenarios of humiliation. A gesture of the host's courtesy becomes instead a serious contest in which their sexual modesty and sense of honor are threatened.

30.2. Shi Chong would always station ten or so slave girls in the privy, all in lovely clothes and makeup. They would have a complete supply of onycha paste, aloeswood lotion, and the like. They would also change the visitor's clothes for him before letting him leave. As a result, all of Shi's guests were too bashful to go to the privy. When Wang Dun would go, however, he would take off his old clothes and put on the new, his expression remaining disdainful all the while. The slave girls all said to one another, "This guest would certainly make a good rebel!" (468)

At first we see Shi Chong's daring employment of the bondservants as a seeming joke on propriety: an act of symbolic violence, like the destruction of Wang's coral tree. Yet the anecdote's focus is different here, and attention to it shows us the sources of Shi's eventual destruction. "This guest would certainly make a good rebel!" say the bondservants; the early Chinese reader draws satisfaction from hearing this historical prophecy from the mouths of ignorant slaves, who were originally only tools of a practical joke. Their comment tells us that Wang's indifference is not *yaliang*, but something more frightening—an indifference to the social rules, a hardened sensibility that can endure what others cannot. Was Shi aware of the implications of Wang's seemingly boorish behavior? Could he see that Wang represented a social world outside his own, in which the intimidation of wealth could be brushed aside with the self-assurance of the violent man? Or were the slave girls and Wang the only ones outside the cultural field within which such gestures as Shi's actually signified?

In another account of Shi's profligacy, the most unpleasant and violent of them all, the slave girls are no longer permitted an objective evaluation of the guests' behavior.

30.1. Whenever Shi Chong invited guests to a banquet, he would always have beautiful women serve the wine. If anyone would not drain his cup, Shi would have a servant take the girl out and behead her. One day Wang Dao and Wang Dun both went to visit Shi. Dao was never much of a drinker, but he forced himself on this occasion until he was completely intoxicated. But when the cup was brought to Dun, he would deliberately not drink in order to see what would happen. Even after three women had been beheaded, his expression remained unchanged; he still refused to drink. When Dao reprimanded him, he said, "If he kills people from his own household, what's that to you?" (467–68)

Shi's horrific contest is matched by another contest on Wang Dun's part. We are not told that he was incapable of drinking the toast or that he did not value the slaves' lives enough to attempt to drink the toast; rather, he did not drink at all "in order to see what would happen." A test of wills is involved here. As Wang Dun calls Shi's bluff, we not only see the consequences of literati excess but also become aware of how such behavior can bring waste and destruction to the literatus himself. Shi's bondservants are members of his own family. He hurts only himself through their execution. Wang's position as outsider to literati culture enables him to see that he, the callous guest, can exploit the weaknesses of others through refusing to be moved.

As we become uncomfortably aware how bizarre and fantastic Shi's behavior looks from outside the literati codes of honor, we also acknowledge how fragile his power really is—that a world exists that cares nothing for his exaggerated version of literati tactics of intimidation and symbolic acts of destruction. *Worldly Tales* relates his own downfall in conjunction with that of his contemporary's, the talented poet Pan Yue.

36.1. Sun Xiu 孫秀 [d. 301] already resented Shi Chong for not giving him Green Pearl 綠珠, and he also hated Pan Yue for treating him uncivilly.

The story of Green Pearl is not related in the collection itself. However, the commentary of Liu Jun fills in the details and provides the most famous legend associated with Shi.

Shi Chong had a singing girl named Green Pearl who was beautiful and had a talent for playing the flute. Sun Xiu sent a messenger to ask for her. Shi was at his country estate at the foot of Beimang 北邙 Mountain. Just as he had climbed up to Coolness Lodge and was sitting by the clear river, the messenger arrived with his message. Shi called out several dozen of his slaves and concubines to show him. "Take whomever you want." The messenger said, "The order I received specified a certain Green Pearl—but I don't know which one she is?" Shi burst out, "Green Pearl is the one I love. She cannot be given!" The messenger replied, "You, Sir, are well versed in events past and present, and know the world well, both near and far. I suggest you reflect." Shi still would not agree. The messenger even came back to ask again, but Shi still refused. (493)

There is irony here. For the first time Shi is unwilling to give something away, even though he could ensure his own friendship with the powerful by doing so. As has been pointed out, when he says "Green Pearl is the one I love," he is expressing not romantic or sexual passion but the covetousness of the possessor: the verb *ai* 愛 here also means "to begrudge."

The *Worldly Tales* account continues:

Afterwards, when they arrested Shi Chong and his nephew Ouyang Jian 歐陽建, they arrested Pan Yue as well.

(Liu Jun's commentary explains that Shi had participated in a plot to overthrow the warlord Sima Lun 司馬倫, the patron of Sun Xiu. When the plot failed, both he and Pan were arrested.)

Shi was escorted to the market first [for execution]. Neither he nor Pan knew that each other would be there. Later when Pan arrived, Shi said, "Ah! You, too, are here." Pan replied, "One might say, 'We will die together when our hair is white.'" The poem Pan contributed to Shi Chong's *Verse Collection of Golden Valley* had the line, "I cast my fate with my friend Shi, / We will die together when our hair is white." Now this prediction had been fulfilled. (493–94)

Oddly enough, Shi is reunited in death with the literati culture he embraced and intimidated. A line of a poem, composed in amity, becomes a prophecy that allows two friends to perish together: tragic, but not without dignity. However, Liu Jun gives another execution ground story—a grim contrast to the touching reunion of poets:

When officers came to arrest Shi at his estate in Hebei 河北, he said to himself, "I'll probably only be exiled to Jiao 交 [Vietnam] or Guang 廣 [Canton]." Only when they brought him by cart to the eastern market did he realize the truth. He sighed. "Those rascals only want to benefit from my wealth!" The man who had arrested Shi said, "If you knew your wealth would bring you harm, why didn't you give it away long ago?" Shi could not answer. (493–94)

That is not quite the end of the Shi Chong legend, however. The *Worldly Tales* compilers and Liu Jun say nothing of the fate of Green Pearl, but the Jin dynastic history (comp. seventh century) adds the following semilegendary account:

Shi Chong was holding a banquet within his mansion. When the arresting officers arrived at his gate, he turned to Green Pearl and said, "I have committed this offense because of you." Green Pearl replied, weeping, "I will now die in your presence." She then threw herself from the top of the mansion and perished.[53]

Stephen Owen has suggested that Green Pearl's action enables her to obtain full status as a human:

By committing suicide, she renounces the safety of her status as possession (to be confiscated by Sun Hsiu [Xiu] after the arrest, along with Shih Ch'ung's other household effects) and seizes for herself, if only for a brief moment as she falls from the tower, the status of human being. By her own choice she enters the deadly economy of feudal relations, based on the acceptance and repayment of obligations. Green Pearl declares herself to be the good retainer, and like many good but poor retainers in early Chinese stories, she pays her debt with the currency of her life. A human being suddenly appears behind the surface of lovely object; the body that falls from the tower is a person, not a thing.[54]

This is undoubtedly true; there are too many other accounts of retainers' loyal suicides (as Owen hints) for us to be able to take it any other way. Yet read in the light of Shi's representation in *Worldly Tales*, we can see her death as yet another example of the trail of ruined property that the literatus leaves behind him in the name of his prestige: broken coral trees, executed stewards, beheaded serving girls. As we have seen, the one thing equal or even superior to the giving of a gift is the willful destruction of property that *could have been given* as a gift; even more so if the destruction prevents the giving of a gift under compulsion. "If you knew your wealth would bring you harm, why didn't you give it away long ago?" In Green Pearl's case, the answer becomes clear. Sun Xiu did not want largess from the hand of the wealthy; he wanted one woman and felt he had the power to force Shi to give in. Green Pearl's final act of suicide grants Shi his dignity and honor: he would rather toss her away, break her on the ground than see her confiscated by another. It is a perverse form of victory; but like Shi's bittersweet reunion with Pan Yue on the execution ground, it represents the odd triumph of literati codes of conduct in the face of a more brutal reality—and in the face of extinction itself.

With the *Worldly Tales* portrayal of Shi Chong, we arrive at a powerful demonstration of literati codes of conduct in both their power and their impotence. Shi's own represented motives are clear, even if they partake of the neurotic and the obsessive: Chinese elite society as it developed in the third and fourth centuries created the forms of cultural expression and competition that allowed for his peculiar manipulation of wealth and status. Although it may seem trivial that two well-educated literati would scheme over the best duckweed pickle recipe, the rules by which they played helped dictate their prestige within literati circles. Yet it is equally clear that such codes of conduct were often worse than useless in relation to other forms of social and economic power. Although literati culture attempted to impose its values on outsiders (and sometimes succeeded), it was just as likely to be a self-contained and self-referential refuge from a more brutal system of values.

Return to Poetry

I have noted that poetry retreated somewhat as a serious form of social exchange at the very time that Chinese elite society was formulating its codes of conduct. We have seen that other forms of cultural expression during this time acted as the glue in literati relations—most prominently, philosophical speculation and "pure conversation." In fact, poetry was used primarily in the fourth century as a medium to express philosophical insights. Yet at the very time that *Worldly Tales* was being edited, poetry was undergoing a tremendous revival, and pure conversation was disappearing from the scene.[55] *Worldly Tales*, as it turns out, is both normative and nostalgic—codifying the forms of elite behavior from a slight distance, at the exact moment when that behavior is entering a new phase.

We cannot hope to exhaust the reasons for this change, nor can we arrive at a definitive explanation of the role that poetry played in it. However, it is clear that one poet embodied this change and led poetry into its greatest revival since the Jian'an era: Xie Lingyun 謝靈運 (385–433). A brief examination of his participation in the social circles of his age and the role his poetry played in them may give us some insight into the next phase of literati literature—and to the development of "court poetry," the genre that paved the way for the full flourishing of Tang dynasty verse.[56]

Xie was born in 385, into a family whose fortunes were declining. That same year saw the death of his talented great-great uncle, Xie An; control of the feeble Jin regime had already passed into the hands of Sima Daozi 司馬道子 (364–402). However, the Xies were still prominent cultural and political arbiters in the capital; by the time the poet had reached his teens, he found himself part of a cultivated circle of kinsmen led by his uncle Xie Hun 謝混 (d. 412). Even in his early years, he was subject to the aesthetic competitions and judgments of fourth-century literati society; one of Xie Hun's poems renders *Worldly Tales*–style judgments on his relatives. He comments on Lingyun's talent as a writer and then adds that additional polish to his style will perfect his work.[57]

As Lingyun grew older, he found his cultural endeavors blending imperceptibly with his political ambitions. From 406 he became advisor to Liu Yi 劉毅 (d. 412), the ambitious younger brother of Liu Yu, a military strongman who had gradually taken over control of the Jin after saving it from the rebellions of Huan Xuan and Sun En 孫恩 (d. 402). Since Xie Hun was a supporter of Yi, it is likely that he placed his talented nephew in Yi's entourage in order to solidify ties of alliance. When Liu Yi attempted to usurp his brother's position, Yu mounted a campaign against him in 412 that resulted in Yi's death; however, Yu recognized Xie's talents and transferred him to his own faction. There he remained, holding a series of literary posts until Liu Yu deposed the last Jin ruler in 420 and assumed the throne; he is known to history as Emperor Wu 武 of the Liu Song 劉宋 dynasty (420–79).[58]

Although Liu Yu was one of the stronger and more capable political figures of the time, he could not prevent the continuing formation of cliques that fought for control of the imperial succession. Ambitious to find a place among the kingmakers, Xie tied himself to a sort of literary-political faction in support of Yu's second son, Yizhen 義眞 (407–24), the Prince of Luling 廬陵; it included the writer Yan Yanzhi 顏延之 (384–456) and the Buddhist monk-philosopher Huilin 慧琳 (fl. 420s). Most prominent in the opposite camp were Xu Xianzhi 徐羨之 (364–426), and Lingyun's own cousin, Xie Hui 謝晦 (390–426), who planned to make Liu Yu's oldest son ruler. When Liu Yu died in 422, Hui's faction seized control and placed its own candidate on the throne (Emperor Shao 少); Lingyun and his allies were exiled.

It was at this point that Lingyun began the literary career that won him his elevated place in the tradition. His most famous early poems describe his journey into exile (he was sent as an administrator to Yongjia 永嘉 in Zhejiang, about three hundred miles south of the capital). He gives details of the landscape he passes through and his increasing depression over leaving friends behind. After stopping briefly at his estate at Shining 始寧 in Guiji (where he continued to write), he passed on to Yongjia. However, depression and illness compelled him to resign his post in 423; he chose to enter into an enforced if leisurely retirement at Shining.

At this juncture one may stand back and consider the factors in the verse production in this period that resulted in Lingyun's popularity. The standard answer given by most critics is simple: landscape description. This is not unreasonable, since commentators and literati living shortly after Xie's own life comment particularly on his ability to capture natural phenomena.[59] Modern scholarship builds on this and attempts to trace the roots of the interest in nature among literati, the origins of describing natural objects and scenes, and the impact of Buddhist and Taoist thought on such descriptions. I do not intend to add to the various arguments on this complex theme, which may be called the dominating issue of Xie Lingyun criticism.[60] Rather, I would suggest that skilled landscape description alone, even as a manifestation of intellectual and cultural history, cannot wholly account for his poetry's popularity. Just as significant is the way the texts position themselves as the products of a member of literati society, who perceives the landscape not merely as an object of beauty or as a representation of philosophical truth, but also as an emblem of "outsideness," of potential exclusion from literati social relations.

Compared to earlier poetry of nature, Xie's poems introduce distinctive rhetorical and stylistic gestures that point to his outsideness. First, his landscapes are rarely static: the poet moves through them, and his representation of them is altered by changes of perspective and by the sheer physical effort it takes to pass through them. Second, this restlessness accompanies a "connoisseur's attitude" to his surroundings: the poet is hunting for interesting experience, and this in turn is simulated by the new, elaborately complex parallel vocabulary that delineates the subtleties of natural phenomena. Third, there are almost always moments when Xie breaks away from description and considers possible "lessons" his traveling may hold or meditates on the emotions that it produces. These vary, but tend to follow certain set patterns: (1) references to the consolations of a "hermit" existence, to Taoist philosophical views of life and death, and to alchemical practices of prolonging life; (2) regret that his closest friends are not with him to enjoy the scene; and (3) an undefined melancholy evoked emblematically through the language of the *Nine Songs* and *Encountering Sorrow*: the poet projects himself as the shaman looking in vain for the god/goddess to keep his/her tryst. Two typical examples may be cited—translated here somewhat clumsily in order to preserve the gnarled, complex nature of Xie's language.

Ascending the Lone Islet in the River 登江中孤嶼

South of the river I have grown tired of passing through sightseeing;	江南倦歷覽
North of the river I have broadly proceeded in circles.	江北曠周旋
I cherish the new, but my road gets longer and longer;	懷新道轉迴
I seek the distinctive, but my time is not protracted.	尋異景不延
A chaotic current rushes to the upright sandbar;	亂流趨正絕
The lone islet beautifies the central stream.	孤嶼迷中川
Clouds and sun shine against each other;	雲日相輝映
Air and water are clear and fresh together.	空水共澄鮮
Manifestation of spirits—no one to appreciate them;	表靈物莫賞
Hiding the true—who can transmit knowledge?	蘊眞誰爲傳
I imagine I'm making out the form of Kunlun;[61]	想像崑山姿
Distant and far from the ties of the central realm.	緬邈區中緣
For the first time I trust the art of Anqi;[62]	始信安期術
I thoroughly succeed in nourishing the years of my life.[63]	得盡養生年

Although these lines incorporate Xie's trademark landscape parallelism, the poem is "about" something else: the need to find "new sights" that do not bore him; the discovery of mysterious spirits who not only hide from human eyes but also possess a secret truth he alone is privy to; the imaginary projection of himself into mythical landscapes, where the little islet becomes Kunlun and he is far from the social strains and stresses of the capital; and the full embrace of alchemical Taoism. To take this message "seriously," however, is to suggest that these discoveries and commitments are indeed acts of conversion, that he is writing to/for himself and that the act of versifying these events will henceforth become unnecessary and superfluous. But Xie is a social poet; he keeps in correspondence with his friends in the capital, and it is highly likely that he sent his poems there to be circulated and copied, as his dynastic biography suggests: "Whenever a poem of his reached the capital, both noble and low-born competed in copying it out."[64] Xie's poems are postcards or travelogues for public consumption; they invite readers to see him as model recluse and yet as an exiled member of their own society. As a consequence, his meditations oscillate ostentatiously from declarations of self-sufficiency to the consolations of nature in exile. "For the first time I trust the art of Anqi," but in poem after poem he repeats this "new" breakthrough, this nature-inspired resolve. The discovery of the spirits here has an ambiguous tone: is he boasting of his discovery of them—something that makes him exceptional—or is he lamenting that their truths cannot be shared with his friends and his social clique? Either way, the poems constantly circle about the issue of being alone and outside. The poet either gazes toward the capital or ostentatiously turns away from it, but the capital is always the conspicuous absence in the landscape that he frenetically and laboriously traverses.

Spending the Night at Stone Gate Cliff 石門巖上宿

In the morning I pick an orchid in the garden;	朝搴苑中蘭
I fear its withering beneath the frost.	畏彼霜下歇
At dusk I return to spend the night at cloud's edge;	瞑還雲際宿
Taking pleasure in the moonlight on the stones.	弄此石上月
Birds cry—so I know they are nesting at night;	鳥鳴識夜棲
Leaves fall—I know the wind is rising.	木落知風發
These different sounds are all the best that can be heard;	異音同至聽
Distinctive echoes, all clear and outstanding.	殊響俱清越
Marvelous things, but no one can appreciate them;	妙物莫爲賞
Fragrant wine—but to whom can I show it off?	芳醑誰與伐
The lovely one never comes;	美人竟不來
By a sunlit vale I dry my hair in vain.[65]	陽阿徒晞髮

Here Xie more explicitly ties his travels to the rhetoric of the *Nine Songs* and so invites the cultivated reader to read his melancholy loneliness as an emblem for political exile. For his fans back in the capital, Xie's delicately contrived landscape descriptions must have brought a particularly vivid power to the old tropes of exile, exclusion, and lament: although allusions and rhetoric dictated how one articulated these themes, they are combined here with detailed description of an individual poet's movements and of what he sees within a defined passage of time. The title sets the specific place. He tells us "in the morning I did this; in the evening, that." He draws conclusions from natural phenomena. He is Qu Yuan in exile, but he is also Xie Lingyun, the poet one knows personally. When he sends poems like this to his friends in the capital, he reintroduces a powerful role for lyric poetry, one that the Jian'an authors had gestured toward but never fully developed: the verse-diary. This verse only makes full sense within the confines of social exchange—as an outgrowth of the relations between men active in the public world and in politics, who could empathize with the complexities of Xie's position "out there."

There are other lessons to be learned from Xie's last years. In 424, Xu Xianzhi's faction succeeded in disposing of both Emperor Shao and the

Prince of Luling, Lingyun's former patron. However, their new imperial choice, another son of Liu Yu's who succeeded to the throne in September (Emperor Wen 文), did not prove as easy to manipulate as the kingmakers had hoped. Soon he created a rival faction of statesmen, and by 426 he managed to overthrow and execute the most prominent men in Xu's clique. As a gesture of reconciliation, he then pardoned Lingyun's own group; Lingyun, Yan Yanzhi, and Huilin were invited to return to the capital. At this point, Xie's life led him into another role that prefigures later cultural developments: he became Emperor Wen's pet poet—famous for his skill, but not for his political importance:

> Lingyun's poetry and calligraphy were both exceptional. Whenever he composed a piece, he would write it out in his own hand. Emperor Wen called them "twin jewels." Xie himself thought that because of his fame and background he should participate in the administration of the time. But he had arrived at the point where he was accepted only for his literary abilities. Whenever he served at an imperial banquet, the ruler would merely chat with him or discuss artistic matters.[66]

That Xie should become less influential politically is in line with a perceived decline of the greater families during the fifth century—although the reasons for this decline have yet to be fully traced.[67] Of course, as a member of the once-great Xie family, Lingyun was not pleased with his new role. His sensitive pride and his failure to recognize the decline of his own position made his remaining years depressingly predictable. In 428 he requested a leave of absence due to illness and returned to his Shining estate. He took with him a circle of aristocratic friends (including his young cousin Xie Huilian 謝惠連 [379–433]). Although he earned a reputation for riotous debauchery during this period, his true downfall came from another direction. He engaged in vast land reclamation projects that came to alienate local officials, who considered him to be poaching on public land. In particular, he earned the enmity of the local magistrate, who seems already to have quarreled with the poet over Buddhist doctrine. Here a revival of Jin dynasty vendettas coincided with the emperor's need to keep an eye on the cliques, and the magistrate succeeded in convincing Emperor Wen that Xie's actions interfered with the public welfare. As a result, some violence seems to have broken out between Xie's servants and the local inhabitants, and the magistrate memorialized the throne, accusing Xie of attempting to rebel.

It is striking here that Emperor Wen decided to take the high road and limit Xie's punishment to exile in a provincial post rather than execution.

Perhaps we see the "benign" gesture of a ruler who knows he can fear little from a decayed family. However, Xie's wounded pride did not allow him to accept these conditions. He ignored his bureaucratic duties, and when another personal enemy attempted to have him arrested, he launched a futile revolt that was quickly put down. The emperor, evidently still reluctant to order the death of a former companion, commanded exile to the distant south. There he seems to have attempted yet another revolt; he was finally captured and executed, in 433. In spite of the high regard the court held for him and his literary fame, his pride led him into a series of foolish and rash decisions.

If Xie is the last of a certain kind of literatus, however, contemporaries embraced him immediately as a striking literary innovator, setting a new standard for poetic composition. At the same time that *Worldly Tales* was being compiled, literati began to write works specifically on poetry—thus elevating it to the first literary genre to earn its own autonomous criticism.[68] It is also impossible not to detect Xie's influence in the landscape poetry of the fifth century—beyond the more obvious point that landscape poetry would not even have been written were it not for his example. Perhaps this would not have come to pass were it not for Xie's reworking of the poet's relation to his social companions and to his political position—that powerful exploration of exile and reclusion, of the attempt to reject social status while reaffirming its power at the same time. For later poets, the landscape would be seen in the same way—a peripheral orbit along which they move, ever facing inward to literati society, especially in the capital. They will gradually avoid the more clumsy, pretentious, and gnarled aspects of Xie's style and integrate more fully the evocation of landscape with the emotional and intellectual reactions the landscape produces, but there will always be the elite society, as well as the elite community of readers, behind that landscape.

CHAPTER FOUR

Spectator Sports

"Lovely that red writing brush"—let there be none who malign it!
—Xu Ling

Ban Gu's *History of the Han* (*Han shu* 漢書) tells us of a certain Lady Ban 班婕妤, a virtuous concubine of Emperor Cheng 成 (r. 32–7 B.C.). Unlike many women in her position, she relied on the ruler's favor not to gain power for herself but to serve him and lead him along the path of virtue:

Once Emperor Cheng was disporting himself in the harem and wished Lady Ban to ride in the same sedan chair with him. She refused: "I have looked over pictures of bygone days and have noticed that the sagely rulers of old always had renowned ministers by their side. On the other hand, the last rulers of the Three Dynasties had their favorite by theirs. If I wanted to ride in the same sedan chair with you, then wouldn't I be like them?" The emperor was impressed by her words and withdrew his invitation. When the empress dowager heard this, she said with delight: "In the past there was Lady Fan 樊, but now we have Lady Ban!"[1]

Lady Ban was not merely a paragon of virtue; she was a lady of literary accomplishments as well, possessed of a considerable erudition and knowledge of the rites. "She could chant the Odes, as well as such works as *Female Modesty*, *Models for Virtue*, and *A Guide for Women*. Whenever she saw anyone or sent them correspondence, she used the ancient ritual forms as her model."[2] However, this erudition was of little consequence when the emperor became enamored of the unscrupulous Zhao 趙 sisters.

In the third year of Hongjia 鴻嘉 [18 B.C.], Zhao Feiyan 趙飛燕 slandered Empress Xu 許 and Lady Ban, claiming that they were using witchcraft to put a curse on the women of the harem and that their imprecations extended to the ruler himself. Empress Xu was immediately deprived of her position. When Lady Ban was interrogated about the accusations, she replied, "I have heard that life and death depend on fate, and wealth and noble station lie with heaven. Even one who cultivates correct behavior may still not be favored—so what would I have to hope from wickedness? If the spirits and gods have prescience, they would not heed the complaints of a disloyal subject; and if they do not have prescience, then what advantage would there be in complaining to them? Consequently I could not have done these things!" The emperor was impressed with her reply and had compassion for her. He rewarded her with a hundred catties of gold. But the Zhao sisters continued in their arrogance and jealousy. Lady Ban feared that her position had been precarious for too long, so she requested to go and serve the Empress Dowager in the Changxin Palace 長信宮. The emperor permitted this.[3]

Although the cycles of fate cannot be avoided, Lady Ban uses her considerable talents and wisdom to avoid the worst of disasters. Her defense most likely wins admiration from the emperor not merely because of its impeccable ethics but also because of its sophisticated use of logic. Later, when Ban realizes that her days are numbered, she withdraws voluntarily. Yet how does the virtuous court lady deal with her failure? The biography concludes with a short rhapsody composed by Ban in her self-imposed exile entitled "Lament for Myself" ("Zidao fu" 自悼賦). The verses begin with an outburst that in many ways resembles the southern, Chu-style lament of the slandered minister (particularly the *Li sao*), although here the speaker is unambiguously female:

I spread out before me paintings of women, took them for my mirror; 陳女圖以鏡監兮
I heeded my governess, asked her about the *Odes*; 顧女史而問詩
Saddened by the warning of the woman of dawn, 悲晨婦之作戒兮
Lamented the vices of Baosi.[4] 哀褒閻之爲郵
I praised Sage Shun's spouses, Huang and Ying, 美皇英之女虞兮
And gloried in Ren and Si, matriarchs of the house of Zhou. 榮任姒之母周
Although I am foolish and base and cannot come up to these, 雖愚陋其靡及兮
How could I dare abandon my thoughts and forget them? 敢舍心而忘兹

. . . .

Suddenly the sun has set aside its light, 白日忽已移光兮
And I am lost in the darkness of dusk. 遂晻莫而昧幽
But still I am borne up by His generous virtue; 猶被覆載之厚德兮
I have not been abandoned for my wickedness. 不廢捐於罪郵

I do my service in this eastern palace,	奉共養於東宮兮
Entrust myself to the lowest ranks at Changxin.	託長信之末流
Together we sweep the curtains and drapes,	共洒埽於帷幄兮
And only death will bring an end to it.	永終死以爲期
May my bones find their place at the hill's foot,	願歸骨於山足兮
Resting in some place by pine and cypress.[5]	依松柏之餘休

Lady Ban's toilette does not consist in self-absorbed attention to surface as she sits before a mirror; rather, hers is an "ethical" toilette. She sits before the paintings of noble women of the past and sculpts her conduct in their likeness. With this moral training, she is prepared for the loss of favor and her relegation to simple housewifely tasks in her palace of exile, where duty is more important than appearance. This is a high-toned and potentially arrogant claim of virtue, after the Qu Yuan model, but it is followed by a more lyrical envoi that turns the poem in new directions. To start, we have one of the first uses of the "desolate and lonely chamber" topos that would become so common in later poetry:

Hidden in this dark hall, secluded and cold,	潛玄宮兮幽以清
With main gates barred and inner doors shut tight.	應門閉兮禁闥扃
The bright halls are now dusty, the jade stairs covered in moss.	華殿塵兮玉階苔
The courtyard overgrown, where green weeds flourish.	中庭萋兮綠草生
The large rooms are filled with shadows, and the curtains are dark.	廣室陰兮帷幄暗
The windows are empty, and the wind is bitter.	房櫳虛兮風泠泠
It stirs my curtains and skirts, fanning the red gauze.	感帷裳兮發紅羅
They rustle and whisper, their white silk murmurs.	紛綷縩兮紈素聲
My spirit is lost afar in some secret place;	神眇眇兮密靚處
My lord does not come—for whom do I make myself lovely?	君不御兮誰爲榮
I look down at the red tiles,	俯視兮丹墀
Recalling the echo of his shoes.	思君兮履綦
I look up and see the cloudy eaves,	仰視兮雲屋
And two streams of tears come down.	雙涕兮橫流

However, Lady Ban now turns against her self-pity and takes a more philosophical—even more cheerful—view of events:

Then I look left and right and grow less sad;	顧左右兮和顏
I pour from a winged goblet to melt away cares.	酌羽觴兮銷憂
Oh, human life in this world	惟人生兮一世
Floats by so quickly!	忽一過兮若浮

I have already enjoyed my share of eminence,	已獨享兮高明
Have had the best of a life of leisure.	處生民兮極休
I will force myself to delight my spirit to the utmost,	勉虞精兮極樂
Since good fortune and high rank are not dependable.	與福祿兮無期
"Green robe" and "White Blossom"	綠衣兮白華
Have been with us since ancient times.[6]	自古兮有之

Lady Ban's injunction to herself to "delight my spirit to the utmost" is not as hedonistic as it sounds; it is more likely a court lady's adapting for her own situation the official's high-minded withdrawal from public service and his return to aesthetic and moral self-cultivation. She thus turns away at the last minute from the emotional and political dangers of the lament and instead finds consolation through reason (a rhetorical move already found in Jia Yi's 賈宜 [200–168 B.C.] rhapsody "The Owl").[7] This final reliance on prudence and moderation must have impressed Ban Gu, who, like many other Eastern Han intellectuals, was skeptical of the emotional and self-destructive excess of Qu Yuan and his admirers.[8] There may be some family pride here, too; Lady Ban was the historian's great-aunt.

However, all these anecdotes are overshadowed in the tradition by one short poem that is not even mentioned in the biography: Lady Ban's "Song of Resentment" (*Yuan gexing* 怨歌行):

Newly cut, the gleaming silk of Qi,	新裂齊紈素
Fresh and pure as frost or snow.	鮮潔如霜雪
Trim it to form a fan with acacias pictured,	裁爲合歡扇
As round and full as the bright moon.	團團似明月
It frequents the folds of My Lord's robes,	出入君懷袖
Trembles to bring forth a light breeze.	動搖微風發
But it always fears that autumn comes,	常恐秋節至
When cool gusts disperse the blazing heat.	涼飆奪炎風
Then it will be cast aside in your box—	棄捐篋笥中
Your favor cut off in its prime.[9]	恩情中道絕

Here, the philosophical and rhetorical elaborations required in the rhapsody are reduced to one simple lament and one single image. Yet how complex this image becomes in the course of ten lines! The linking of the speaker with moon, fan, and purity; the sexual suggestiveness of the acacias, whose name in Chinese (*hehuan* 合歡) means "joyous union"; the hint at sexual intimacy as the speaker penetrates the lover's robes and trembles in his hand; the fear of an inevitable "natural" process that will bring dismissal (Does autumn signify aging? The lover's cooling passions? A much more powerful and significant rival?). This is no longer a poem of someone who is in control and capable of asserting moral worth and authority; it is lover as utile object, lover as dependent, lover as pure emotional response. There are no compensating gestures here, no intellectual and rhetorical consolations.

Most modern scholars are skeptical that Lady Ban authored this poem. Since there are no traditions before the fifth century that tie the text to her, and since many of the images turn up as part of the shared conventions of anonymous poems on sexual rejection, they are probably right.[10] Yet in the fifth and sixth centuries, when most literati looked back at the development of poetry and attempted to trace its origins, they believed in her authorship and had a high opinion of her. Zhong Rong 鍾嶸 (ca. 465–518), in his *Shi pin* 詩品 (Evaluations of the poets), ranked her among the eleven greatest poets of the tradition, for this poem alone.[11] For literati, she deserved respect because they realized the importance of her situation. For them, the political and social context that led to her poem (slander and loss of favor) was a significant theme in poetic composition in general. The fan poem appealed not because it confirmed some male concept of women as necessarily weak and fragile creatures, but because Lady Ban's predicament could be their own.[12] In turn, the biographical background to the fan poem (the life of Lady Ban) dignifies the text, which might have seemed too emotionally excessive in and of itself. It sanctions the poem by legitimizing the sentiments that provoked it; it allows for the element of pathos and of sexual longing.

Yet there also exists another, more cynical tale of a Han court lady, in which virtue and sincerity are called into question:

> Empress Chen 陳, a consort of Emperor Wu, for a time received his favor. But since she was rather jealous, he sent her away to the Tall Gate Palace 長門宮. She was melancholy and depressed there, filled with sad thoughts. She heard that a man from Chengdu (the capital of Shu), Sima Xiangru, was the most skilled in letters throughout the world. Offering him a hundred catties of yellow gold so that he and [his wife] Zhuo Wenjun 卓文君 could purchase ale, she took the opportunity to ask him to compose a piece to relieve her grief. Xiangru composed a poem to awaken the ruler, and as a result Empress Chen again obtained Wu's personal favor.[13]

So runs the preface to the "Tall Gate Rhapsody" 長門賦, as it appears in the *Literary Anthology* (*Wenxuan*). Scholars have debated the authenticity of the piece; after weighing the evidence, David Knechtges suggests that although the preface was obviously composed much later, there is no compelling evidence for rejecting the attribution.[14] Be that as it may, it stands enshrined in the anthology, providing another, alternative, and less pleasant genealogy for court ladies' laments.

In this alternative world, women are thought to be incapable of expressing themselves, and their motives are quite possibly insincere. Moreover, money is exchanged, and art is commissioned; one could easily see that a charge of emotional insincerity might be leveled against the author. The composition would then become an exercise in pure poetic talent and "projection"; as Sima stands on the outside, envisioning the abandoned empress, he projects her both as desiring and as an object of desire—part of the object of the poem is to make the emperor "soften" and turn his sexual attention to the lady once again. Unlike Lady Ban's "proper" rhapsody, "Tall Gate" brings a new type of passion and desire to the woman's lament, and the voyeuristic participation of the male reader is emphasized by the text's seamless move from an exterior description to the court lady's monologue.

In summation, Lady Ban's rhapsody supposedly affirmed the scholar's dignity, as well as the court lady's, because her original goals lay precisely in moral remonstrance, not in imperial favor. Although she saw that "women had their place," she (like Mencius' mother) had a right to speak and advise. Not only that, but her integrity and education gave her the competence to speak. Her passions are, ultimately, "correct." With Empress Chen, we are asked to take vicarious pleasure in a woman's despair. Ban is the true "wife," Chen, the concubine; their status clarifies the two ways of evoking the abandoned female. Needless to say, Lady Ban's remained the ideal path for the poet who would "allegorize" his own position at court in a "proper" way; Empress Chen's was the improper path, one that suggested the pleasures and profits of composition, rather than its ethical purpose. In practice, however, such distinctions became impossible to maintain.[15]

Canonizing the "Feminine" Voice

From at least as early as the *Songs of Chu,* the feminine voice of lament has been explicit in the tradition—and implicitly political. As literati in the fifth and sixth centuries A.D. looked back on their literary genealogies and traced sources for their literary heritage, this mode of expression remained an ever-present possibility, one that was sanctioned by tradition. In the preface to his critical evaluation of poets, the *Shi pin,* Zhong Rong discusses the emotional inspiration for verse writing:

Springtime breezes and springtime birds; autumn moon and autumn crickets; summer clouds and muggy rains; winter moon and sharp cold—these elements of the four seasons inspire poetry. At banquets we exchange poetry to delight each other; when parted from our companions, we use poetry to lodge our resentment. When a minister of Chu leaves his home or when a Han concubine departs from the palace; when bones strew the northern plain or when souls pursue the flying tumbleweeds; when spears are carried at distant outposts or when killing vapors grow strong on the borders; when frontier travelers have but thin clothing or when tears are exhausted in the widow's chambers; or when the minister unties his seals of office and leaves the court, giving no heed to returning once he's gone, or the lady enters into favor with her expressive brows, her second glance overturning the state—all these situations stir the heart and soul. If it were not for poetry, how could these meanings be expressed? If it were not for verse, how could these feelings be released? Thus it is said, "Poetry can be used to be social, and poetry can be used to express resentment."[16] Nothing is better than poetry for bringing ease to the poor and afflicted, for relieving the gloom of those in retirement. For this reason it is universally loved by those literati who compose it.[17]

After commenting on the influence of the seasons, Zhong lists a number of human situations that inspire poetry. Some are tropes associated with *yuefu* poetry, and so sanction the use of personae. These descriptive phrases are parallel, and in most cases each parallel phrase describes two phenomena that are the same or similar. Hence, "when bones strew the northern plain or when souls pursue the flying tumbleweeds; when spears are carried at distant outposts or when killing vapors grow strong on the borders" are two pairs of descriptive phrases that summarize the warlike poetry of the frontier.[18] They are flanked by two other pairs: "When a minister of Chu leaves his home or when a Han concubine departs from the palace" most likely alludes to Qu Yuan and to Lady Ban; their linkage here makes almost explicit the identification of slandered minister with slandered lady.[19] "When frontier travelers have but thin clothing or when tears are exhausted in the widow's chambers" is not explicitly political, since it alludes to a popular poetry tradition in which men lament the hardships of frontier service and women mourn the absence or the death of their soldier-husbands. Although this is a husband-wife pairing rather than a pairing of equivalence—the same poetic situation seen from two different perspectives, rather than two different poetic situations that are equivalent—it does reinforce the presence of the female voice in serious poetry. Finally, "when the minister unties his seals and leaves the court, giving no heed to returning once he's gone, or the lady enters into favor with her expressive brows, her second glance overturning the state" creates a rhetorical pairing of greater sophistication; the general meaning is "poetry is inspired by the situation of those who are leaving/going into exile (either voluntarily or compulsorily), as well as that of those who are coming into favor." Again, the minister / court lady link here is self-evident. Thus, in spite of an almost exclusively male writing community in Zhong Rong's day, distinctively female situations figure in three out of ten—and two of these are linked to the "serious" matter of politics. As if to emphasize this fact, he finally regroups poetic composition under the elite male umbrella: "For this reason it is universally loved by those literati who compose it."

Zhong Rong's statement is part of a lofty preface on serious poetry; it means to emphasize the canonical status of composition, its prominent role in literati cultural construction. One has only to observe most traditional commentaries on the Nineteen Old Poems to see how male/female relations are interpreted as relations between ruler and minister—many of them dating as early as the seventh-century Li Shan commentary, with roots going even earlier. For example, the fifth poem in the series describes a man(?) hearing and empathizing with the lonely song of a widow living in a high tower. At the end he transfers the empathy to an image of desire: "I wish we were a pair of calling cranes, / Who could spread their wings and fly away high."[20] Li Shan comments, "This poem describes a person of lofty talent who has yet to succeed in obtaining office, because few know him."[21] I do not believe that in the fifth and sixth centuries such interpretations resulted from an attempt to make "unimportant" themes "morally significant," to "rescue" the tradition for "Confucianism"; Zhong Rong's preface places too much emphasis on such themes for that to be true.[22] Rather, the male/female and ruler/minister readings exist in the text side by side, mutually reinforcing each other, playing upon one's interests and emotions depending on how one takes them at any moment. Li Shan chooses to emphasize this meaning, I suggest, for the same reason that the Mao commentaries emphasize the historical contexts of the Odes: multiple meanings are always present, but for the literatus the one that has the greatest ("canonical") significance is the political one.

However, these canonical moves are constantly called into question by the ambiguity the lyric poem often presents to the reader and the impossibility of being sure about the context. In a number of the Nineteen Old

Poems, for example, the gender of the speaker cannot be determined with certainty. Are those who express grief at separation spouses, lovers, or friends? Is the solitary insomniac an abandoned woman or a traveler longing for home?[23] In other poems (as in the "Long Gate Rhapsody"), the line between the "lyric" first-person voice of lament and an exterior voice of description is impossible to determine. Such ambiguity may have contributed to later readers' conflation of the tropes of "male" and "female" experience; it also helped erase the line between description and self-expression. For readers, female representation in texts became implicated in complex mechanisms in which a "lucid" reading is possible only through an explicit act of interpretation, of "clarification": a reader explains the poem and gives one reading, but he has no evidence that that is the "right" one. Women can become men, men women; political relations can become those of desire; tropes that suggest anguished self-expression can instead become opportunities for voyeuristic and pseudo-objective description. Some readers may say that this ambiguity is present only as a modern reading problem, and that contemporaries knew how to read the poem. This is true, but only to a certain extent. Within the social occasion that a poem is composed, issues like gender and perspective may indeed have been clear, but (as I shall show) expectations of reading during this period could not prevent the ambiguity of texts once they were read outside their contexts. If a context was not known, only a constructed one could solve all problems. Beyond this, we have the additional issue of accepted allegorical traditions, which made male and female tropes of desire interchangeable. Once men use women as an emblem for the self, there is often no way of determining the difference between observing and becoming.

Even beyond the Nineteen Old Poems, one can detect the implication of these interpretive issues in determining literary "worth." Consider, for instance, the work of Cao Zhi. Cao's popularity and importance are unchallenged in the fifth and sixth centuries; critics and theorists may disagree on the value of every other poet, but everyone agrees that Cao is one of the greatest figures in the tradition since the authors of the Odes.[24] The reasons are manifold: he presides over the "golden age" of the Jian'an era, he has had a biography created for him that is tragic and that suggests his poetry is imbued with a strong sense of lament and emotion, and he provides a strong example for the Chinese literatus who desires to find ways of articulating his social and political concerns.

Later views of the poet differ. In subsequent critical discussions, Cao and the Jian'an poets in general are admired for a style that embodies *fenggu* 風骨 (wind and bone), a quality that resides in a "forthrightness" of emotion, a solid sense of structure and coherence, and a refusal to conceal feelings behind overly self-conscious or artful language.[25] Although in modern criticism, *fenggu* has come to be associated with a discourse of "manliness," in contradistinction to the "effeminate" and "decadent" poets of the southern court, one should not overlook the ways that Cao Zhi's poems construct an identity that partakes of the "feminine."

Cao's *yuefu* and his adoption of *yuefu* "types" and themes are often considered the most significant aspects of his work. Two "types" seem to have attracted him particularly strongly: the abandoned woman and the urban bravo. Both appear in his poems as fully delineated examples of gendered conduct. Their behavior is rendered clearly and unambiguously, often through the description of social activities (hunting, weaving); the verses evoke "manly men" and "womanly women." In this respect, the ambiguity of the Nineteen Old Poems has been replaced by an attempt to create an explicitly gendered poetic language. But would a reader of Cao's time or later see this differentiation as a manifestation only of Cao's "manly" character, a desire on the poet's part to delineate his own manliness even more clearly? If so, we would expect a reader to see Cao exclusively as the (usually heroic) male, while distancing himself from the character of the female. Using Grace Fong's useful distinctions, the reader would interpret the bravo as a "mask," behind which the self is detectable, and the abandoned woman would be a "persona," assumed for an entirely objective interest in aesthetic creation.[26] It is more or less impossible to re-create a reader's response to specific texts at so early a date, of course—a surviving commentarial tradition does not really begin on any large scale until the Song dynasty, four hundred years later than the period we are discussing. However, I think it reasonable to assume that a reader of Zhong Rong's time (the sixth century) would see both bravo and abandoned woman as essential aspects of Cao's personality. Furthermore, he would see the male hero figure to a certain extent as Cao's idealized projection of the self—that is, what he had hoped to be if he had not been constrained by the hostility and persecution of his brother, Cao Pi. In the fifth of the six "Poems on Various Subjects," for example, the speaker explicitly decries the fact that he is not allowed to join the army and fight against the empire's enemies in the south.[27] "A Brown Sparrow in

Wild Fields" 野田黃雀行 presents a striking allegory for the male hero / idealized self:

A sad wind often blows in the tall trees,	高樹多悲風
And the sea waters raise their waves.	海水揚其波
If you do not have a sharp sword at hand,	利劍不在掌
Then there's no point in having many friends.	結友何須多
Don't you see the sparrow in the hedge?	不見籬間雀
It sees the hawk, falls into the net.	見鷂自投羅
The fowler's happy to catch the bird,	羅家得雀喜
But the youth is sorry to see it.	少年見雀悲
He draws the sword, cuts through the net,	拔劍捎羅網
Letting the sparrow fly away.	黃雀得飛飛
The sparrow flies and flies, touches the sky,	飛飛摩蒼天
Then comes back to thank the youth.[28]	來下謝少年

Within the reading tradition, there is an irresistible temptation to see this poem as Cao Zhi's lament over his inability to protect his friends and allies, a number of whom Cao Pi is reputed to have killed. And if the sparrow poem is an exercise in wish fulfillment, then the poems describing bravo life are fantasies of idealized heroism, granting far more space to action and exploit than to emotional expression.[29] However, an equally common tone in Cao's "fictional" personae is the melancholy lament of the abandoned woman. One must remember that for readers of the "tradition" up until Zhong Rong, at least, the explicit lament over rejection is a "feminine" gesture; in turn, passivity or immobility can be considered yin or feminine as well, thus providing the motivation for reading the female lament as politically "serious."

Ballad of the Abandoned Wife 棄婦詩

Pomegranate planted in the front courtyard;	石榴植前庭
Its verdant leaves tremble, flashing blue and emerald.	綠葉搖縹青
Crimson blooms shine fiercely,	丹華灼烈烈
Their jeweled colors bright in their splendor.	璀彩有光榮
Splendor glimmering with the brightness of jade—	光榮曄流離
As if one might sport with their pure magic.	可以戲淑靈
Birds come here and flock,	有鳥飛來集
Beat their wings, sing with sorrow.	拊翼以悲鳴
And why should they sing in sorrow?	悲鳴夫何爲
No fruit has come from these crimson blooms.	丹華實不成
I beat my breast, give out long sighs—	拊心長歎息

A childless one must return to her old home.	無子當歸寧
If you have children, you're the moon ascending the sky,	有子月經天
If barren, nothing but a shooting star.	無子若流星
The moon will have a full cycle of life,	天月相終始
While the shooting star sinks, without a spark.	流星没無精
I lived long with him, but failed in my duty;	棲遲失所宜
Now I've fallen amidst the tiles and stones.	下與瓦石并
Deep worries stir from within,	憂懷從中來
I sigh and sigh until cockcrow.	歎息通雞鳴
I toss and turn, cannot sleep,	反側不能寐
Then go and pace in the front courtyard.	逍遙於前庭
I hesitate—then return to my room,	踟躕還入房
Amid the swishing of my bed curtains.	肅肅帷幕聲
I roll back the curtains, order my sash,	褰帷更攝帶
Strum, pluck the silken strings of my lute.	撫節彈鳴箏
So full of feeling, their lingering sounds—	慷慨有餘音
Magical and marvelous they sound, but sad and clear.	要妙悲且清
Then I hold back tears, give a long sigh—	收淚長歎息
Wondering how I could have offended the Spirits.	何以負神靈
The Dipper Star awaits the seasonal frost—	招搖待霜露
But must I bear fruit only in the warmer months?	何必春夏成
A late harvest brings fine fruit;	晚穫爲良實
I wish My Lord to be patient a while.[30]	願君且安寧

The speaker begins with a landscape that has emblematic significance: the blossoms have attracted the birds with their beauty but cause disappointment when they do not lead to the production of "useful" fruits. So the beauty of the woman, whose own career becomes that of a shooting star; in the absence of a worthwhile contribution to her husband's family, she will be cast aside when her beauty fades. To express her grief, she turns to music—a standard move for the abandoned woman, not only in the Nineteen Old Poems but in the rhapsodies of Sima Xiangru and Lady Ban as well. Because the emotive power of music was seen as analogous to that of poetry, her music here is a reperformance of the text of the poem itself: emotional release and petition for mercy.

I suggest that a literatus would find pleasure in this poem particularly because he might see it as Cao Zhi's own autobiographical lament and interpret Cao's literary precocity as the blossom that has yet to produce fruit and see Cao as begging the ruler (either his father or his brother) to give him the opportunity to prove himself more than just a literary ornament. Again, we are dealing not with provable biographical realities but with certain habits of a reading tradition that help to support the appeal of a poet and make him canonical and influential for the work of later writers.[31] As we have seen and will continue to see, the cultural, social, and literary traditions of the sixth-century literati would continue to encourage and validate this reading and would affect their own poetic practice when employing the female voice.[32]

However, this reading of Cao is significant for more than the way it elevates the "self-expressive" function of his work; it also makes it easier to accommodate Cao within a history of literati poetry writing that stretches into later centuries. In his *Shi pin*, Zhong Rong establishes set genealogies for the better poets: in Zhong's analysis, a poet's work is derived from a specific predecessor, and he in turn becomes the inspiration for a later poet, and so forth. Generally speaking, the closer a poet is to the "great ancestors," namely the *Odes* and *Songs of Chu*, the greater he or she is; as generations pass, there is an inevitable decline. Zhong sees Cao as the inheritor of the *Odes*; and he sees Xie Lingyun as Cao's heir. In fact, Xie is the only poet after the Western Jin dynasty to have such a lofty pedigree. For a modern reader, this may at first seem puzzling: What does Xie, with his rhetorically elaborate occasional travel poems, have in common with Cao's simpler, "nobler" epistles and ballads? There are, I think, three particularly significant answers. First, Xie and Cao share a burden of emotional grief that results from their public failures; this manifests itself in a particularly powerful tone of lament (or so Zhong and his colleagues would read it). Second, in reconstructing a literary heritage, Zhong wishes to link his own age with that of the Jian'an as directly as possible. What better way than to associate the most brilliant "modern" poet—the inventor of descriptive parallelism—with the most distinguished poet of the past? Genealogy compensates for gaps in literary history; the barrenness of the fourth century can conveniently be elided. Finally, if Cao and Xie are seen as part of the same tradition, they can expand the modes available to the poet by combining between them a wide variety of styles and themes while still supposedly partaking of the same roots. Cao's ballads (particularly his emotional evocations of abandoned women) can be reincorporated into the world that created Xie—the world of social interaction and competition examined in the preceding chapter. Cao's work, like Lady Ban's fan poem, canonizes the female lament.[33] As we move from Zhong Rong and his estimations of the tradition, we are confronted with an even more pointed attempt to do this: Xu Ling's 徐陵 (507–

83) anthology of "female"-oriented verse, the *Yutai xinyong* 玉臺新詠 (New verses from a jade terrace).

Many poems in *Jade Terrace* have been dismissed as examples of "immoral" or even "pornographic" art. In understanding these attacks, we must confront some of the ambiguities of the "lament" in the tradition and how its canonicity came to be undermined to some extent through the pressures of poetic practice. Xu Ling's work was a product of the salon of Xiao Gang 蕭剛 (503–51), son of the Liang emperor Wu 武, from 531 until 549 the heir to the throne, and brief ruler of the dynasty (as Emperor Jianwen 簡文) from 549 to 551. Xiao is often dismissed as decadent simply for living in the time that he did: by the time he ruled, the Liang was already crumbling. Its successor state in the south, the Chen, would last only a few decades before all of China was unified under the Sui.

Most significantly, however, history attributes to Xiao Gang and his circle the invention of a kind of poetry known as "palace poetry" (*gongti shi* 宮體詩). The retroactive spread of this term and its application to Xiao are rooted to a large extent in the awareness of the disasters that were to overcome the southern world a few decades later and the desire to attribute those disasters to moral factors. Xiao Gang's biography in the *Liang History* gives one (typical) derivation for the verse: "The emperor once said of himself, 'When I was only six, I was seized by the poetry bug and never outgrew it.' However, his verse was marred by frivolity and sensuality. At the time they called it 'palace style.'"[34] Here, palace verse is termed *qingyan* 輕艷; *qing* means "light," "unsubstantial," hence "frivolous." Under these circumstances, it probably indicates a lack of ethical or political seriousness in composition—a failure to heed the prescriptions of the Great Preface. *Yan* is harder to define. The word seems originally to have referred to a form of southern harem music that became popular after the Han; other uses suggest that it came to be applied to any kind of "new" popular music that was suspect for its immorality—particularly if it evoked sexual attraction.[35] Simply calling such poetry or songs "love poetry" gives an inaccurate sense of its emphasis on its physicality; perhaps calling it "sensual" poetry gives some sense of its meaning.[36] Another, less critical, source attributes the beginning of the style to the influence of Xiao Gang's close associate, the literatus Xu Chi 徐摛 (472–551): "Chi's literary style was distinctive, and everyone at the crown prince's establishment imitated it. What we call the 'palace style' arose from this."[37] Either way, later critics saw in it sensuality combined with a predilection for novelty for its own sake—generally speaking, an unattractive combination.

However, an examination of the attitudes of poetry practitioners reveals a continuing allegiance (on the surface at least) to canonical values. It is true that a passing comment of Xiao Gang's has suggested to some a willingness to embrace novelty and even triviality: "The way of establishing oneself in the world is different from that of literary composition. In establishing oneself, one must be cautious and serious. In literary composition, one must be uninhibited."[38] Guo Shaoyu once argued that the term "uninhibited" (*fangdang* 放蕩) revealed Jianwen's essential immorality and licentiousness; he termed palace poetry "pornographic."[39] However, as Wang Yunxi has pointed out, the quote comes in a letter of short instructions to his son and is not necessarily the place to look for serious statements on the nature of literature. He and others have also suggested that the term *fangdang* would have had a long history of favorable associations for Xiao; it tended to be applied earlier to a form of unrestrained behavior characteristic of, for example, the Seven Worthies of the Bamboo Grove.[40] In a more serious critical document (a lengthy letter to his brother Xiao Yi 蕭繹), Xiao reaffirms poetry's importance but distinguishes it from Confucian scholarly texts. For him, the power of poetry to be self-expressive and to evoke emotions remains unique and unsurpassed.[41] In this he echoes Zhong Rong and other critics of his time.

Of course, we must not create a coherent set of attitudes toward literature in Xiao's writings, as if every poet was a theorist. Xiao never attempted a comprehensive discussion of literature, and his comments are found here and there, mostly in letters. Since context and situation often dictate what he will say, even if he had some set "idea" about what he was doing in his salon, it is ultimately irrecoverable. Nevertheless, it is highly improbable that literati culture, framed as it was by the rigors of the tradition and by a continuing emphasis on cultural capital, would have made a self-conscious claim to be "breaking with" the past and its emphasis on the expressive qualities of verse.

Within *Jade Terrace* itself, there are interesting conflicts between the supposed "immoral" motivations behind "palace poetry" and the claim of following an orthodox tradition. An anecdote in a ninth-century miscellany exposes these conflicts in a rather revealing way:

Emperor Taizong 太宗 [of the Tang; r. 627–49] said to his attendant ministers, "I write 'sensual' poetry for diversion." Yu Shinan 虞世南 [558–638] remonstrated with him. "Although Your Majesty's works might be skillful, the genre is hardly dignified. Those below you will certainly follow you in your likes. I fear that once this type of writing finds circulation it will bring about a weakening in the customs of the land. From now on I must graciously decline any commands [to compose such verse]." The emperor replied, "A minister as honest as yourself must be commended. If all of you were like Shinan, I would have no worries about keeping the empire in order." He then rewarded him with fifty bolts of silk. Earlier, when Emperor Jianwen of the Liang was crown prince, he was fond of sensual poetry. The entire empire was transformed by it, and it spread until it became a custom. They called it "palace style." In his later years he tried to change this development, but it was too late. He then commanded Xu Ling to edit *Jade Terrace* in order to magnify the style.[42]

The author relates the courage and uprightness of Taizong's ministers (the traditional, oft-cited reason for the "greatness" of Taizong's reign) to the implied cowardice of those who did not warn Xiao Gang of his folly. As a result, Xiao must engage in a "salvage operation" on customs. The author may be attempting to reconcile the seeming frivolity of palace poetry with the canonizing effort implied by the commission of *Jade Terrace* itself. How seriously we take this anecdote depends on how big a gap we perceive between the sixth century and the ninth (when it was set down on paper).[43]

Xu Ling's preface to *Jade Terrace,* however, confirms the tradition of the canonized lament. At first, as Xu describes it, the anthology seems to be solely concerned with women authors and readers; he seems to be making the claim that since women are inferior and less "serious" than men, they need their own "less serious" literary tradition—a place where they can see the work of their predecessors and find models for imitation. This is at the very least disingenuous, since almost all the poems are male-authored, and the readership would very obviously have been mostly male. But a closer examination of the preface's themes reveals the literati concerns underneath. It commences with an evocation of palace women:[44]

These pavilions like Crossing-the-Clouds and Sun-Leveler were never seen by the likes of You Yu 由余; these thousand gates and myriad doors were once rhapsodized by Zhang Heng 張衡. Atop the jade terraces of Zhou's king, within the golden chambers of Han's emperor, there are jade trees with coral branches, pearl drapes with tortoiseshell hooks. There are lovely ladies within.

Xu begins with a visual contraction. From the mention of great palaces (You Yu was an emissary from the north who was awed by the splendor of Chinese mansions, Zhang Heng the author of a rhapsody on the imperial capitals), we move inside, away from external glories into "feminine" space; thus, the author suggests, we will be dealing with the interior, the *yin*.

Some ladies fill the side palace rosters from the great clans of Five Barrows 五陵; from fine families of the Four Names they resound in fame throughout the royal lanes. But there are also maids from Ying Creek 潁川, Newmarket 新市, Midriver 河間, and Gazeford 觀津—known from the first as "Lady Charming" or already named "Skillful Smile."[45] Within the king of Chu's palace, they are promoted because of their slender waists, and the beauties of Wei all proclaim their slender hands. They are read in the Odes, and familiar with the Rites, hardly like that Eastern Neighbor who served as her own matchmaker; they are sinuous and lithe, romantic in bearing, not like Xi Shi, who had to be taught.[46] Their brothers serve as court musicians, so they've studied song from their youth; they grew up in Heyang 河陽, so they've learned the skill of the dance. They needn't wait for Shi Chong 石崇 to write new mandolin songs for them, and their various harp songs owe nothing to Cao Zhi. The Yang family transmitted to them the strum of the zither; the Girl of Qin gave them breath for the flute.

When rumor of their favor reaches the Longjoy 長樂 Palace, Empress Chen 陳 grows wary with the news; and when the portraits of these immortals are displayed, the khan's wife grows jealous from afar.

Various kinds of palace women are described here: those from great, prestigious families (Five Barrows was an upper-class suburb of the Han capital, the Four Names were the four families most famous for supplying Han empresses), and those from more modest circumstances. The talents of these women (literary, scholarly, musical) are stressed throughout as a major reason for their favor. Perhaps most significant here is the mention of song and dance: the "court musician" is Li Yannian 李延年, performer for Emperor Wu of the Han; his sister, Lady Li, managed to become court favorite despite her obscure background. Heyang was the home of Zhao "the flying swallow," who replaced Lady Ban in Emperor Cheng's affections. Li and Zhao are the two most famous examples of women whose seductive arts won them an elevated position in the ruler's favor. Meanwhile, the arrival of new recruits inevitably produces jealousy among old favorites. Empress Chen was furious over Emperor Wu's preference for a lowly serving girl, Wei Zifu 衛子夫 (later Empress Wei). Emperor Gaozu 高祖 (r. 206–195 B.C.) of the Han managed to break a siege by the northern Xiongnu tribe by sending the chieftain's wife the picture of a beautiful court lady he was sending to the chieftain as a present; she compelled her husband to raise the siege and depart.[47]

Xu's preface is not merely an attempt to catalogue the various women in a harem; it is pointing to a central issue in harem politics, the competition between favorites. Here is the crux of the preface: Xu is signifying to readers that the court lament tradition is the proper topic here by alluding to the age-old habit of reading the ruler's court as a harem, and literary talents as the attractiveness of the women. Although Xu writes in the ornamented, allusive diction of parallel prose, his desire to introduce "canonical" themes emerges at last—and thus opens up the way for us to read *Jade Terrace* as a more "proper," literati anthology. At the same time, because *Jade Terrace* grew out of Xiao Gang's patronage of Xu Ling, he attempts to bring the poems within the aegis of the court and imperial judgment.[48]

Another long, heavily allusive passage continues the description of the performance skills of the ladies; this is followed by a detailed account of their beauty—but as is typical in such descriptions, beauty is couched in terms of fashionable cosmetics, elegant costume, and gentility of manner. Soon, however, Xu returns to the matter at hand: literary achievement.

> Add this [to their beauty]: their basic natures are open and bright, their untrammeled thoughts clever and flourishing. With a subtle understanding of composition and superior skill in verse, they keep at hand their crystal inkstone cases all day and never once put down their kingfisher brush-holders. Lucid essays fill their baskets—and not just about peonies; newly fashioned works come one after another—not stopping with the grape vine alone.[49] On the Double Ninth they climb high and write poems of feeling for the occasion; the Princess of Myriad Years will have eulogies for her many virtues. Just so beautiful are these ladies; just so, their talent and sentiment.

Beauty and talent become two parallel qualities that may be constructively compared. In parallelistic thinking, comparison invites identity: beauty itself is a form of talent (signified through the tasteful and seductive ornamentation of one's surface); talent is a form of beauty, an ornamentation and manifestation of the beauties within.[50] Moreover, Xu suggests, this literary, "inner" ornamentation (namely, "women's" literature) is versatile and covers a wide variety of subjects and situations. Xu alludes here to Ban Gu's comment, "The one who can climb to a great height and compose poetry will become a great officer."[51] Of course, composing poetry from a high vista, particularly on the festival of the Double Ninth (the ninth day of the ninth lunar month), was a common custom even at this relatively early time, but there is a hint at the idea of literary accomplishment as a ticket to high office.

Afterward, in the winding corridors of Pepper Hall, in the remote corners of Persimmon Lodge (where crimson hall-keys stay solemn at dawn, and bronze door-knockers are silent at noon), since it is not yet the night for an imperial summons, she will not come and serve, quilt held to her chest. Since her day of service is still distant, who will help her do her hair?[52] She has little to occupy her mind in her strolls and too much leisure in her loneliness. She grows tired of the intermittent bells of Longjoy Palace, and she wearies of the slow waterclock in the central hall. . . . Though she *could* act the Jade Maid and play arrow-toss, her attention would go after a hundred rounds;[53] if she were the Qi lady competing at *liubo* 六博, her delight would end with six tallies.[54] Since she has no pleasure in her leisure, she turns to new verse; such works can replace those herbs that mysteriously relieve melancholy and grief.

However, if the famous pieces of bygone times and the crafted work of the present age—writings that have already been distributed in imperial libraries—if they are not collected in a single book, she will have no opportunity to read them. Therefore, burning the lamp, I have copied late into the night; wielding the brush, I have written from dawn. I have recorded here the following "songs of glamour," ten scrolls in all. Although they cannot be placed next to the Elegantiae and Hymns, they do not go beyond the bounds of the poets of the Airs. Just as the waters of the Jing 涇 and Wei 渭 stay apart even though they flow through the same channel, so do I make the same distinction here.

A court lady must be *employed* if she is to fulfill her purpose in life. Any other time must seem useless and filled with ennui. She can try her hand at various hobbies and games, but it is much more practical if she works at her reading and writing. Here Xu hints at the whole system of images associated with the abandoned court lady who waits in vain for the arrival of an imperial visit. Poetry here is not just an occupation. Rather, the reading of other poems describing the predicament of court ladies will "relieve melancholy and grief." In an intriguing reinforcement, the very poetry that they read is meant to reflect the situation they seek to escape from. The poetry speaks "for" them, relieves their inarticulateness—so in one sense, the poetry performs the activity of Sima Xiangru, writing "for" Empress Chen. Xu nowhere explicitly states that these talented literary women are actually the *authors* of the poems in the anthology. Nonetheless, the anthology can serve as a training ground for future articulation: these poems will teach the women how to state their condition in an artful and sophisticated way, perhaps even teach them how to petition the ruler for attention. It will train them to be better competitors in the harem. But if a reading tradition already exists that suggests that "cross-gendered" poetry is in fact the male speaking for "himself," then there is nothing that prevents us from seeing the preface otherwise: *Jade Terrace* tells of generations of male poets turning to lament and petition when they have been removed from office. Their own idleness results in literary production; their capacity at displaying their literary talents and participating in a tradition of lament will relieve their own sorrow and lead perhaps to reinstatement.

Xu continues with a self-effacing comment about his own efforts; the tradition he maps cannot be compared to the more formal of the Odes (a text, incidentally, that women are intimate with, by his own admission). Yet there is a tradition here, nonetheless; he claims that the *Jade Terrace* poems are in the same league as the poems from the more emotional, expressive *feng* (Airs) section. He compares *Jade Terrace* and the Odes to the Wei and Jing rivers—two currents that in time ran into the same passage, with the Wei maintaining its muddiness, and the Jing, its clarity. To assume that this means that the Wei's muddiness indicates immorality or a tradition only for women is to miss the point: rather, Xu argues for two kinds of literature. The river channel is emotional self-expression; the Jing River, the *feng*, is the canonical, normative expression of emotion and hence guides the student to a normative, proper sense of public demonstration of feeling. The Wei River, the poems of *Jade Terrace*, is the expression of emotions within a less lofty, but still significant sphere—harem ladies talking one to another, court poets in the grace and ease of the salon. There is room for both streams. Xu self-effacingly relegates the *Jade Terrace* tradition to the private, the noncanonical, but whether Xu really believes that the poems are truly that unimportant in the lives of their public authors—men who often carried their political feuds and concerns into the salon—is another matter.

After a section describing the beauty of the manuscript itself, Xu turns to its use (like the ladies, the collection is a pleasure to look at and a pleasure to employ):

> Just when she draws aside her black-ox curtains, and that final serenade is still unplayed; just when she stands before her redbird window, while her fresh toilette is still incomplete—then she will unroll these emerald scrolls, undo these scroll ties, peruse the poems for a while by the library curtains, ever turning them with her slender hands. Of course, her labors are hardly like those of Empress Deng 鄧, who

studied the *Spring and Autumn Annals* (a scholar's achievement, so difficult to emulate!); or those of Empress Dou 竇, who gave herself over to Taoist techniques (the art of gold and cinnabar is still unperfected!). Nonetheless, they are far superior to that rich man of Western Shu, who exhausted all emotion in "Lu Palace," or the heir-apparent in his quarters, where fluid chanting ended with "The Flute."

"Charming those maids of Qi"—I hope they will use this to spend their leisure. "Lovely that red writing brush"—let there be none who malign it!

Again, the gesture of the text's unimportance is emphasized—yet Xu continues to demand a place for it as well. The humility before the more "serious" activities of Empresses Deng and Dou is undermined by the suggestion that conventional Confucian and Taoist preoccupations are perhaps too weighty and ideal to task the "practical" lady. Literary variety and versatility are praised in relation to a single-minded obsession with one kind of literature alone. Xu's valedictory closes with a final linkage of beauty and talent. Both quotations come from the *Odes*; the second line is particularly interesting because it refers to a Mao interpretation of "Gentle Girl" (Ode no. 42). For the Mao commentators, the enigmatic "red pipe" of the verses, a gift of a girl to the poem's speaker, is a writing brush wielded by the "women historians" of the Zhou king—a band of secretaries whose recording of the events of the harem acted as an interior counterpart to the male court records of external events.[55] The final line thus refers to female traditions of historical and textual transmission—traditions in which *Jade Terrace* seeks to participate. Again, however, the tradition can be read as simply a male interior tradition—the relative positioning of the courtier as *nei*, "within," speaking as a woman in the context of interior politics. Xu does not need to mention these explicitly; the rhetoric of competition sets up the analogy and the rereading. *Jade Terrace* speaks to women, as well as to men who are refiguring themselves as women.

Subverting the Canon: Structure, Voyeurism, Envy

The *Jade Terrace* preface says other things as well. A reader familiar with the "palace poems" reproduced in its last chapters will realize that Xu Ling uses the very vocabulary with which women are described in such verse to evoke the idealized, objectified women who *produce* verse. We are granted the power of spying on these women artists, of making them the objects of our own desire. At the same point, Xu exhibits grace and skill in composition—so that the symmetrical rhythms of the language will echo the ornamented balance of the ladies in their robes and makeup. Once again, gendered texts begin to fragment into alternative possibilities, in which description, self-expression, and desire become confused. Here, however, confusion is caused not just through the ambiguity of representing gender, as with the Nineteen Old Poems, or through the impossibility of fully knowing an autobiographical context, as with Cao Zhi. Rather, it is caused by an increasing emphasis on the description of sexual attraction and the male game of competitive erotic stimulation. Such evocations work against the "self-expressive" traditions of female lament; they also work within and against the new formal structures of poetry, which gives them expression and power.

I argued in the preceding chapter that after Xie Lingyun poetry derived its popularity from its efficacy as a form of communication and cultural capital in the social circles of the post-Jin world. In addition, Xie developed certain techniques that made it easier for poetry to become a "competitive" art form. Most evidently, syntactic parallelism created a limiting structure for experience. Xie's carefully crafted landscapes, in which journeys are mapped into chiasmic patterns of mountain and water scenery, demonstrated how an experience could be structured and carefully laid out in stages. Poetry became more structured and less likely to ramble.[56] Few poets before him used parallelism with the consistency, artistry, and sophistication evident in his work. Its popularity among the following generations marks a major turning point in the development of Chinese verse, one that far exceeds the mere "literary" or "artistic."

Once Xie provided an introductory model for poetic social exchange, there was an acceleration of poetic activity among literati. Other developments soon followed: the growth of "refined" features in poetic diction; the elimination of vulgar language, as well as inelegantly technical or obscure vocabulary; the substitution of emotional "solidarity" with the group, rather than individual passion; and a concern with the metrical and phonetic qualities of Chinese verse. The last led to the increasing emphasis on "tonality" in poetry: a series of accepted proprieties on how syllables with different sound values should be juxtaposed within a line and within a couplet.[57] The old competitive context of the "clear conversation" is continuing, but now it often moves into the salons, many of them patronized by imperial princes. There contests are held, and judges are selected to determine the best verse, based on an increasingly rigid conception of versification. As with "clear conversation," a literati "game," strongly marked by sophisticated wit and insouciance, actually conceals a serious competitive determination of power and control.

The increasing number of metrical restrictions accompanied by the ritualization of poetic composition within a salon setting also led to what Kang-i Sun Chang has called "the interior turn of the landscape."[58] Poems tended to become shorter, to give poets a respectable yet reasonable length for verses, which often had to be composed under pressure within a set period of time. A regularization occurred: most poems tended to be four, five, or six couplets long. The opening couplet usually set the scene; the middle couplets give descriptive details, in fairly rigid parallel structures; and the last couplet is a witty or emotional response by the poet to the situation.[59] Subject matter became more fragmented, small-scale, and interior. A large vista might be replaced by a focus on one detail—a single flower, the effect of moonlight on a building. Artificial objects could become subjects of a poem, and palace women, a particularly striking example of the closed interior, were an attractive subject for many poets.

Here we have a new genealogy for palace poetry, one independent of the female lament, in which self-expression in its conventional sense is no longer as important as the "playful" description of palace women. In fact, in *Jade Terrace*, one sees a turning away from the "lament" tradition in some of the later chapters. Almost all poets up until the mid-fifth century write in the voice of abandoned women, with a dozen exceptions—some mourning poems for wives, a few husband-wife exchanges, or a paean to a goddess. However, after this point (especially after Shen Yue 沈約 [441–513], one of the innovators of the court style), laments yield to other, less lugubrious verse—elaborate descriptions of court ladies or entertainers, evocation of women's possessions, even drinking verses. Laments of abandoned women still make up well over half of the selections, even in the last chapters, but lamentation is no longer the only theme.[60]

Even as the lament canonically suggests emotional honesty in fourth- and fifth-century verse, there is alternatively an intellectual and social exercise in progress. Both the need to appear in control, while seemingly indifferent to "serious" things (*yaliang*), and the "natural" display of talent form essential elements in the cultivation and maintenance of cultural capital. These poems are forms of social exchange: thus, the laments with titles such as "Six Poems Matching Secretary Wu," "Spring Evening: To Princely Command," "Respectfully Matching the Prince of Xiangdong's Poem, 'Winter Morning, to Princely Command,'" and "Grasping My Brush, I Playfully Compose." Many of these exchanges are actually *yuefu*; thus, the titles explicitly remove *yuefu* personae from the realm of social or personal critique to that of social conversation. The erotic is a major part of this world because the "creation" of women in the text can act as a particularly potent form of social exchange: men are the judges of female beauty, and such judging forms a bond between men. Poems circulated on the topic also act as a form of symbolic bonding in the homosocial community, a potent focus for competition and ownership. And, of course, the mastery of sensual language in evoking the erotic is a particularly striking test of the poet's skill in self-control: the pimp remains unmoved even as he challenges the customer to remain so. Sexual desire is evoked, harnessed, displaced into the serious competitions of the social occasion.[61] It was partially a recognition of this social construction of the sensual that made later readers reject palace poetry as a canonical contribution to the "female lament."

However, it would be a mistake to analyze palace poetry as having only one "message"; the tradition has become complicated, and all the factors come to bear in a sensitive reading. Again, gender and desire shift from one position to another, as poets simultaneously engage in acts of voyeurism, cross-dressing, and competition. A poem by Liu Xiaochuo 劉孝綽 (481–539) supplies an excellent introduction to this world:

From a distance I saw that the host in a neighboring boat had tossed an article into the water; all his singing girls competed for it. A guest asked me to compose a poem on the subject. 遙見鄰舟主人投一物，衆姬爭之，有客請余爲詠

The river flows, so level and smooth;	河流既浼浼
The river birds cry *guan guan*.	河鳥復關關
Falling blossoms float out from the banks,	洛花浮浦出
Flying pheasants cross the islet, then return.	飛雉渡洲還
On this day, the singing-house girls	此日倡家女
Compete with their charming peach and plum complexions.	競嬌桃李顏
Their fine lord, although he begrudges a fine earring	良人惜美珥
Wishes to trade it for fragrant rushes.[62]	欲以代芳菅

The first eight lines set the scene for the competition mentioned in the title: a serene natural landscape for the boating party. However, the host and master of the singing girls throws this "natural" scene into commotion by sacrificing an earring. By throwing it overboard, Liu wittily says he exchanges the valuable earring for the worthless water rushes, but he alludes also to the lines of Ode no. 139:

In the Pool by East Gate, 東門之池
One can soak and soften the rushes; 可以漚菅
That lovely Shuji— 彼美淑姬
I can chat with her knee to knee.[63] 可與晤言

The host thus uses the earring as a way of buying the best maid among the group, the one he hopes "to chat with." Formerly, the women were competing only in their "natural" beauty, in a passive way, as part of the scene. Now they are thrown into an active struggle for predominance. In the next few lines, clichés of female competition are catalogued:

The new favorite with her fine silks suspects the old one with her plain silks, 新縑疑故素
And Zhao Feiyan as she flourishes has contempt for the fading Lady Ban. 盛趙蔑衰班
Plain gauze dragged on the stream vies with fine gauze that restricts; 曳綃爭掩縠
With trembling pendants they seize the jingling ring. 搖珮奪鳴環

Women are described in *Jade Terrace* either as allusions or types (as here with Zhao Feiyan and Lady Ban) or by what they *wear* or *decorate themselves with*—here, all sorts of silk apparel. There is a real lascivious joy in this scene—sweaty girls wrestling. But these ornamented objects struggle here for yet one more ornament that is the most important of all—the earring, because it holds the key to favor.

So far, so good. Yet in the final couplet the poet draws attention *back* to the event itself, to the poet's own identity, to *male* competition and jealousy:

The heart of the guest is shaken in vain, 客心空振蕩
For the highest branches cannot be grasped. 高枝不可攀

This is a polite compliment: it is left to the host to pick the best of the singing girls. Although the poet, the guest in the other boat, may marvel and admire, he cannot pick the highest branch, the woman who finds the earring. Her acquisition of the earring is not matched on his part by an equally symbolic acquisition. However, writing the poem does give something to strive for, as the poet participates in a competition of his own. His text suggests that the struggle between the singing-girls is re-enacted at the level of the verse—one in which Liu himself is invited to produce a witty, sophisticated poem on the subject that will be judged by his peers and by the host. In writing, he must struggle to find the best, most effective words for the verse: he must try to "find the earring" before others do; he is one of the singing girls, not one of the hosts. (The "host" of the title [*zhuren* 主人] can be a "host" in many ways: Liu Xiaochuo's host, assuredly, but also the "master" or even "employer" of the singing-girls.)

Obviously some complicated issues are being worked through in some texts—issues where "outside" description and identity are merging into a complex thread that cannot be easily unwound. Even in the simplest laments this confusion of desire and identification occurs through the absence of pronouns and the impossibility of determining when male description of a woman ends and her words of lament begin. In Cao Zhi, the boundary laments are relatively clear. In a typical "palace poem" by Wang Sengru 王僧孺 (465–522), however, the boundaries are not so self-evident. In the following translation, I have avoided inserting pronouns except those present in the text.

Matching Secretary Sima 與司馬治書
"I Hear a Neighbor Woman Weaving in the Night" 夕聞鄰婦夜織

In a remote chamber, the breeze has already sharpened;	洞房風已激
On a long veranda, moonlight clear again.	長廊月復清
Vague and dim—the night-darkened courtyard is broad;	藹藹夜庭廣
Fluttering about—dawn curtains light.	飄飄曉帳輕
In a confusion, hear the brooding noise of a hundred cicadas;	雜聞百蟲思
One-sidedly saddened by one bird's song.	偏傷一息聲
The bird song never seems to end,	鳥聲長不息
My [female] heart will never cease.	妾心復何極
Still I fear you [probably male] have no clothes;	猶恐君無衣
Night after night, weave by the window.[64]	夜夜當窗織

The complexities commence with the title. This poem explicitly disavows that it represents an actual event—rather, it imitates another poem, one that supposedly derived from the event. Secretary Sima heard the neighbor woman and so wrote the poem. Or is that also fictional? The woman in Wang's poem turns out to be weaving clothes for her husband on the frontier, a stock folk-song trope. At any rate, the text now becomes part of a competition, to see who can "evoke" the situation better. The two men both stand, listening to the woman.

How much do they see? And how much do they hear? Reading Chinese poetry is largely a reconstruction of points of view, of working out for oneself who is speaking, and what she/he is doing.[65] In this poem, however, two equally viable scenarios suggest themselves. First, as Wang and Sima listen to her weaving, they project a speech for her, a conventional lament for an abandoned wife. Lonely in her empty room, she is sensitive to the increasing cold of autumn. Unable to sleep, she decides to weave instead, through the whole night (the parallel mentions of dark courtyard and morning drapes are meant to suggest the passage of time). The next couplet repeats that passage of time, and so the poem continues logically:

> At night I heard a throng of cicadas,
> But now at dawn I hear only one bird's song.
> How that bird song never ends!
> My heart, too, will never cease.
> Always I fear you have no clothes,
> And so night after night I've worked by the window.

As she works through the night, the cicadas, with their multitudinous, sociable racket, seem hardly worth attention. But at dawn she hears one single, solitary bird. It is (to be as sentimental as the poem) the lonely cry of her heart. It annoys her because it will not stop chirping, just as her heart annoys her for not stopping and ending her misery. Here we have a fairly conventional lament poem in the tradition of Cao Zhi and his successors—a poem of supposed "emotional honesty," which, even if not written from the perspective of the poet itself, presents the poet as a man who empathizes with her situation—possibly because her loneliness and longing for attention echo his own political dilemma.

There are other correspondences as well. Although the woman is not subject to the same insecurities as the palace lady is (and thus cannot become a fairly obvious correlative for the courtier), the poem nonetheless plays on the palace lament themes of anxiety, insecurity, fear of abandonment. The crickets (harbingers of winter and cold weather) bring on the call of the bird, who "self-expressively" gives way to emotion. Self-expression is then routed from bird to woman and eventually to the poet himself—textile becomes text, as he engages in rhetoric, fearing that no one will ultimately benefit from his own "weaving"—weaving carried out for his distant master.

Of course, this is only one possibility. In the second scenario, the men move in on the neighbor's house, stand outside her chamber. They stand listening all night—and at some point they hear the woman speak. They detect her voice, like the song of a bird, lost at first amid the chirping of the cicadas—a woman's voice and not a bird's, because what bird sings at night with the cicadas?

> Although everywhere we hear the throng of cicadas,
> We alone are moved by her solitary bird-song,
> And that song of hers never seems to end.
> "My heart will never come to an end!
> Always I fear you have no clothes,
> And so night after night I weave by the window."

This voyeuristic scenario is meant to enhance the woman's desirability: lonely, vulnerable, perhaps ready for a new love, the site for an erotics of pity. Poem after poem in *Jade Terrace* lavishes attention on this attractiveness of lonely women.[66] The elaborate descriptive powers of social poetry (an inheritance from Xie Lingyun's landscape verse) play on this inclination by evoking the object in sensually attractive language. But who will succeed in entering the chamber? Perhaps the poet who is thought to be the most successful in the theme, whose own version of the woman will displace the others. Wang seeks not only to replace the soldier-husband, but Sima as well; to take the woman symbolically means to best his fellow poet.

The French critic and philosopher René Girard has created a scenario for human desire when it is not "simple" and "reciprocated." In what he calls "triangular desire," an external mediator that desires the same object comes between the lover and that object. This mediator stirs complex emotions within the lover: reverence and love, in some cases, because the mediator provides an example of successful, fulfilled desire that the lover can aspire to; hatred in others, because the mediator is in the way. One might add that there is also the desire to become and imitate the rival and so replace him in the relation with the desired object. For Girard, this mechanism is not typical simply of nineteenth-century fictional narratives but runs through much of human experience.[67]

One might add that it also functions particularly well as a model for social "sensual" poetry. If anything, the introduction of a semi-explicit physical desire through the creation of a language evoking the sensuality of the woman's body doubles the triangulation until an even more complicated structure is revealed. The poet begins in a competitive situation with potential rivals within the salon; there is always someone who is potentially in greater favor than himself with the patron, because salon competition induces oversensitivity and the constant reading (and misreading) of others' motives. Versification is "idealized" and elevated through the poetry competition, while at the same time it is repressed: competition is supposedly "playful," a witty form of entertainment. The relation with the rival varies: desire to befriend him, to gain his trust and alliance, to have him help one gain access to the beloved (the patron); hatred of him, as a competitor; and a desire to be him or replace him, to write the poem that *overwrites* his and wipes it away. Every poetry exchange has the potential to become a replacement.

But the text points in two directions: to the rival and to the woman it describes and mimics. Here, too, there are different attitudes toward the woman. In most palace poems, the woman is court lady or hired entertainer. She has access to the privileged body (host, patron, emperor). The poet desires her, and in this sense he desires to replace both the other poets who similarly desire her and potentially the master himself; the possession of the woman is the greatest proof of male supremacy. Yet there is also the desire to *be* her, because the lady has special access to techniques that the poet is either incapable of or ostensibly disdains: the sexual arts of dance and song or (more demurely) the language of political lament associated with the court lady. Becoming a woman allows the poet either to seduce or to petition the patron.

The poem is the medium that articulates his desire for the lady and the medium through which he *becomes* one. Language is the ornament of the poet, and seductive beauty can be represented in literati culture only through ornament, exterior adornment. Of course, as we have frequently seen, one claims that exterior ornament is the index to interior beauty, but it is also the only way to make that beauty manifest. The text labors to substitute for heterosexual male desire, to displace the superiority of the poet-rival, to act as a verbal representation of the poet's ornament / inner virtue and thus to replace the lady and speak as/for her, to act as a seduction upon the ruler, and to express the poet's lament in a culturally sanctioned way. Of course, the movement into the woman allows for a seemingly *fictional* reconstitution of triangular desire, along the simplified lines of harem politics: the woman of the poem is to the other harem women as the poet is to his court poetry rivals. The patron is now the main, simple object of desire, unmediated by a rerouting through other systems of meaning.

All these readings can be implicit, explicit, consciously worked out, or merely present in the text as a possibility.[68] "Sensual" poetry was written in many different social situations, by poets of many different social ranks. An emperor or imperial prince can participate in these games and add his voice to gestures of voyeurism or competition: since a discourse of linguistic display and "distance" has been worked out, his participation is not precluded by the "literati lament" element. At the same time, we will never be able to reconstruct the exact situation that produces a "sensual" poem or pinpoint more accurately the precise social motivations for its composition. The construction of the female as well as the poets' identification with her allows for spectator to become spectacle, to mingle in a realm in which conventional boundaries are down while still operating ostensibly within a privileged space defined by the rules of literati competition.[69]

Aside: Court Poet or Poet at Court?

Unfortunately, we may never know the poets' real attitude toward the court and the emperor. Is the ambiguity between voyeuristic distance and female role-playing a genuine one, or merely an effect of not having a context? Does a comparison with other court societies offer us any insights?

One of the best attempts to construct a possible context for late Six Dynasties poetry has been carried out by Cynthia Chennault, who argues that these poets lived in a time when rulers seeking greater autocratic authority deliberately created "court societies" in order to centralize and make impotent their troublesome aristocracies. She emphasizes the relations between this more "centralized" court and literati:

> In conditions that in many ways parallel the cases in Europe, emperors of the Southern Dynasties shed the feeble position they held during the Eastern Jin, of being *primus inter pares* among members of the top social echelon. As they exercised a rulership that grew increasingly autocratic during the Liu-Song and Qi, and were able to concentrate in their own person the disposition of political and economic resources, men who hoped for prominence needed to connect themselves with that center. In China, as in Europe, webs of patronage that involved royal relatives and high officials were a result.[70]

And later:

> The impromptu recitation of many *yongwu shi*, along with poems "written upon royal command" and other *ad hoc* displays of skill, may have owed their popularity at court to reasons other than an interest in frivolous diversion. To the extent that acquired abilities were becoming a factor of position, the political arena was thrown

open to anyone who could learn the areas of knowledge which were valued. A means of narrowing the field was to demand *ex tempore* proofs of accomplishment. Likewise, as lyrical practice became more rigidly conventionalized and aesthetic standards were codified in guidelines, the act of spontaneous and seemingly effortless composition may have defined an exclusive in-group among salon members who were able merely to satisfy the rules.[71]

Chennault's main source for a description of European "court society" comes from Norbert Elias's *Court Society*, a work largely completed (in the 1930s) before his wider analysis of European cultural "modernization" in *The Civilizing Process*. Elias dealt mostly with Louis XIV's court. Although he sought to introduce analogies from other cultures, it remains self-evident that his interest lay in the role of a court society in the rise of early modern European absolutist states. Roger Chartier's summary of Elias's argument is essential for redefining for the contemporary reader some of the author's sociological models.

The figuration of court society is indissolubly linked to the construction of the absolutist state. In such a state, the sovereign has a double monopoly: a fiscal monopoly that centralizes tax-gathering and gives the ruler the possibility of rewarding his faithful supporters and servants in coin rather than in land, and a monopoly of legitimate violence that attributes military force to him alone and thus makes him master and guarantor of the pacification of the entire social area. This fiscal and military monopoly, which dispossessed the aristocracy of the traditional foundations of its power and obliged its members to live near the sovereign, dispenser of rentes (annuities), pensions, and gratifications, was the result of a dual process that Elias studies in detail in *Power and Civility*, the second volume of *The Civilizing Process*.

Chartier continues:

The affirmation of the power of the absolute sovereign marked the outcome of a competition of many centuries' duration, opposing, in the same area, several units of domination.[72]

In describing the rivalry between the *noblesse d'epée* and *de robe*, Chartier emphasizes the practical institutionalization by the ruler of their antagonistic roles:

The antagonism that existed between the dominant social groups was thus, in the first place, the result of a differentiation of social functions that had reinforced the power of a bourgeoisie of office-holders and administrators that paralleled the traditional landed and military aristocracy. But this rivalry, the very condition of absolute power, could be and had to be perpetuated by the sovereign, who, playing one group against the other, reproduced the "balance of tensions" needed to create the personal form of the monopoly of domination. This meant, first, the parallel strengthening of the monarchical state and of the bourgeoisie of the robe, who took on the charges of justice and finance to counterbalance the claims of the nobility. Next, this meant that the royal will both protected and controlled the aristocracy as an indispensable counterweight to the power of the office-holders. The court became an institution essential to accomplish this end. . . . By preserving the aristocracy as a socially distinct group as well as making it subject to the ruler, the court constituted the principal mechanism that permitted the French kings to perpetuate their personal power. Fiscal monopoly, military monopoly and court etiquette were thus the three instruments of domination that together defined this original social form that was court society.[73]

Chartier also emphasizes the appropriateness of the ruler's lack of charisma as a contributing factor to the routinization of aristocratic/bourgeois tensions. Thus, Elias created a model of considerable complexity to analyze the early modern French court. Whatever one's opinion of Elias's model, the question remains: Does the Chinese case fit?

As I have noted above, it has been argued successfully that the great families of the Eastern Jin began to decline precipitously during the fifth century; it is also true that emperors attempted (as emperors are wont to do) to undermine the status of great families by encouraging the careers of minor gentry (sometimes termed *houjin* 後進) or even "commoners."[74] However, scholars who have noted these attempts have also stressed the general failure of southern ruling families to carry out any significant centralization in comparison with the northern courts or even the Tang (and of course even the Tang's success in this respect is often debated as well). Imperial interest in promoting the underprivileged does not guarantee a successful creation of an absolutist polity. Charles Holcombe, normally a scholar quick to suggest the literati's continuing dependence on the emperor for recognition of status, is nonetheless skeptical:

Southern dynasties emperors had tried for years to achieve a restoration of imperial authority. The power of the court probably sank to its lowest ebb during the Eastern Chin dynasty, in the fourth century. When the dynasty collapsed into rebellion, beginning as early as 399, peace was forcibly restored by military men from the Northern Palace Army. After their leader, Liu Yu, established his own dynasty, the Sung, in 420, it is possible to detect a subsequent strong imperial resolve to exclude great literati families from positions of real influence (as distinct from important-sounding sinecures) in the government, both to escape the powerful influence of the literati and because great literati made such miserably ineffective civil servants. . . .

If active administrative posts in the imperial government were increasingly staffed by men of humble origin, however, great literati families continued to dominate Southern dynasty society for another two centuries. Such men could even afford to ignore the mundane details of actual civil service because their position was so secure. Southern dynasties attempts at imperial restoration were at best only partially successful, and in the end it was the heavily non-Chinese Northern empire, where literati *pouvoirs intermediaires* had never been as intransigent, that triumphed where the Southern dynasty courts had failed in restoring imperial autocracy.[75]

And Dennis Grafflin concurs:

Not only did the Eastern Chin super-elite fade rapidly, but also there was no replacement of it by an equivalent Southern Dynasties super-elite of great families. From the rise of Ssu-ma Tao-tzu 司馬道子 in 385 to the extinction of the Ch'en dynasty in 589, aristocratic preeminence and its institutional fossil, the Nine Rank 九品 system, were short-circuited and ignored in the political arena by a variety of court and army factions drawn from sub-aristocratic sources.[76]

Although Grafflin does agree that the earlier aristocracy had vanished, his use of the term "factions" here suggests the continuing widespread violence of those centuries and of the complete inability of the emperor to assert any real authority over his state; at best he could set up his own faction and contribute to the quarrel. If poets of the time tended to congregate in the salons, it was more because of the personal ties they developed to certain specific princes.

Moreover, the lack of clarity in this move toward absolutism and its impact on literature is represented well by Chinese scholars' continuing questions and disagreements on the subject. Yan Caiping stresses the power of the emperor as critical judge of literati works, whereas Yang Donglin sees increasing imperial prestige as a motivation for aristocracies *to fall back on* their clan cultural skills and to emphasize their supremacy in this field. Yang points to the continuing literary prestige of the great clans and also notes that the royal families themselves competed as equals in boasting of their clan traditions of learning (a very different attitude from Louis XIV's deliberate avoidance of "charisma"). He also notes that a decline in clan political prestige did not mean a decline in the economic power of families.[77]

Thus, the question here is not whether the emperor was *trying* to centralize—but whether there is enough evidence to believe that such a centralizing move had an effect on the literary compositions of literati surrounding the emperor in the way that Louis XIV's court literature was affected by his centralization program. Did the *houjin*, or minor gentry, see themselves as more dependent on the ruler? And if so, were they influential enough on the literary scene to affect the cultural atmosphere of the times? I think it is possible that they did, but we simply do not know enough at present to be sure. Throughout my discussion here, I have assumed that competition and anxiety among poets was often more decentralized and focused on many different centers of power.

Paintings and Mirrors

Lady Ban sits looking at the portraits of noble ladies and bases her conduct on them. The painting is based on an ideal constructed for court ladies: "Be like her." Not all models are models of virtue, though; in fact, models are much more reliable guides to the outside surface than to inner worth, to ornament (*wen* 文) rather than substance (*zhi* 質)—just as a literary text (*wen*) provides the possibility of feigning substance. If one models oneself perfectly on the source, then who can tell the difference between the ideal and the response? Paintings can compel the viewer to compare him/herself to ideals of ornament, to change and refashion appearance.

Xiao Gang: A Court Beauty Beholds a Painting 詠美人觀畫

On the palace wall a goddess is painted;	殿上圖神女
From the hall emerges a splendid lady.	宮裏出佳人
It's lovely that both are painted,	可憐俱是畫
But who can detect the fake and the true?	誰能辨僞眞
I can discover clear eyes and brows in both,	分明凈眉眼
And the same kind of slender waist for each.	一種細腰身
Here's the way of telling them apart—	所可持爲異
Which one of them is in a good mood?[78]	長有好精神

Xiao has placed the painting in a public place; now he and his fellow poets hide behind the doorway and watch the fun. As a witty exercise in voyeurism, the poem's meaning is clear—from the emphasis on surface in determining a woman's worth to the equation between painting and the application of cosmetics, to the joking ambiguity of the last line, which leaves it open to the reader to decide if the painting is meant or the "real" woman. Yet the confusion here suggests not only the easy manufacture and replacement of women but also the imitation of one to another—the painting is also a mirror in which the woman sees another version of herself. This is true even if the portrait is not a portrait of her. Yet the painting is not merely the double/rival of the woman; it provides the model. She freezes before it, morphs into its twin.

Xiao Gang (the inventor of palace poetry, the master of "sensual" verse) presents a poem to his poets, who are invited to respond with their own verses. How does your work compare with this? The poet's response is highly charged: a sympathetic delight with surface and its articulation (the creation of another painting), an attempt to ornament the self so as to outdo the painting (standing before it, making distinctions between self and painting). One response is by Xiao's most talented poet-companion, Yu Jianwu 庾肩吾 (487–550)—father of the even more famous Yu Xin 庾信 (513–81):

Yu Jianwu: To Imperial Command
"A Beauty Sees Herself in a Painting" 詠美人自看畫應令

If you wish to know the skill, the cleverness of painting,	欲知畫能巧
Summon the real and set her off against it.	喚取眞來映
Both issue forth as if one body divided,	並出似分身
Behold each other as mirrors reflecting.	相看如照鏡
They steady their hairpins, spaced the same;	安釵等疏密
They wear their collars neat and tidy.	著領俱周正
If the Xiongnu hadn't raised the Pingcheng siege,	不解平城圍
Then who could have vied with this work of art?[79]	誰與丹青競

The identity of the woman is undermined by implying a splitting of the woman in two. No longer does the poem labor at the effort to determine which is real; for Yu Jianwu, there is no difference of one to the other—unlike Xiao, he does not claim to judge differences. The third couplet rhetorically restates this: the parallel structure makes each line a mirror of the other line in the couplet (each more or less says the same thing with different words), and the attractiveness of woman and painting is found in their orderliness and neatness—every ornament is in place, like the well-chosen words of a parallel couplet. The "real" woman steps out of the mirror the painting has become, as model and imitator fuse perfectly; or rather, she becomes a mirror as well, as mirror reflects mirror, reproducing a vertiginous multitude of identically perfect women. The poet himself becomes trapped within a paradox—the more he strains to attain the ideal model provided, the more he becomes indistinguishable from others. There lie the roots of the "sameness" of palace poetry, its infinite repeatability.

The last couplet refers to a semi-legendary story, mentioned above, surrounding the first Han emperor, Gaozu. Besieged in the town of Pingcheng by an army of nomadic invaders, the Xiongnu, he can think of no stratagem of escape. His clever advisor, Chen Ping, sends an emissary to the Xiongnu khan's wife with a painting of a beautiful woman. "The emperor has many such women in his harem. He has sent for one now, whom he plans to present to the khan, in the hope that he will raise the siege." Out of jealousy, the khan's wife immediately persuades her husband to let the emperor escape.[80] Jianwu's poem is ambiguous, yielding two possible meanings: if the chieftain had taken the city, his wife could not possibly have competed with the real beauties he would find within; or if he had taken the city, he would soon have discovered that the women within did not come up to the beauty of the painting. The perfection of ornament now becomes a political stratagem, deliberately eliciting jealousy through its own perfection. But is Yu pointing to the ultimate success of modeling—the true ladies who could replace the khan's lady? Or is he still claiming that there is some mark that distinguishes the real and suggesting that the painting is still an impossible ideal, leading to disappointment for the khan who pillages the city looking for reality and substance?

Paintings aspire to the state of mirrors in this poem, but mirrors are far more interesting in other poems.[81] As an element of a woman's bedroom, it is ordinary enough, and for Flying Swallow, the favorite of Emperor Cheng, a woman confident of her favor and yet conscious of the need to remain fashionable, the mirror is an indispensable guide, the surface that allows one to check and maintain one's own surface:

Fei Chang (fl. 510): Reflecting in the Mirror	費昶：詠照鏡
Dawn sunlight reflects off the apricot-wood eaves	晨暉照杏梁
As Flying Swallow rises for her morning toilette.	飛燕起朝妝
With a careful mind she spreads broad the mascara;	留心散廣黛
With a light hand rubs on flower-yellow powder.	輕手約花黃
She straightens hairpins, at times checking her image;	正釵時念影
Brushes with powder puff, treasures the scent.	拂絮且憐香
But she now dislikes these emerald colors,	方嫌翠色故
Suddenly feels her jades lack luster.	乍道玉無光
"In the city, everyone paints brows halfway,	城中皆半額
And would scorn the length I paint my own."[82]	非妾畫眉長

The touch of anxiety at the end is hardly a ripple on the front of the reflected brow; we have a warrior of the boudoir, who can contemplate her needs with the best of judgment. Competition requires this quality of distance, of self-evaluation, as well as sensitivity to change—ever ready to follow circumstances. But with the loss of the patron's favor, the mirror becomes useless; the subject is outside the competition, deprived of the judge who validates worth.

Gao Shuang (fl. 502): A Mirror　高爽：詠鏡

When I first climbed the Phoenix Stairs,	初上鳳皇墀
This mirror reflected my mothlike brows.	此鏡照峨眉
It reflected our keeping company together,	言照常相守
But will not reflect my longing for you.	不照常相思
Empty of hopes, I never use it now—	虛心會不采
But I still preserve its bright purity.	貞明空自持
Do not say the mirror's an old and useless thing,	無言此故物
For once again it will reflect our new trysts.[83]	更復照新期

"Empty of hopes, I never use it now"—"use" here is *cai* 采, "to select," what the ruler does when he picks his companion for the night. "But I still preserve its bright purity"—these adjectives are often applied to the human character as well. "Disillusioned, now that I am never picked, it's useless for me to preserve my bright purity." The double reading completes the fusion of woman and mirror here, and she projects her own actions onto the emperor in an act of sympathetic magic. Although the mirror and I grow older, we will not be discarded, for we may still be of use.

This is why abandoned court ladies in palace poetry debate whether to abandon their mirrors, put them aside in a dusty case or hold on to them in hope of change. But are mirrors true friends, revealing a fantasized meeting in the future? Or do they remind one constantly of aging?

Xiao Gang: Grieving in a Bedroom, Reflected
in a Mirror　蕭剛：愁閨照鏡

I've grown pale and thin since you left–	别來憔悴久
Others find my features strange.	他人怪容色
Only the mirror in my case	只有匣中鏡
Knows me at once when I hold it up.[84]	還持自相識

She has not only become unimportant with the departure of her lover—she has become unrecognizable. Made doubly alone by this change into another being, she must depend on the mirror to show her "what she is." There is an act of trust here, the belief that a mirror will always say "This is you." It is only that belief that can help, since she can see herself only with the mirror. If she did not believe that mirrors would tell her this truth, she would no longer know herself either. Yet is this something she desires, when the awareness of the self no longer matters, when no one else knows who she is? The mirror now promotes an act of self-identifying—it can no longer guide her in the social necessity of presenting surface. Yet there is consolation there as well; a relation has been created between woman and mirror that circuits them outside social evaluation. Xiao inadvertently suggests a role for culture outside social competition.[85]

In the end, however, mirrors are implicated in all the factors that produce social "sensuous" verse: they participate in the evaluation that grants poets prestige; they are perfect voyeurs, taking pleasure in the woman's appearance while spying on them unawares. Above all, mirrors are echoes made visible. If ladies use them to feel the way around terrain they cannot see, those echoes can also become the enemy, the companion that merely doubles one's words, mocking one's solitude. In the competitive community, everyone is striving to be like everyone else: to be like the woman in the painting, to be like each other. Mirrors may seem to be the only confident and unchanging echo of oneself as one paints the attractive surface or utters the attractive word. But in the land of doubles, they may be the enemy, the Other preying on your resemblance.

Xu Ling: Written for a Lady of Yang Kan's Household,
In Reply to a Gift of a Mirror 徐陵：爲羊袞州家人答餉鏡

Your letter came, and the gift of a mirror.	信來贈寶鏡
How tall it stands, like the round moon!	亭亭似圓月
But as a mirror ages, it gets more bright—	鏡久自愈明
And as I age, feelings fade away.	人久情愈歇
I'll take the mirror, hang it on an empty stand,	取鏡掛空臺
But will never uncover it from today.	於今莫復開
Haven't you seen that lone *luan*-bird?	不見孤鸞鳥
And how can a lost soul ever return?[86]	亡魂何處來

Xu Ling speaks for someone else—again, the assumption is that she cannot respond on her own, and the discourse is rerouted around the woman, becomes poet confronting poet as they enjoy the woman and her emotions. Did a poem accompany this mirror, thus calling for a poem in response? If so, it would have politely complimented Yang Kan's lady, suggesting perhaps that the mirror would allow her to survey her true beauty. Xu Ling typically takes her response in a different direction, speaking instead of the fear of aging and of the loss of favor (by the rules of palace-poetry reading, "feelings fade away" must refer to the cooling passions of the master as she ages and becomes unattractive). Yet the mirror as one side of the exchange makes it the dangerous, aggressive gift, like the poem that accompanied it: the rival/seducer infiltrating the lady's bedroom, the competitor. It was generally believed that mirrors became clearer with age; so as time goes by, the mirror (in its typical role of rival/solipsistic delineator of self-worth) will have increased capacity to show her every sign of aging. Rather than an unchanging record of reality, it grows strong on her diminishment. The woman gains revenge by covering it and refusing to let it bear witness. After all, the uncovered mirror is a bringer of death:

Once upon a time, the King of Jibin 罽賓 went off to snare birds on a steep mountain and caught a *luan*. The king cherished it greatly. He wanted it to sing, but could not find a way to bring it about. He graced it with a golden cage, fed it with rare delicacies, and treated it with increasing kindness; yet for three years it did not sing. His wife said, "I've heard that a bird will sing only when it sees its own kind. Why don't you hang a mirror for it so that it will see its reflection?" The king followed her advice. When the *luan* saw its reflection, it cried out sadly, and the mournful echoes rose into the sky. Then it gave a leap and expired.[87]

We are tempted from our heterosexualized viewpoint to see the *luan* bird's singing as a mating ritual, but in fact Chinese lore says only that birds like to sing in company—they are sociable. The placement of the mirror is an act of deception, giving the bird a partner supposedly different from itself and encouraging it in an act of false conversation. But the bird reads the reflection as the record of its own decline and perishes with the realization. Yang Kan's lady recognizes the deceit of the intruder-mirror and protects herself—but in the process cuts herself off from the Other, the community of rivals and fellow ladies. She potentially reduces herself to a silence that can only be broken by a death song.

Short-circuit?

In a few distinctive places, women poets are recognized in *Jade Terrace* and given voice. Their poetry does not grandly reconfigure the male games of competition: most resituate themselves as female speakers, constructing a female persona as ambivalent and complex as the males; after all, the court lady is indeed in the same position as the court poet. Perhaps one difference may be that the court lady can "cross-gender" herself as a man simply by writing as a woman: if speaking "as a woman" is a male right, then the act of writing poetry can allow the woman to break out of female restraints and reconfigure herself. The women poets represented in *Jade Terrace* seldom exploit this move in interesting directions. But there are a few moments in the women authors that demonstrate some awareness of the complicated constructions that poetic rhetoric is becoming.

Liu Lingxian (fl. ca. 500): In Reply to
My Husband 劉令嫻：答外詩二首之二

"The eastern neighbor is a perfect beauty."	東家挺奇麗
"Most delightful face in all the south."	南國擅容輝
"In the evening moonlight she appears like a goddess."	夜月方神女
"In dawn mists she becomes the Luo River Maiden."	朝霞喻洛妃
Yet when I look at my face in the mirror,	還看鏡中色
I know your foolish similes are frauds,	比艷自知非
Your descriptions are only clever flattery,	摛辭徒妙好
Your comparisons far off the mark.	連類頓乖違
Although you, my wise husband, think me lovely,	智夫雖已麗
I would never dare bring walls down with my looks![88]	傾城未敢希

Liu strips off the accretions of cliché and demands a certain honesty in address: my mistress's eyes are nothing like the sun. Yet we will never know whether this is an attempt to remove rhetoric from the exchange between husband and wife, to deny the male gaze (since articulating desire in poetry involves bringing in so many other factors seemingly alien to it), or whether it is simply a witty way of being modest.

CHAPTER FIVE

The Textual Life of Savages

From their horses' flanks hung the heads of men,
While their women were carried behind.
—attrib. Cai Yan

In 200 B.C., the Han emperor Gaozu 高祖 launched a punitive expedition against the Xiongnu 匈奴, an alliance of northern tribesmen who had been disrupting Chinese settlements north of the Yellow River. Sima Qian gives the details in chapter 110 of his *Historical Records*:

At that time, the Han had just brought peace to the empire. [The general] Han Xin 韓信 was transferred to Dai 代, with his capital at Mayi 馬邑. The Xiongnu launched a major attack on Mayi and besieged it. Han Xin surrendered to them. After the Xiongnu obtained his services, they led troops southward past Mount Juzhu 句注, attacked Taiyuan 太原, and advanced to below Jinyang 晉陽. Gaozu himself led troops to go and strike them. Just then the winter was bitterly cold and snow fell. Two or three out of every ten soldiers lost fingers from frostbite. [The Xiongnu khan] Modu 冒頓 feigned flight in order to lead on the Han forces.[1] They pursued to the attack. Modu concealed his elite forces and presented his weaker men. The entire Han army came after them, accompanied by 320,000 infantry.

The emperor reached Pingcheng 平城 ahead of the infantry. Modu then unleashed his cavalry, 400,000 strong, and surrounded the emperor at White Slope 白登 for seven days. Those outside the encirclement could not provide assistance to those within. . . . The emperor secretly sent an emissary to bribe the khan's wife.[2] She then said to Modu, "Why should you two create difficulties for one another? You may have taken Han territory, but you'd never be able to live there. Besides, the Han have their own protective deities. You should consider the matter carefully."

Modu had previously agreed to link up with Han Xin's generals, Wang Huang 王黃 and Zhao Li 趙利, but neither of them had arrived with his troops. He suspected that they were plotting with the Han, and so now he took his wife's advice and opened one corner of the encirclement. The emperor commanded his officers all to face outward with crossbows ready and go straight through the break, and he was finally reunited with the main army. Modu then led his army away. The emperor likewise brought away his troops and discontinued the campaign. He then sent Liu Jing 劉敬 to arrange an alliance with the Xiongnu by marriage (*heqin* 和親). (*Shi ji* 110.25–27)[3]

Gaozu's attack on Pingcheng is the opening volley of a dramatic border war that continued on and off for the duration of the Han. The outline of events is well known: the initial weakness of Han forces in the wake of Xiongnu raids; the costly though somewhat effective military campaigns of Emperor Wu 武 from 130 to 90 B.C.; the Xiongnu civil war that resulted in an improved Chinese position by 30 B.C.; a resumption of border conflicts that finally culminated in the north-south split of the Xiongnu in the mid-first century A.D.; and their increasing weakness in the face of new nomadic confederations during most of the following two centuries.[4] For this account, we are almost totally dependent on the official Chinese histories: Sima Qian's 司馬遷 *Historical Records* (*Shi ji* 史記), Ban Gu's 班固 *History of the Han* (*Han shu* 漢書), and Fan Ye's 范曄 (398–445) *History of the Later Han* (*Hou Han shu* 後漢書). Modern scholars can only speculate and hypothesize in an attempt to fill in the gaps created by an official and Chinese-biased historiography.

Yet scholars have said little concerning the way the official histories present the Xiongnu. Granted, we assume that they portray non-Chinese peoples less than positively and that the Chinese worldview is "ethnocentric," to say the least. But Chinese attitudes toward this foreign "other" create a subtler discourse of the alien as well, one that is intertwined with the historical narrative and continually surfaces in various ways; certain motifs emerge that both question and emphasize Chinese cultural distinctiveness. Of course, we can attribute this language in part to the obvious culture clash of Xiongnu and Han: the ancient historians do not invent a people *ex nihilo* merely for the purposes of creating edifying contrasts. But the "historical reality" (whatever that might ultimately be) is still refashioned into important ideological constructions.[5]

Even within the brief account of the Pingcheng campaign, Sima Qian manages to incorporate many aspects of this ideology. Gaozu must go on campaign because of a Xiongnu alliance with a rebel (and in fact the early Han emperors often faced the possibility of renegade Chinese generals and leaders joining forces with an external enemy). When Gaozu finally launches his expedition, the desolate and inhospitable north wreaks havoc on his soldiers; this is no place where Han Chinese would willingly live. Later, the Xiongnu lure the Han forces into a dangerous position. Chinese failure is thus attributable more to enemy perfidy and Han carelessness than to the inability of Han fighting men. Gaozu's and Modu's armies are originally comparable in size (though these numbers are probably greatly exaggerated for effect);[6] but a stratagem allows a small Han force to be surrounded by a horde of barbarians—thus creating a situation in which the tide of savagery threatens to swallow the small forces of "civilization." Finally, the Han, unable to cope with barbarian perfidy, manage to escape only through clever negotiation and a willingness to swallow their pride. Pingcheng will later become a symbol of the tendency of the early Han emperors to maintain the peace not through military might but through one "humiliating" treaty after another.

Gaozu's escape plans introduce a woman into the picture. Sima Qian's account in chapter 110 sounds at least possible, if not terribly plausible; it may reflect Han prejudices concerning non-Chinese women that they are meddlesome and far too politically active for the good of the state and society. However, Sima develops this part of the story in a more fanciful manner elsewhere:

Chen Ping 陳平, acting as a colonel in the guard, accompanied the emperor when he went to attack the rebel Han Xin in Dai. When the troops reached Pingcheng, they were besieged by the Xiongnu. For seven days they went without food. The emperor employed a marvelous plan of Chen Ping's and sent him to the khan's wife. In this way the siege was lifted. After the emperor escaped, he kept the plan secret, and no one in the world has managed to learn of it. (*Shi ji* 56.14)

Here, in the biography of Chen Ping, Sima seems to follow a different tradition, since *Historical Records* 110 makes the plot fairly explicit: the khan's wife was simply bribed. The Later Han philosopher Huan Tan 桓譚 (43 B.C.–A.D. 28), however, elaborated spectacularly on this "unknown plan"—and so created a classic Chinese myth of race and gender:

Someone said, "Chen Ping raised the siege of Pingcheng for the emperor. Afterward it was said that the affair was kept secret and that no one in the world has managed to learn of it. It must have been kept secret because it was marvelously clever and exceedingly excellent. Can you speculate as to the nature of it?" I replied, "In fact, their scheme was uninspired, clumsy, and poor. *That* is the reason it was not leaked to outsiders. After the emperor had been besieged for seven days, Chen Ping went to persuade the khan's wife, and she in turn persuaded the khan, who then let the emperor escape. From this we can fathom what he must have done to persuade her. Chen Ping must have said, 'The Han is filled with beautiful women, their appearance surpassing that of anyone else in the world. Now that the emperor is in difficulties, he will have already sent a swift messenger to bring one back. He will present her to the khan. If the khan sees this woman, he will be smitten with her; and if he does, you will fall daily into greater disfavor. It would be better to let the Han forces escape before she arrives. If they escape, they won't bring the woman.' The khan's wife had a jealous nature; she must have hated this possibility and thus made sure that this would not happen. Chen Ping's plan was a simple one; but after it was used, he wished to make it seem strange and mysterious and thus concealed it so that it would not leak out." (*Shi ji* 56:14)

Ying Shao 應劭, the second-century commentator on the *Han History*, elaborated on this version by suggesting that the envoy to Modu's consort showed her a portrait of a beautiful Chinese woman to clinch his argument (*Han shu* 1B.63).[7]

Huan Tan's speculation may spring from some Chinese advisors' contempt for the Han policy of *heqin*, the peaceful and renewable negotiation of treaties with the Xiongnu that was usually sealed not only with "bribes" to the chieftain but with a Chinese princess in marriage as well. In this limited sense, *heqin* means not merely "creating harmonious and friendly ties" but, more specifically, "bringing peace to one's kin"—that is, reasserting the friendly relations that should exist between relatives by marriage.[8] By projecting a form of *heqin* as the shameless and desperate scheme of a besieged emperor and his minister, Huan manages to allegorize an aspect of Han/Xiongnu relations that reflects badly on the imperial order and on imperial policy. Not only does Han military might give way to a tawdry deception, but a woman is used to make the deception work. The empire is saved in this case by a woman's jealousy (just as in other cases it is wrecked by it); the internal and the *yin* are allowed to influence public matters. Beyond this, however, the Han itself is subordinated and feminized. The emperor plays a passive role within the siege, with no active or forceful method of escape. The remedy comes not through action, but through words, persuasion, false rhetoric. Finally, Liu Jing—the man who is credited with developing the *heqin* system—negotiates the treaty. The surrender of a Chinese princess means the marriage of the Han to the Xiongnu; although later rulers might claim that they are the "elder brothers" of the tribesmen, they are also symbolically their wives.

The role of gender in the construction of Xiongnu/Han relations allows us to discuss it within the framework of our larger argument; for the non-Chinese Other creates a new field within which to introduce the same gendered polarities analyzed above: husband/wife, ruler/minister, superior/subordinate. In this chapter, I examine early historical accounts of the Xiongnu as the basis for such gendered constructions; although historians have examined them repeatedly in an attempt to reconstruct Xiongnu history, I am more interested in the cultural attitudes they imply. We will then move to other depictions of the savage by discussing texts and poems concerning Wang Zhaojun 王昭君 (also known as Mingfei 明妃, "The Brilliant Consort"), the first-century B.C. court lady who left China to become a bride of a Xiongnu chieftain.

Constructing Aliens

The primary texts for Xiongnu history and ethnography are chapter 110 of the *Historical Records* and chapter 94 of the *History of the Han*; the latter duplicates Sima's work with minor changes and continues it up until the fall of the usurper Wang Mang 王莽 in A.D. 23. In a reflection of the importance of the Xiongnu and the role they played in Chinese literati consciousness, these narratives grant them considerably more space than any other "barbarian" people. Throughout, subtle and not-so-subtle descriptions and anecdotes build up an image of this enemy of the Han. This commences with Chinese conceptions of Xiongnu genealogy. *Historical Records* 110 begins not with the confederation's founding in the third century B.C. but with an account of northern barbarians' influence on earlier Chinese matters. First, it identifies the Xiongnu ancestor as a member of the Chinese cultural world, a descendant of the Xia 夏 royal house by the name of Chunwei 淳維 (*Shi ji* 110.2). This is a typical move: traditional historiography tends to trace non-Chinese peoples back to a Chinese ancestor, although generally the more "alien" or geographically distant they lie from the core of Chinese civilization, the more remote the ties of kinship. The text passes immediately to a discussion of the Rong 戎 and Di 狄 tribes, who appear often in the early Classics. Sima Qian goes out of his way to emphasize the rather significant role the Rong and Di played in the most prominent events of early Chinese history:

barbarian incursion or participation serves as a catalyst for domestic affairs. An ancestor of the Zhou line, Gong Liu 公劉, lives among the Western Rong, adopts their customs, and founds a city; later, his descendant, Danfu 亶父, flees their attacks and thus founds the Zhou state itself, at Mount Qi 岐 (*Shi ji* 110.4).[9] A Zhou military expedition against barbarians anticipates the Zhou assault on the Shang itself. Meanwhile, the first "tributary system" is established, in which barbarians are forced to live in an area beyond the Chinese ecumene and are compelled to attend court periodically and present gifts to the Zhou ruler (*Shi ji* 110.5).

Next, Sima relates the fatal loss of western China, and the transfer of the capital to Luoyang is brought about by a pact between Marquis Shen 申, a disaffected nobleman, and the Quan 犬 barbarians (*Shi ji* 110.5–6). A major Chinese setback is attributed to an alliance between a renegade and a foreign threat. From this point on, many of Sima's incidents, both positive and negative, demonstrate the consequences of barbarian-Chinese mingling. At the time of the loss of the west, one of the wilder feudal states, Qin 秦, begins its climb when Duke Xiang 襄 drives the barbarians westward. Qin thus takes on the role of peacekeeper and aggressor on the western frontier. A little later, the great Duke Mu 穆 of Qin (r. 653–21) continues his attacks on the Rong; the duke's policies as Sima Qian describes them elsewhere prefigure the uneasy shifts of alliance that will become typical later:

> The king of the Rong sent [his advisor] You Yu 由余 to Qin. You Yu's ancestors were Jin 晉 men who had fled to the Rong, and he could speak Jin dialect. The king of the Rong had heard that Duke Mu was a worthy ruler, and so he sent You Yu to observe Qin customs. The duke showed You Yu his palaces and halls, piled one atop the other. You Yu said, "If you had spirits build them for you, then you must have tired them out. But if you had the people build them, you caused them much suffering." (*Shi ji* 5.7)

You Yu goes on to defend the "barbarian" way of life. Sima employs the rhetorical strategy of using the Other to define the faults in one's own society; but ultimately it is You Yu's naïveté and honesty that bring about the downfall of the Rong when Duke Mu successfully suborns him:

> The duke seated You Yu close to him on his mat and fed him from his dishes. He inquired after his country's topography and the disposition of its forces. He learned everything. Then he commanded a court clerk to send the Rong king two sets of eight female musicians each. The Rong king accepted them with delight and did not

return them even after a year. Qin then sent You Yu home. You Yu remonstrated with the king repeatedly, but the king would not listen. The duke also secretly dispatched emissaries to invite You Yu to return, and You Yu then surrendered and went over to Qin. The duke treated him with courtesies befitting a guest and inquired of him how to attack the Rong. (*Shi ji* 5.8)

Rong virtue is undermined by the *shengse* 聲色 (women and song) of Chinese culture. As a result, You Yu's own moral standards compel him to abandon the people his clan had adopted.[10]

Other parts of China also see foreign incursions merge with internal politics. One king of Zhou marries a barbarian queen in order to enlist her people's help in attacking the state of Zheng 鄭; his later abandonment of her results in a barbarian invasion and the king's temporary exile (*Shi ji* 110.7). However, as the feudal period proceeds, many of the border states begin to beat the tribes back, either through military efforts or through political intrigues. The initial stages of the "Great Wall" are built after an incident in which nomads meddle in the internal and sexual politics of a state:

During the reign of King Zhao 昭 of Qin, the king of the Yiqu 義渠, a tribe of the Rong, carried on an affair with Xuan 宣, the queen dowager. They had two children. She then killed the barbarian king through a deception at the Sweet Springs Palace. Afterward she raised an army to attack and devastate the Yiqu. Qin then took possession of Longxi 隴西, Beidi 北地, and Shangjun 上郡. They constructed a long wall to keep out the Tartars. (*Shi ji* 110.11)

The permeability of sexual boundaries results in an attempt to erect a more solid border. A similar mechanism occurs when King Wuling 武靈 of Zhao 趙 succeeds in driving the tribesmen north after he makes his people adopt their dress and modes of warfare. In Yan 燕, farther to the east, the general Qin Kai 秦開 is sent as hostage to the tribesmen and puts his knowledge of their ways to use when he returns; as a result, he expels them from the northern borders of his own state. In both Zhao and Yan, these acts of cultural intermingling end with the construction of barrier walls—walls that will also contribute eventually to the Great Wall (*Shi ji* 110.12–13).

The narrative repeatedly suggests that the actions of non-Chinese peoples have constantly changed and transformed Chinese history. Although rulers attempt to create boundaries to keep them out, both sides constantly cross them in order to achieve their goals. All of this provides an ironic counterpoint to the idealized system of Chinese-barbarian relations presented in the "Tribute of Yao" (*Yao gong* 堯貢) chapter of the *Book of Documents* (*Shang shu* 尚書). Here, the state is viewed as a series of five demesnes nestled one within the other. At the center is the king and the area under his immediate control; he is surrounded by fiefdoms granted to his immediate relatives, which are surrounded in turn by fiefdoms held by less closely allied noblemen. In the fourth and fifth areas dwell allied and non-allied barbarians, who rarely come to court to present tribute. The goal of this system is to provide a series of protective barriers around the central Chinese core; if enough distance is kept between the truly alien barbarian and the ruler, the ruler can essentially carry out his duties while paying little or no attention to barbarian matters.[11] As Yang Lien-sheng has suggested, this created at the very least an idealized conception of diplomacy, whereby Chinese and non-Chinese could each mind his own business—and historians such as Ban Gu praised such a policy as the only one that could assure peace.[12] For him (and perhaps for Sima Qian as well) the interference of barbarians in internal Chinese politics was a sign of the deterioration of the Chinese polity from the more ideal conditions of antiquity, although it is highly unlikely that the *Book of Documents* model of concentric polities ever existed in reality.

If Sima's early history of the northern frontier suggests a view of the barbarians as a dark presence ready to interfere at any moment and to exploit the discontent of prominent Chinese at the very core of the state, worse was yet to come. In this early period, as Sima points out, the tribes "were scattered, dwelling in small valleys each with its own lords and leaders. Now and then they would join to form groups of a hundred or so, but no one was able to unify them" (*Shi ji* 110.10). All this changed when these tribes began to use horses and developed a culture of nomadic pastoralism in the fourth and third centuries B.C. Soon cavalry replaced the foot soldier; barbarians became considerably more deadly, and their raids almost impossible to defend against. They eventually developed into larger political units, until the western barbarian leader Modu forged tribes throughout Mongolia into the Xiongnu confederation about the time of the founding of the Han. The new Chinese empire was now face to face with another empire, capable quite often of collective action and sustained warfare—or at least capable of a more consistent policy than the disunited tribes had ever been.

Thomas Barfield has argued that the formation of the Xiongnu is not coincidental. Rather, as the Han polity itself unifies, the nomads living on its edge find a peaceful and prosperous nation more profitable for raiding. Drawing on recent theories that pastoral nomadism depends economically on such raids for its stability, Barfield concludes that a strong Chinese empire will inevitably produce a strong nomadic empire in turn. When the Chinese empire disintegrates, the loose alliance of tribes that makes up a nomadic empire crumbles as well, since the nomad chieftain can no longer promise his followers a high level of success and profit from raiding a no longer prosperous land.[13] Whether Barfield's model is accurate or not I leave to historians to judge. His interpretation of the Xiongnu is based on data drawn chiefly from Sima Qian. As a result, he may be noticing not so much a historically accurate account of events as the Xiongnu's impact on the Chinese imagination. In *Historical Records* 110, the Xiongnu confederation takes on the role of the Han's dark Other—a competent, organized polity that competes with the Han for territory and allegiance. As Barfield himself notes, one of the consequences of Xiongnu unity is to limit the political options a borderman has: he can either cast his lot northward with the Xiongnu or southward with the Han.[14] The result is a polarization of possible identities.

The Xiongnu empire thus repeatedly serves as the dark mirror of the Han in Sima's account, most prominently in his rhetorical attempts to define the Xiongnu as the radical opposite of the Chinese. At the very beginning of his account, before he discusses the genealogy of the tribes, the historian gives an overall evaluation of Xiongnu character. In this, he repeats a model he used in the personal biographies, whereby he often briefly summarizes the subject's character before commencing the main narrative. Most of this evaluation emphasizes the difference between Xiongnu habits and those of the Chinese:

> They move about in pursuit of water and pasturage, and they have neither walled cities, fixed abodes, nor fields to till. Nevertheless, each group has its appointed territory. They have no system of writing and depend on verbal agreements. The boys can ride sheep and go out to shoot birds and mice with bow and arrow. When they are slightly older, they shoot foxes and rabbits and eat the meat. Thus all the men can use bows, and all participate in their armored cavalry. Their custom is to follow their flocks when at peace and make a livelihood at hunting; but in times of crisis they practice warfare and make raids. This is their nature. Bows and arrows are their long-range weapons, and they use knives and spears at short range. When the advantage is on their side, they advance; otherwise, they retreat, and they do not find it shameful to flee. They will take advantage whenever they can, giving no heed to

the proprieties. Everyone from the rulers on down eats meat taken from their herds; they dress in hides or in felt robes. The able-bodied eat the best, and the old live off the scraps. They honor the young and healthy and scorn the old and weak. They marry their stepmothers, and when a brother dies, one of the siblings marries the widow. They have personal names but no taboos in using them, and they have no surnames or "polite" names. (*Shi ji* 110.3–4)

One might analyze this brief ethnographic description exclusively as a set of polarities that would be self-evident to the Chinese reader of the time: migratory/sedentary; oral/written; military education / civil education; war as marauding / war as self-defense; cowardice/courage; self-advantage/righteousness; heavy meat diet / heavy grain diet; leather, fur, felt / silk; respect for the young / respect for the aged; impropriety in marriage / propriety in marriage; disregard for name proprieties / indiscriminate naming. Not all these polarities portray the Xiongnu custom as negative, but the structure of them/us is so pervasive here that the ultimate effect is to render *all* Xiongnu customs as inherently non-Chinese and explicitly or potentially immoral. Note, for example, the last two items in the list: it is easy to see that a Chinese reader would find the custom of widow remarriage to a brother-in-law or stepson an abhorrent violation of the rites, but this disregard for the integrity of family identity produces in turn a disregard for the customs of naming that go far to identify the individual within a network of social proprieties. Quite strikingly, almost the only statement that does not contribute to a polarity is the one concerning the Xiongnu polity: "Nevertheless, each group has its appointed territory." It is this one concession to political organization that make the Xiongnu a coherent threat to the Han order.

Of course, Sima Qian's analysis here is most likely not as pernicious as much nineteenth-century imperialist writing was, but we cannot blind ourselves to the fact that our main historical sources on the Xiongnu are saturated with this language.[15] Although it is offset to some extent by the arguments of a figure like You Yu, the use of outsiders as mouthpieces to criticize Chinese customs only emphasizes their essentially alien nature.

The Xiongnu inversion of Chinese cultural practices emerges occasionally elsewhere. One may note, for example, the dramatic story of Modu's founding of the confederation—one that involves the ultimately unfilial act of parricide (*Shi ji* 110.15). Although this account may reflect the Xiongnu's own founding legends (the Oedipal drama of assuming authority through killing one's father is common enough), a Chinese reader is more likely to feel some revulsion.[16] Another inversion has more explicit political consequences, however: the barbarians, simply by being barbarians, ought to be subordinate to the superior Han cultural order. But in the second century B.C., the policy of appeasement and *heqin* determined that they were not—at least to some policy critics. Here we see the beginning of the debate over *heqin*, in which failure to keep the nomads in line is related to an inversion of the "way things ought to be." Thus Jia Yi 賈宜 (200–168 B.C.):

> The situation of the world is as if it were turned upside down. Generally speaking, the Son of Heaven is the head of the empire. Why? Because he is on top. The Man 蠻 and Yi 夷 barbarians are the foot of the empire. Why? Because they are on the bottom. Now the Xiongnu arrogantly invade and carry out raids. This is the ultimate in disrespect. The disasters in the empire have no end, and yet every year the Han sends them gold and silk floss and dyed fabrics. Now the Yi and Di issue commands, and this should be the right of the ruler; the emperor submits tribute, and this is the ceremony of the subject. The feet perversely occupy the top, and the head the bottom. We hang inverted like this, and yet no one can release us. Can it be that there are no men who can help the state? (*Han shu* 48.2240)

Not only have the Xiongnu inverted all the sacred values of Chinese culture, they have threatened to invert the political order as well, destroying the proper relations between ruler and vassal. Although Sima's and Ban Gu's views of the Xiongnu are not unrelievedly negative (they level a considerable amount of criticism at the waste, dishonesty, and violence of Han foreign policy), this essential analysis of the Xiongnu's significance continues to hold true through their discussions of them. The Xiongnu are not like us, and they continue to pose a threat to the world order.

Men on the Border

However, there is another, partially contradictory theme going on within the historians' writings. As I have suggested above, Sima knows well that boundaries are continually being crossed, in spite of ideal structures that would relegate the barbarians to an outside realm. Owen Lattimore commented on this long ago from a practical, geographical perspective. He suggested that a border did not exist between Han and Xiongnu; rather, they were separated by a *frontier*, a broad belt of territory in which Han lifestyles gradually shaded into pastoral nomadism. The changes are imperceptible, movable, and permeable.[17] Ultimately, the construction of a system of walls during the Qin was intended to make borders more permanent, to keep barbarians out and to keep Chinese in. However, as Arthur Waldron has recently suggested, these walls were largely useless as a military or defensive barrier. Rather, territorial walls were symbolic boundaries that allowed the Han polity to define a solid boundary between Han and Xiongnu. In fact, agreements between the Han emperors and the Xiongnu during the second century B.C. often mentioned the walls as defining foci for treaties and their provisos: the Xiongnu have a right to everything north of the walls, the Han to everything south.[18] In this way, the government attempted to ignore the existence of the large transitional frontier and of the existence of hybrid ways of life and of hybrid bordermen.

As Sima Qian had already pointed out, however, northern and southern cultures often intermixed; their political histories influenced each other. Han Xin was only the first in a series of Han renegades who negotiated with the Xiongnu, invited their incursions, or deserted to them in the capacity of advisors. On the other side, particularly during the campaigns of Emperor Wu, Xiongnu chieftains and generals surrendered and took up Han arms against their former comrades—and sometimes deserted again and returned to the north. It is thus not surprising that agreements also included provisos insisting on the Han and Xiongnu right of dealing with their own people.[19] This constant interchange across the frontier (and the threat it posed to cultural stability) serves as a subtext to Sima's narrative, just as much as the portrayal of Xiongnu as Other, un-Chinese. One of the most striking examples of this is the attention Sima gives to the career of Zhonghang Shuo 中行說:

> Emperor Xiaowen 孝文 [r. 179–157 B.C.] again sent a princess of the royal house to be the khan's wife, and he sent the eunuch Zhonghang Shuo, a man of Yan, to attend on her. Shuo did not wish to go, but the Han forced him. Shuo said, "My going will certainly be a misfortune for the Han!" After he arrived and granted his allegiance to the khan, the khan treated him with great intimacy and favor. (*Shi ji* 110.33)

As a eunuch, Zhonghang Shuo has already crossed one boundary; now he is forced to cross another one as well. He has surrendered being a man *and* a Chinese. It is no coincidence that his importance begins with a *heqin* agreement: he becomes a proxy for the princess herself and thus creates a paradoxical fertile hybrid of Chinese and Xiongnu culture. This is, ironically, a fatal reversal of the positive goals of hybridity and miscegenation that Liu Jing introduced in his invention of the *heqin* policy after the disaster at Pingcheng. Claiming that ties of kinship ultimately transcend political enmity, Liu had assumed that once the khan became the son-in-law of the emperor, Chinese authority would prevail. He not only imagined the genetic triumph of the Han but also saw *heqin* as a wedge that would allow for the introduction of Chinese moral reasoning as well as the softening influence of the material products of Chinese civilization (*Shi ji* 99.8–9). Jia Yi, in spite of his generally hawkish attitude toward Xiongnu relations, also saw the export of Chinese civilization as a successful method of weakening Xiongnu belligerence and martial spirit—just as Duke Mu's gift of the female musicians weakened the Rong.[20] Zhonghang Shuo, by bringing Chinese know-how to the Xiongnu, recognizes the danger of such cultural pollution:

Before this, the Xiongnu had a liking for Han fabrics and foodstuffs. Zhonghang Shuo said, "Although the Xiongnu are numerous, they cannot match even one district of the Han in population. But their strength lies in the fact that their food and clothing are different and that they have little esteem for the Han. But now you, khan, have changed customs and have become fond of Han goods. If no more than two tenths of your goods were of Han origin, then you would all lose your freedom to the Han! If you simply put on Han fabrics and wear them as you gallop through the fields and brambles, your robes and trousers will become torn and tattered. That will prove that they cannot compare with the durability of your felt and fur. When you obtain Han foods, refuse them in order to show that they do not possess the convenience or soundness of your kumiss." (*Shi ji* 110.33–34)

On the other hand, Zhonghang's role as cultural hybrid allows the Xiongnu to adopt the right kind of cultural habits:

He taught the khan's officials recordkeeping in order to make a record of his people and herds. Whenever the Han sent the khan a letter, the wooden slip bearing the message was a foot long and an inch wide. It would read, "The emperor respectfully inquires after the health of the khan. The gifts we have presented are as follows," etc. Zhonghang Shuo had the khan send the Han letters on slips a foot long and two inches wide and had him enlarge the size of the seal. He would make the message more overbearing: "The Great Khan of the Xiongnu, born of heaven and earth and established by the sun and moon, respectfully inquires after the health of the emperor. The gifts that we have presented are as follows," etc. (*Shi ji* 110.34)

Later on, Sima comments: "Day and night he taught the khan how to exploit his opportunities" (*Shi ji* 110.37). Throughout these passages, Sima suggests that the Xiongnu needed a figure like Zhonghang to deal with the Chinese; since they are so completely "opposite" from their cultural opponents, only a selective adoption of certain Chinese habits will allow them to communicate and compete with their southern enemies. Zhonghang teaches them bureaucratic account-keeping and Chinese ceremonial practices. The result is ominous.

Most striking of all, however, is the speech Sima creates for Zhonghang in which he virulently defends the Xiongnu way of life and code of ethics:

One of the Han envoys once said to him, "It is the Xiongnu custom to scorn the aged!" Zhonghang berated the envoy. "Is it not the Han custom that when soldiers first set out for the border posts, their elders surrender warm clothing and fine foods in order to see them on their way?" The Han emissary assented to this. Zhonghang continued, "The Xiongnu make fighting their business. Since the old and weak cannot fight, they give the best of their food to the young and healthy; since these latter can thus act as their defenders, both fathers and sons are successfully preserved. So how can you say the Xiongnu disregard the aged?" The emissary said, "But among the Xiongnu both fathers and children sleep in the same tent; and when the father dies, son marries stepmother, and when a brother dies, all his siblings marry the widows. They do not possess the ornament of cap and sash, and they lack courtly rituals." Zhonghang replied, "It is the custom of the Xiongnu to eat the flesh of their herds and to drink their milk and wear their hides. Their herds feed on the grasses and drink water, moving with the seasons. This is why in times of crisis they know how to ride and shoot, and in times of peace they are happy with no required business to attend to. Agreements are lightly held and easy to carry out. Since the relations between ruler and subject are simple, ruling a state is as easy as ruling one man. The reason why they marry the wives of their fathers and brothers is because they hate the loss of a clan. This is why they keep families alive even in times of chaos. Now in China, even though they are cautious not to marry their own kin, relations grow increasingly alien from each other, and it ends with them killing one another. The change of ruling houses comes from things like this. Moreover, when propriety is in decline, those above and below resent each other, and the people's strength is weakened through excessive building projects. Men plow and engage in sericulture with all their effort merely in order to find food and clothing; they build city walls to protect themselves. For this reason the people are untrained in the military arts during times of crisis, and in times of peace they weary themselves in their occupations. Well, you house-dwellers shouldn't chatter so much. Of what use are all your ceremonial hats?" (*Shi ji* 110.34–36)

In this extraordinary passage, Sima allows Zhonghang to defend almost every cultural practice that had been implicitly criticized at the beginning of the chapter.[21] Moreover, Jia Yi's and Liu Jing's arguments on the "softening effects" of Han civilization are now employed to attack the weaknesses of the Chinese themselves.

Sima Qian is noted for his ability to balance many points of view by projecting the conflicting speeches of his dramatis personae; he rarely simplifies or explains away the complexity of historical events. As David Schaberg has noted, Zhonghang's speech does not quite elevate the Xiongnu as an alternative to a corrupt Chinese society. Rather, it is Zhonghang's status as Chinese eunuch turned renegade that allows him to create a plausible pro-Xiongnu anti-Han argument within the boundaries of Chinese discourse itself.[22] The Xiongnu are not defending themselves; no doubt most of them would be largely indifferent to Han attacks on their lifestyle. Zhonghang's rage is the rage of a man involved in an act of vengeance on the society that has rejected him. Most important, however, Zhonghang knows how to use the Xiongnu as the instrument of his revenge: Sima suggests that the Xiongnu become dangerous only when Chinese renegades and marginal men allow them to compete successfully within the Han cultural sphere.

In spite of Zhonghang's warnings against Chinese goods, however, Sima's account makes it clear that the Xiongnu were eager to engage in trade when they were at peace, and so the Chinese continued to sell to them—for their own profit, as well as in the constant hope that the Xiongnu would be softened by the exposure and thus become proper subordinate vassals. The policy of *heqin*, then, continued to be rooted in an ideological paradox: the princess, symbol of an effeminate Chinese civilization, would be the Trojan horse introducing that civilization to the Xiongnu. The subordinate, the emperor, would thus undermine the supremacy of his temporary superior, the khan. It may have been this continuing ideological subordination (not to mention its ultimate failure to obtain its aim) that made Chinese ministers so uncomfortable with *heqin* agreements. Zhonghang Shuo could show how *heqin* could benefit the Xiongnu by making them stronger, not weaker.

This is illustrated further by the marriage proposal Modu sent to the Empress Dowager Lü 呂 during the period she controlled the government as regent (195–180 B.C.). Granted that the empress was in control of the empire (thus inverting the gender hierarchy of the government), his insulting missive comes as a comic parody of the *heqin* system and its implicit subordination of China to the barbarians:

> I am a lonely and fragile man. I was born in swamp and marsh, and I grew up on the plains, in the realm of ox and horse. Often I have traveled to our borders, for I wished to visit the central land. Your Majesty now reigns on your own; lonely and fragile yourself, you dwell apart. Since we two rulers are both joyless and have no way to make ourselves happy, I wish for us to trade what we have for what we have not. (*Han shu* 94A.3754–55)

The last line surely mimics the language of trade agreements. Liu Jing had advised Gaozu: "On a yearly basis you should ask after the Xiongnu ruler's health and make presents of what the Han has in excess and what they lack" (*Shi ji* 99.8). The khan recognizes the essential link between *heqin* and trade, between miscegenation and material exchange.

Chinese advisors became increasingly hostile to the *heqin* system and the dangers of cultural interaction; the result was the outbreak of hostilities at the beginning of Emperor Wu's reign, followed by forty years of protracted fighting. Yet if these campaigns were started in the hopes of driving the Xiongnu beyond the frontiers of the Chinese cultural sphere, they ended by producing yet another "man on the border," the Chinese campaigner. Sima and Ban Gu pay particular attention to these men, often granting them their own biographies, and they in turn have become some of the most popular figures of Han history. Li Guang 李廣, Wei Qing 衛青, Huo Qubing 霍去病, Li Ling 李陵, Li Guangli 李廣利—all had their lives and military campaigns heavily dramatized in the histories. Yet almost all of them were compelled by the vagaries of their campaigns to become to some extent "marginal men," caught between the demands of Han politics and the assaults of their Xiongnu enemies. Readers of the *Historical Records* and the *Han History* are made aware that these commanders were threatened not so much by Xiongnu attacks as by the ruler's imperious demands that they produce results. Defeat (or even failure to follow commands to the letter) could result in cashiering or execution. Both Li Ling and Li Guangli were compelled to surrender to the Xiongnu, and both became renegades. Li Guang had a reputation for personal courage and wiliness in dealing with the Xiongnu, tempered with bad luck that served to infect those who came in contact with him. When in the end he is not allowed to participate in a major campaign for fear that he will bring failure with him, he commits suicide out of shame.

All these men respond to two severe and unrelenting forces; they live out much of their lives in the no-man's-land of the frontier and attempt to achieve wealth and fame and avoid death at the hands of either side.

Many of these social factors (and the continuing ambivalence Chinese felt about their own identity and that of the barbarians) emerge in the famous joint biographies of Li Ling and Su Wu 蘇武 in the *History of the Han*. Li Ling, the grandson of the hapless Li Guang, continues his family's history of disasters when he is cut off by a Xiongnu force deep in enemy territory and is eventually forced to surrender. Ban Gu describes this failure carefully: a commander was generally expected to kill himself rather than surrender, and there was the additional horror in this case of renegadism, of joining the "other," alien culture and thus losing one's own humanity. In compensation, the historian stresses how Li Ling held out as long as he possibly could: opposed even before the battle by personal enemies within the Han army, he must assume an almost impossible task, which he nearly accomplishes with brilliance and panache. When he is finally surrounded, he carries on a ruthless, week-long battle with a force that far outnumbers his own. Nor does Emperor Wu completely condemn his defeat: after his initial rage, he relents and sends forces to assist the men in Li Ling's army who had managed to escape (*Han shu* 54.2455–57).

Ban Gu also tells us that Li Ling, although feted by the Xiongnu, who greatly admired his courage, refused to aid his captors by joining them in their campaigns or supplying information to aid their assaults. This situation changes through a tragic misunderstanding:

After Li Ling had been with the Xiongnu for a little over a year, the emperor sent the Yinyu 因杅 general Gongsun Ao 公孫敖 to lead troops deep into Xiongnu land and meet up with Ling. Gongsun's troops returned without success. He said, "I captured someone alive, and he reported that Li Ling is teaching the khan military tactics in order to help him withstand the Han armies. That is why I had no success." When the emperor heard this, he put Li's entire clan to death. Mother, brothers, wife, and children were all executed. . . . Later, when the Han sent an emissary to the Xiongnu, Ling said to him, "I commanded a force of five thousand infantry in a sortie against the Xiongnu, but because I went without relief forces, I was defeated. How could I have betrayed the Han so badly that my entire family should warrant execution?" The emissary replied, "The Han heard that you were teaching military tactics to the Xiongnu." Ling said, "That was Li Xu 李緒, not I." Li Xu was originally a frontier commander stationed at Xihou 奚侯 city. The Xiongnu had attacked him, and he had surrendered. The khan treated him as a guest, and he always occupied a place above that of Li Ling. Li Ling was upset that his family had been executed because of Li Xu, and so he sent a man to assassinate him. The khan's mother wished in turn to kill Ling; so the khan hid him in the north. He only returned after her death. The khan thought Li Ling a brave man and married one of his daughters to him. (*Han shu* 54.2457)

Li Ling is thus first undone by a general forced to concoct an excuse to save his own skin; then ruined again by a renegade that shares his name. Li Xu appears nowhere else in the *History of the Han*; one wonders if someone invented this dark dopplegänger of Li Ling's to save the hero's reputation. Yet it is the loss of his family (and hence, symbolically and practically, the breaking of any ties he had with Han society and culture) that finally drives Li Ling to surrender to the khan's advances. In return, he joins not only the khan's family but also the violence and political intrigues of the Xiongnu tribe as well.

Ban Gu perceives Su Wu as a foil to Li Ling, a man who refuses to surrender his Chinese nature. Sent as an envoy to the Xiongnu, he is detained by the khan when a member of his staff becomes embroiled in a conspiracy at the khan's court. A renegade Chinese, Wei Lü 衛律, attempts to win Su over, but he adamantly refuses to surrender his loyalty. Eventually the Xiongnu send him to the distant north as a shepherd:

They then moved Su Wu to an uninhabited region on the Northern Lake and made him herd rams, telling him that when his rams yielded milk he could return. They then separated the officials in his retinue and sent them to different places. After Wu arrived at the Northern Lake, his government food supplies had not arrived. He dug up and devoured the supplies of weeds and nuts the field mice had hidden away. He used the Han insignia staff as a shepherd's crook, keeping it at his side while sleeping and awake, until the insignia's yak-tail pennants had fallen out. (*Han shu* 54.2463)

Taunted with the promise of a return dependent on a sex change in his sheep (perhaps a sign of his own refusal to assume a subordinate position), Su adamantly clings to his staff of office, a symbol of cultural potency as well as the last shred of native Han ornamentation. To lose his staff would mean becoming as Zhonghang Shuo, a eunuch deprived of his Han identity. When a Han emissary to the Xiongnu attempts to persuade Li Ling to return to the Han and give up his renegade status, he replies, "I already wear barbarian dress" (*Han shu* 54.2458). The outer surface betrays the change within. For Su Wu, the staff (now used ironically as a shepherd's crook) maintains its powerful symbolic function. He finally returns to the Han as an old man but unbowed in his allegiance to his ruler.

If Su Wu is the Han minister who keeps the borderlines clear, however, the possibility of interbreeding occurs even in his story:

Su Wu had grown old, and his son had been executed [he had become involved in a conspiracy shortly after Su's return to China]. The emperor felt sorry for him. He asked his courtiers, "Wu was among the Xiongnu for some time. Did he perhaps father a son there?" Wu confessed to the Marquis of Ping'en 平恩: "Formerly, when I was finally preparing to leave, my Tartar wife had a son, Tongguo 通國. I have heard news of him, and I'd like to send a messenger with gold and silks as a present for him." The emperor agreed. Later, Tongguo returned with the messenger, and the emperor made him a palace attendant. (*Han shu* 54.2468)

The son's name, Tongguo, means literally "passing through countries" or "facilitator of countries." Yet it is not clear how we should interpret this final act of miscegenation. Is Su Wu's status as a Han minister unsullied because he fathers a boy on a Xiongnu woman rather than become the "female" himself? And is the boy's final "sinicization" a sign of his final benignity? Or is it simply one more instance of how impossible borders are to maintain between two cultures that depend on each other for their sense of identity?

The Dilemma of Surrender

These cultural prejudices and ideologies emerge in the case of a woman noted in both history and literature, whose career and subsequent literary reputation differ from those of Wang Zhaojun.[23] Cai Yan 蔡琰 (b. ca. A.D. 178), daughter of the prominent literatus Cai Yong 蔡邕 (133–92), was kidnapped by a Xiongnu tribe and lived with them for twelve years, giving birth to children first by a chieftain and then by one of his sons. She was later ransomed from captivity by the warlord Cao Cao, who remarried her to his courtier Dong Si 董祀. By the time Fan Ye compiled the *History of the Later Han* in the fifth century, two poems describing her experiences were already attributed to her; both poems are entitled "A Song of Grief and Resentment" ("Bei fen shi" 悲憤詩).[24] Although scholars are undecided whether she authored them, they generally agree that they were composed close to her own era.[25]

One of the more interesting aspects of the Cai Yan poems is the way they demonstrate how deep-rooted the Chinese assumptions about the Xiongnu had become by the early third century A.D. Politically, the situation had changed: the Eastern Han was much less militarily aggressive than its predecessor, and the Xiongnu themselves had splintered into northern and southern alliances, the latter of which largely remained loyal to the Han. Cai Yan was captured by southern Xiongnu mercenaries employed by the usurper Dong Zhuo 董桌 (who had seized control of the last Han emperor and moved the capital from Luoyang to Chang'an) in his fight against an alliance of opposing officials and warlords. Evidently, Dong Zhuo gave his mercenaries carte blanche to raid and capture native Chinese inhabitants in their area of operation. Despite these different conditions, the Cai Yan poems reflect the perspectives manifested earlier in Sima Qian and Ban Gu.

The more interesting of the two, a long narrative poem of 108 lines, gives a straightforward account of her experiences. After a brief description of the background (Dong Zhuo and the turmoil caused by his ambitions), the narrator evokes her captivity in vivid imagery:

On the plains the folk were weak and fragile, 平土人脆弱
The troops that came were Tartars all. 來兵皆胡羌
They hunted the fields, besieged the towns, 獵野圍城邑
All perished wherever they rode. 所向悉破亡
They slashed and killed, left nothing living, 斬截無孑遺
And the bones of the dead were heaped in piles. 尸骸相掌拒
From their horses' flanks hung the heads of men, 馬邊懸男頭
While their women were carried behind. (ll. 11–18) 馬後載婦女

Here, the sedentary "people of the plains" are contrasted with the non-Chinese aggressors. The gender contrast makes the sexual nature of the assault explicit and powerful: the same horse carries the head of the husband, the body of the wife.

Ten thousand numbered the captives: 所略有萬計
Yet they would not let us camp together. 不得令屯聚
At times, shared flesh and bone would meet, 或有骨肉俱
And wished to speak, yet did not dare. 欲言不敢語
If we let down our guard for an instant, 失意幾微間
They would say, "Kill the prisoner dogs! 輒言斃降虜
We should settle you with our blades! 要當以亭刃

We will not preserve your lives!"	我曹不活汝
How could we care for our lives any more?	豈復惜性命
For we could not bear their shouting and cursing.	不堪其詈罵
Some used the rod on us as they pleased,	或便加棰杖
Pain came to us all alike.	毒痛參并下
At dawn we trudged, wailing and weeping.	旦則號泣行
At night we sat, moaning in grief.	夜則悲吟坐
We wished for death, but could not gain it,	欲死不能得
We wished to live, but they would not grant it.	欲生無一可
What grudge did Heaven hold for us	彼蒼者何辜
That we should endure this horror? (ll. 23–40)	乃遭此厄禍

The resemblance to colonial American captivity narratives, in which forced marches were combined with constant beatings and threats of violence, is striking. The narrator now steps back and comments (as one might expect) on the inhospitable nature of Xiongnu lands. The environment literally violates her personal (Chinese) space:

The frontier wilds are so different—	邊荒與華異
Their customs lack decency and order.	人俗少義理
They live buried in frost and snow,	處所多霜雪
And a Tartar wind rises in summer and spring.	胡風春夏起
It flaps my robes about,	翩翩吹我衣
And roars as it enters my ears. (ll. 41–46)	肅肅入我耳

Since the southern tribes of the Xiongnu did not inhabit such barren territory at the time of the raid, Hans Frankel cites the description here as evidence that the poem was written not by Cai Yan but by somebody unfamiliar with the concrete details of her life.[26] However, that is not the point: the rhetorical power of the situation commands this description of desert and wilderness, and any poet, including Cai, might alter reality for the effect.

Rather than speak directly and in clear chronological sequence of her twelve years in captivity, the narrator chooses to address first her desire to return and her hope that ransomers will arrive. Only after she has her wishes fulfilled does she describe her Xiongnu children:

Times I'd brood, thinking of parents,	感時念父母
Then my painful sighing would have no end.	哀歎無窮已
Whenever a traveler came from beyond,	有客從外來
I'd hear of his coming with joy.	聞之常歡喜

I'd greet him, hear the news he'd bring,	迎問其消息
But he never came from my home.	輒復非鄉里
But by chance my hopes were fulfilled,	邂逅徼時願
My own flesh came to claim me.	骨肉來迎己
And though I could escape at last	己得自解免
I had to leave my sons behind.	當復棄兒子
Though Heaven tied our hearts together,	天屬綴人心
I knew, once parted, we'd never meet again.	念別無會期
In life or in death forever parted—	存亡永乖隔
And so I could not bear to leave.	不忍與之辭
My sons clung to my neck in front,	兒前抱我頸
Asked me, "Mother, where do you go?	問母欲何之
People say you are going away,	人言母當去
Can you ever come back again?	豈復有還時
Oh mother, you were always so good!	阿母常仁惻
How can you now be so cruel?	念何更不慈
We haven't even grown up yet—	我尚未成人
What shall we do if you don't take care of us?"	奈何不顧思
To see this—I crumbled within!	見此崩五內
I grew dizzy, grew mad from it.	恍惚生狂癡
Wailing and crying, I stroked them gently,	號泣手撫摩
About to go, then holding back. (ll. 47–72)	當發復回疑

Since the narrator is recounting her misfortunes in a poem entitled "Song of Sorrow and Resentment," it would be inappropriate thematically for her to dwell on her children until compelled to leave them. Yet this striking distortion of chronology has another purpose as well: it emphasizes the degree to which she is forced to choose between two cultures and is not allowed to belong entirely to one or the other. The lament for her family in China leads immediately to a lament for the family she must now leave behind. Unlike Su Wu's son, who came to China and was accepted, Cai Yan's children must be abandoned. Being Chinese and being a mother cannot be reconciled. It is this split that makes the poem so successful; it is not so much about the hardship of life with the Xiongnu as it is about the tragedy of being forced to put down roots in two contrasting and opposing worlds.

The psychological desolation of this condition creates the environment for the remainder of the poem. Here the narrator moves decisively away from biography and instead draws on Han-era *yuefu* and *gushi* tropes—most strikingly, that of the lifetime campaigner who returns home to find his village deserted:

Away, away, cut off from my loves,	去去割情戀
Rushed on my journey, every day farther.	遄征日遐邁
So remote! A thousand miles!	悠悠三千里
When shall we ever meet again!	何時復交會
I thought of those sons of my womb,	念我出腹子
And my breast felt crushed because of it.	胸臆爲摧敗
When I arrived, my family was gone—	既至家人盡
Neither father's side nor mother's.	又復無中外
Town walls turned to mountain woods,	城郭爲山林
In the courtyard grew brambles and weeds.	庭宇生荆艾
White bones—who knew whose—	白骨不知誰
Lay unburied, scattered around.	從横莫覆蓋
I came out of the gate, no human sounds:	出門無人聲
Jackals and wolves barked and howled.	豺狼號且吠
All alone I faced my single shadow,	煢煢對孤景
Heart and liver shattered in grief.	怛咤糜肝肺
I climbed a hill, gazed far away;	登高遠眺望
At once my soul went flying off.	魂神忽飛逝
But just when it seemed my life had ended,	奄若壽命盡
Those nearby showed me grace.	旁人相寬大
For their sake I forced myself to live	爲復強視息
Though life had no point for me.	雖生何聊賴
I entrusted my life to a new man;	託命於新人
Tried my hardest to carry on.	竭心自勗厲
But since my wanderings have cheapened my worth,	流離成鄙賤
I fear I'll be cast aside once more.	常恐復捐廢
How long does a human life last?	人生幾何時
I will brood on my sorrows till the day I die. (ll. 81–108)	懷憂終年歲

The narrator's statements parallel the conventions of the first of the Nineteen Old Poems: "Going on and on, and on and on, / parted from you while alive." Yet here the grief is for her Xiongnu family, and by traveling back to Han China, she actually goes "far away." When she returns, she encounters a situation like that found in another anonymous "old poem" (*gu shi* 古詩):

At fifteen I followed the army on campaign,	十五從軍征
At eighty I finally came home.	八十始得歸

On the road I met a man from my town: 道逢鄉里人
"Is there anyone left at home?" 家中有阿誰
"Gaze far over there—that's your house—" 遙望是君家
Pines and cypress grew thick over the mounds. 松柏冢累累
Rabbits entered by the dog doors, 兔從狗竇入
And pheasants flew from the rafters. 雉從梁上飛
In the courtyard wild grains grew; 中庭生旋穀
On top the well, wild mallows. 井上生旋葵
I boiled the grain to make gruel, 烹穀持作飯
And plucked the mallows to make soup. 采葵持作羹
Once the soup and gruel were cooked, 羹飯一時熟
I had no one to serve them to. 不知貽阿誰
I went out the gate, gazed to the east— 出門東向望
Tears fell and soaked my clothes.[27] 淚落沾我衣

Here, frontier tropes associated with male tribulations are rerouted to a woman's experience. Although it is doubtful that the historical Cai Yan confronted such desolation upon her return, the psychological situation is valid: she now has neither her Chinese nor her Xiongnu family. Although she eventually finds support, she is now subject to continual anxiety about the stability of her new position: "I fear I'll be cast aside once more." The ultimate message of the poem is clear: the two worlds of Xiongnu and Han cannot be brought together, and she who must live in both will find that she can live in neither. Perhaps the anxiety of miscegenation that ran through the narratives of Sima Qian and Ban Gu has hardened by the end of the Han. Even if complex social and cultural interactions between China and its neighbors continued to determine reality, in the world of texts the boundaries were drawn much more clearly.

The Rejected Courtier Wang Zhaojun

Cai Yan's poems were well known and frequently read in traditional China, and during the Tang they inspired a series of eighteen "Songs of a Nomad Flute" (胡笳十八拍) composed in Cai Yan's voice. However, Cai Yan did not become the quintessential figure of racial and sexual alliance in Chinese literature. It may be that her story was still too ambiguous for authors seeking dramatic distinctions between nomad and Chinese: after all, she was kidnapped by Xiongnu forces under command of a Chinese leader, during a period of civil war; and perhaps she expressed too much regret over the loss of her Xiongnu family. Maybe her realization that her wanderings had "cheapened her worth" came to color some views of her as well. But I suspect that the single most important factor is that her story did not quite allow for literati identification; that is, her own position could not become an allegorical representation of dilemmas faced by the literatus who might read and imitate her story. It may well be for this reason that Zhong Rong chose not to evaluate the poems, when he was so strongly affected by the fan poem of Lady Ban. It would take another lady—Wang Zhaojun—to supply a role to which literati could truly relate.[28]

The only fact we know of any reliability concerning Wang Zhaojun occurs in the Xiongnu chapter of the *History of the Han*. After the friendly khan Huhanye 呼韓邪 emerged victorious in a civil war with his brother, he solidified his Han alliance with a grand visit to the court in 33 B.C. "The khan expressed a desire to become an in-law of the emperor's and thus become part of the imperial family. Emperor Yuan 元 [r. 48–33] gave him the hand of a young woman from his harem, a child of a good family, named Wang Qiang 王嬙, polite name Zhaojun. The khan was so delighted that he guaranteed in writing the peace and security of the passes from Shanggu 上谷 to Dunhuang 燉煌 in the west" (*Han shu* 94B.3803). Later, her descendants (first through marriage to Huhanye, later through marriage to his son, in keeping with Xiongnu law) played a prominent role in Xiongnu politics. According to the *History of the Later Han*, one of her grandsons founded the line of "southern khans" who continued to maintain an alliance with the Han royal house—and who commanded the troops that were to capture Cai Yan two centuries later.[29]

Almost nothing was known about Wang, then or now, and yet she nonetheless became the object of endless speculation. This is in part because she was the last significant participant in the *heqin* system during the Han and so served as a focus for the anxiety literati had felt about such exchanges from the beginning. Three early texts, all likely composed by the fifth century, demonstrate the way the legend had developed. The simplest is the one found in the *History of the Later Han*. In describing certain central events in Xiongnu politics, it fills in some details of Zhaojun's career. Here, disgusted by Emperor Yuan's failure to recognize her great beauty, she volunteers to go to Huhanye. When the emperor sees her beauty at the time of her departure, he regrets the decision but feels obliged to keep his word. Later, when the death of Huhanye results in her marriage to her stepson, she sends a letter to Emperor Cheng 成 asking to return to China. The emperor refuses her request.[30]

Two anecdotal works contain more dramatic accounts: the *Miscellaneous Records of the Western Capital* (*Xijing zaji* 西京雜記, attributed to Liu Xin 劉歆 [50 B.C.–A.D. 23] but undoubtedly later), and *A Manual for the Harp* (*Qin cao* 琴操, attributed to Cai Yong—Cai Yan's father—but also probably later). The first associates Zhaojun's departure with the execution of a number of palace painters:

Since there were many women in the rear palace of Emperor Yuan and he could not see them often, he had painters paint their features, and he would summon his ladies based on the portraits. All the palace women bribed the painters—at the most, some one hundred thousand cash, and not less than fifty thousand. Only Wang Qiang, confident in her own beauty, was unwilling to give anything. The artist then portrayed her as ugly; so she failed to appear before His Majesty. After this, the khan came to court, seeking a palace lady to make his wife. Thereupon the emperor consulted his pictures and selected Zhaojun. When the time came for her to leave and he summoned her to appear, she proved most beautiful of those from the rear palaces, clever in conversation, graceful and elegant in deportment. The emperor regretted his actions, but the selection was already set. Just then since he considered his faith with other countries important, he did not exchange her for another. Afterward, he investigated the matter thoroughly. All the painters were taken to the marketplace and executed. A record was made of their possessions, and they amounted to many tens of thousand.[31]

This is supplemented by the more elaborate record in *A Manual for the Harp*. To explain the origins of a popular melody, "Longing in Resentment and Loneliness" 怨曠思惟歌, the author gives the following account:

Wang Zhaojun was the daughter of Wang Rang 王穰 of Qi 齊. When she was sixteen, she was bright and pure in beauty, famed throughout the land. Rang saw that she was upright and virtuous and gracefully lovely and that she would never show herself at gate or window. And so, because she was so different from others, he would not betroth her to those who came seeking her hand. He presented her to Emperor Yuan; but because she came from a distant area, she was not favored by the ruler. She dwelt in the rear palace for five or six years; growing resentful and lonely, she would make a show of not adorning her face and form. Each time Emperor Yuan passed through the rear palace, he would not visit her residence, giving it short shrift.

Later, the khan sent an emissary to present congratulations at the court. Emperor Yuan put on a show of singers and dancers and then commanded the ladies of the rear palace to come forth in their finery. Zhaojun had been angry and resentful for some time, and since she could not obtain a position among those attending on His Majesty, she adorned herself even more with her fine cosmetics and abundant ornaments. Her form and figure shone forth brightly. When they had all taken their seats in their ranks, Emperor Yuan said to the emissary, "How does the khan prefer to take his pleasure?" The emissary replied, "He has supplied himself enough with precious, marvelous, and unusual things. Only his women are ugly and vulgar and do not come up to those of the central kingdom." The emperor then addressed his palace ladies, "I wish to give a woman to the khan. Let anyone willing to depart arise." Then with a sigh Zhaojun came forward over the banquet mat. "I have fully received the good fortune of the Emperor's favor in the rear palace. But because I am ugly and vulgar I could not please Your Majesty's desires. I truly wish to depart." At the time the khan's emissary was present. The emperor was greatly astonished; he regretted the affair, but could do nothing to stop her. After a long while, he sighed and said, "I have already committed an error." He then gave her over.

Zhaojun arrived among the Xiongnu. The khan was delighted, imagining that the Han had been generous with him. He gave himself over to drinking and making merry, and he sent an emissary to return presents to the Han: a pair of white jade disks, ten swift horses, and various pearls and treasures of the northern lands. However, Zhaojun resented that the emperor had originally paid no attention to her, and she brooded and was unhappy in her heart. As she thought of her native land, she composed a song, "Longing in Resentment and Loneliness":

Autumn trees grow thick,	秋木萋萋
Their leaves grow sere and yellow.	其葉萎黃
Birds stop upon them,	有鳥爰止
Nesting in the clumps of mulberry.	集於苞桑
Looking after their feathered wings,	養育毛羽
Form and features glow with light.	形容生光
Then they rise into the clouds,	既得升雲
Where they attend within the curtains.	獲倖帷房
The separate palace is cut off and lonely,	離宮絕曠
My body is broken and buried.	身體摧藏
My will sunken and desolate.	志念抑沉
I cannot soar away.	不得頡頏
Though I have food for my hunger,	雖得餧食
My heart stumbles and hesitates.	心有徊徨
Oh, what have I done?	我獨伊何
I've departed, changed from the norm.	改往變常

Fluttering, the swallows 翩翩之燕
Gather far away among the Western Qiang. 遠集西羌
Lofty hills, so steep and tall! 高山峨峨
Yellow River, how it flows and floods! 河水泱泱
Oh father! Oh mother! 父兮母兮
The road is far and long! 道里悠長
Oh alas! Oh alas! 嗚呼哀哉
My worrying heart is sad and broken. 憂心惻傷

Zhaojun had a son named Shiwei 世違. When the khan died, Shiwei inherited from him. Generally, among the barbarians, when the father dies, the son marries the mother. Zhaojun asked Shiwei, "Are you a Han? Or a northern barbarian?" Shiwei replied, "I prefer to be of the northern barbarians." She then took poison and died. The khan buried her. The grass of the northern lands is mostly yellow, but her tomb alone is always green.[32]

These anecdotes make up the principal details of the legend of Zhaojun. Already we have a simple item of diplomatic history changed into a tragedy—a tragedy of misused talent. The roots of Wang's fate lie in the failure of the emperor to recognize her extraordinary beauty, combined with Zhaojun's resentment over his behavior. Literati self-identification emerges once again: since the *Li sao*, physical beauty and outward adornment have been the external manifestation of virtue. The anecdotes are more compelling, however, in the way they work out this theme. In the *Manual for the Harp*, the emperor's initial disregard for Wang provokes an action of resentment (*yuan* 怨) on her part: her refusal to adorn herself and to make herself presentable leads him to continue to be unresponsive to her. Thus, an element of pride not completely unlike hubris enters into her actions. This becomes clearer when she willfully makes herself attractive for the banquet with the Xiongnu emissary and then volunteers to go to the khan. In this she serves as a cautionary example for the literatus/courtier: he must be patient with his lord and not let resentment cause him to behave spitefully and against his own self-interests. The account in the *Miscellaneous Records* was more influential on later traditions, however. Here, the emperor must depend on an intermediary—the court painter—to supply him with information about his ladies. The painter thus holds the important role of flatterer/slanderer so central to post–*Li sao* rhetoric: on the one hand, he is the matchmaker needed for any true congress between husband and wife, ruler and minister; on the other, he is the obstructer, the one who conceals true ministerial worth from the ruler or who promotes worthlessness for the sake of profit. The emperor's book of paintings is a form of selection, just as the examination system will become in later ages, and tampering with such a system threatens the well-being of the entire state. Zhaojun is willful in this version as well, although her willfulness incorporates a strong sense of virtue in her refusal to join the other court ladies in offering a bribe.

In both versions, however, there is a general stress on Wang's motives, on the reasons for her behavior. For better or for worse, she is ultimately the author of her own fortunes, not simply the unwitting tool of court politics. Despite her self-destructive pride, the *Manual for the Harp* shows us the actions of an imperial servant willingly accepting exile. Her willingness to do so comments on the emperor's failed ability to recognize talent and restores autonomy to her actions. *Resentment* becomes the emotion through which she establishes this independence and autonomy, and so (in good literati fashion) resentment becomes the musical mode through which she expresses herself (in the Tang encyclopedias, Zhaojun's story appears as the principal example of *yuan*). It does not matter so much that her poem is not particularly good; what is much more significant is that she should sing her grief.

In the biography of Su Wu, Su refused to give up his allegiance to the Han ruler. As a consequence, the khan sent him to the far north to herd sheep. Su Wu's allegiance temporarily disrupted the fluid border dynamics of Han-Xiongnu relations. In a world in which frontiers were permeable and generals and court ladies traveled back and forth between the two cultures, Su drew a clear line; for him, there was the land of the Han or there was exile. Herding sheep by Lake Baikal, he saw himself not as a Xiongnu herdsman but as a Han Chinese far from home. Cai Yan, on the other hand, came to identify with her captors and to accept her role as mother of Xiongnu children; although she started out as a mere captive, she came to accept the new life so thoroughly that return to China became a new form of exile. The anecdotes around Zhaojun make her reject Cai's alternative and embrace that of Su: she rides off to the north not to become the wife of a chieftain but to become an exiled Chinese subject. Unlike Su, she does submit to the khan, but only because imperial politics and the bidding of her lord require it; in an odd way, she continues to be a Chinese imperial consort. When the khan dies, however, the terms change, because she is now confronted with a decision that will unalterably change her cultural identity—just as Li Ling did when he donned barbarian dress. In the *History of the Later Han,* she writes a letter to the emperor, begging that she be allowed to return rather than marry the stepson of her husband. When Emperor Cheng commands her to remain, she does so and continues to produce heirs to the Xiongnu throne. It may be that those imperial orders allow her to maintain the illusion that she is a loyal minister in exile, rather than merely a woman passed from father to son. However, the wholly legendary account in the *Manual for the Harp* provides a more unpleasant story. Zhaojun's question—Are you barbarian or Chinese?—conceals the key to her own fate: for this incarnation of Zhaojun, to cease to be Chinese is to cease to be. The text emphasizes this by misinterpreting the Xiongnu custom: here, sons marry their own mothers, not their stepmothers, and thus the horror and repulsiveness of symbolic incest is made literal. When she learns that her son, in spite of his mixed blood, has not been truly sinicized, she kills herself rather than become a barbarian. Her own blood has betrayed her. Like the Cai Yan poems, this version argues for the tragic consequences of miscegenation and cultural exchange, but unlike the Cai Yan poems, its take is more uncompromising and more xenophobic. Here the savages are savages, and a Han court lady can only be their captive, never their queen.

Erasing the Body

Unlike Cai Yan's long narrative poem, the four-character lyric lament preserved in *A Manual for the Harp* did not become famous; the poetic representation of Wang Zhaojun's life was in the hands of literati from the beginning. The earliest literati imitation preserved today (and hence generally accepted as the beginning of the tradition) was composed by Shi Chong 石崇, the aristocrat whose excesses are traced in chapter 3:

Song of Wang Mingjun 王明君辭

Wang Mingjun's original name was Wang Zhaojun (it has been changed because of a taboo on Emperor Wen's personal name). When the Xiongnu were at the height of their power, they requested a bride from the Han. Emperor Yuan sent as bride the daughter of a fine family from the rear palace. Earlier a princess had been married to the Wusun 烏孫, and she had commanded that the musicians play the lute on horseback in order to comfort her as she traveled with her melancholy thoughts. It must have been the same when Mingjun was escorted. The new melody that I have composed is filled with grief and resentment. I hereby commit it to paper.

	I came from a family of the Han;	我本漢家子
	Now I go to the khan's court.	將適單于庭
	Before I can bring my parting to an end,	辭訣未及終
	The vanguard horsemen raise their standards.	前驅已抗旌
5	The charioteer's tears fall in torrents,	僕御涕流離
	The chariot horses neigh in grief.	轅馬爲悲鳴
	Grief and melancholy wound my very vitals,	哀鬱傷五內
	Tears soak my crimson hat-strings.	泣淚沾朱纓
	On and on, daily, farther I go,	行行日已遠
10	Until I reach the Xiongnu's city.	遂造匈奴城
	They lead me within his yurt dwelling,	延我於穹廬
	Grant me the title of queen.	加我閼氏名
	But I cannot find rest with an alien people;	殊類非所安
	Though ennobled, it's not to my glory.	雖貴非所榮
15	I have been disgraced by both father and son,	父子見凌辱
	I face them in shame and terror.	對之慚且驚
	Indeed, it is not easy to kill oneself,	殺身良不易
	So I silently get on with my life.	默默以苟生
	But how can I plan to get on with my life	苟生亦何聊
20	When filled with deep brooding and constant rage?	積思常憤盈
	I wish I could use a flying swan's wing,	願假飛鴻翼
	Ride it and travel far away.	棄之以遐征
	But the flying swan pays no heed to me,	飛鴻不我顧
	And I stand long in my vacillation.	佇立以屏營
25	Once I was jade kept within a case;	昔爲匣中玉
	Now I am a blossom fallen in ordure.	今爲糞上英
	A morning blossom's joy is not good enough—	朝華不足歡
	I prefer to join with the autumn grass.	甘與秋草并
	I send word to later generations:	傳語後世人
30	Marrying far from home brings grief hard to bear.[33]	遠嫁難爲情

Notes

Preface: A Han princess was sent to the Wusun in an attempt to create an ally on the far side of the Xiongnu domains. A poem attributed to her survives in the *History of the Han* (96.3903):

My family marries me to one corner of the sky.	吾家嫁我兮天一方
Sent afar to an alien land, to the King of Wusun.	遠託異國兮烏孫王
His yurt will make my dwelling, his felt my walls;	廬爲室兮旃爲牆
Meat will be my food, and kumiss my drink.	以肉爲食兮酪爲漿
I dwell forever in home-thoughts, my heart is hurt within;	居常土思兮心內傷
I wish I were a brown swan and could fly back home.	願爲黃鵠兮還故鄉

The "lute" is the *pipa* 琵琶, a stringed instrument of central Asian origin. In a move typical of early ethnographic thinking, writers came to believe that the Wusun princess (and later, Wang Zhaojun) introduced it to central Asia and that it returned from there to China.
l. 15: There is disagreement among commentators on this line. I believe it refers to Zhaojun's revulsion over incest practices and that her husband and her second husband, his son, have disgraced her by marrying her. Others believe that the line refers to the shame she feels before her own father and brothers. However, since line 16 suggests that she is in the physical presence of the khan, I believe the first possibility more likely.

Shi Chong shows little awareness of the legendary accretions found in *A Manual for the Harp* and the *Miscellaneous Records* and, in fact, may not have read them (if they were even in existence at the time). There is no mention of the painting and no mention of Wang Zhaojun's willful decision to accept the khan's hand in marriage. He instead centers almost exclusively on the sorrow of parting and the hardship of living in the wilderness; at times he employs imagery found in the Cai Yan poem. Nonetheless, unlike the Cai Yan poem, these verses do not attempt to tell a story: they move away from narrative, as literati poetry tends to do.[34]

There is also little of the symbolic import of the Wang Zhaojun story as represented in the brief narrative texts. Shi makes little use of the language of the exiled or rejected courtier. Only one couplet seems to anticipate certain themes that will recur in later Wang Zhaojun poems:

Once I was jade kept within a case;
Now I am a blossom fallen in ordure.

Here the imagery of exposure and pollution that characterizes the *Li sao* is suggested within the contrast of a parallel couplet. By and large, however, Shi Chong is content to emphasize the role his poem plays in a musical tradition, a *replacement* text for some lost song Wang may have sung on her way to exile (again, he may not have known the four-character poem found in *A Manual for the Harp*). In this, he speaks for Wang, as many later literati would do.

In fact, pre-Tang poems on Wang Zhaojun make many of the same moves that poems on the feminine do in general: Wang was considered either a *yuefu* topic that called for literati impersonation of the female voice or a subject for palace poetry, in which eroticism was muted through the rhetorical elaboration of certain recurring aspects of her legend. Parallelism is frequently employed to contrast a fragile woman and the harsh environment, as in these lines of Shen Yue, in which the savage winds rip through clothes and assault the body beneath it:

Daily I see the flying sand rise,	日見奔沙起
Gradually feel the tumbleweeds increase.	稍覺轉蓬多
The Tartar winds violate flesh and bone—	胡風犯肌骨
They do not halt at harming silk and gauze.[35]	非直傷綺羅

This assault is constantly mediated by the typical language of image, mask, and illusion we saw in the last chapter. This is particularly obvious by the fifth century, when allusions to the legends surrounding Wang, most typically the painting story, begin to enter poems. Since we have already seen poets' use of the painting trope to illustrate the paradox of surface appearance and inner worth, this should hardly seem surprising. In one couplet by a sixth-century female poet, the sister of Liu Xiaochuo 劉孝綽, Shi Chong's jade case is paired with the painting in order to bring the poem into the interior space of the boudoir:

The reds and greens [of painting] have lost their original ornament,	丹青失舊儀
And the jade in the case becomes an autumn weed.[36]	匣玉成秋草

This last line makes little sense beyond its suggestion that a sheltered item becomes worthless when exposed to the elements. In other poems, the painting is linked with a mirror—the mirror predictably becomes a sort of self-actuating painting, in which the signs of physical decline can be observed (in other words, Wang begins to resemble the portrait that had unfairly disfigured her).[37] Most interesting, however, is a quatrain by the female poet Shen Manyuan 沈滿願 (fl. 530):

If I had trusted at first to the skill of paint,	早信丹青巧
I would have bribed that Luoyang artist—	重貨洛陽師
A thousand in gold buys cicada tresses,	千金買蟬鬢
A million to sketch moth eyebrows.[38]	百萬寫蛾眉

Familiar with the artificiality of any court reputation, Shen cynically suggests that beauty is useless without a well-paid promoter.

Although for obvious reasons the theme of Wang Zhaojun, her suffering in the desert, and her disfigured painting would have appealed to the Six Dynasties poet, there is little in poetry from this period beyond the dynamics we have already traced in palace poetry. This does not mean that poets disregarded the link between Wang Zhaojun and the suffering courtier; rather, the exiled lady and the exiled courtier could share the same kind of suffering without the need for explicit equation. In the Tang, this gradually changed. As we shall see in the chapters to come, women became increasingly differentiated from the male conception of the self and instead took on a human identity independent of obsessive male concerns. Allegories in which Wang stands in for the frustrated literatus will continue, right up until the Qing, but Tang poems on Wang Zhaojun are more striking for the way they explore central problems of Chinese identity, express anxieties over nomadic pressures on the culture, or (more lightheartedly) create a Chinese perspective through which the "romantic mysteries" of the savage can be viewed. In Tang society, where popular music (especially music from central Asia) displaced native traditions, Wang's association with the lute (*pipa*) emerged as an important image. Poets also became more intrigued with the beginning and ending of Wang's career. On the one hand, they viewed her as a significant representative of the Han people, with her own family and her own origins. Her supposed native village in the Yangtze Gorges seems to have become a tourist site, inviting on-site meditations on the social injustices of palace lady recruitment. On the other hand, poets wrote increasingly about the Verdant Mound (*qing zhong* 青冢), Wang's burial tumulus that supposedly stayed green amid the arid desert. Oddly enough, Wang is most often present through her erasure—either by death, as a result of suffering the rigors of a desert existence, or by her replacement by other rhetorical concerns such as historical evaluations of her significance, or evocations of a landscape where she once dwelt.[39]

Perhaps the earliest poems to show this new perspective are a series of quatrains by Chu Guangxi 儲光羲 (707–60). Here the routine rhetorical elaboration of Wang's suffering is replaced by the frontier vignette, in which specific moments of nostalgia or homesickness are evoked through vivid, often ironic scenes of Xiongnu life:

Four Songs of Mingfei 明妃曲四首

I

Bright-colored riders, pair on pair, led on my jeweled carriage;	彩騎雙雙引寶車
Tibetan pipers two by two performed on Tartar fifes.	羌笛兩兩奏胡笳

How could I bear to leave and take the road
past the Wei River Bridge, 若爲別得橫橋路
And not pine after my palace's
jade branch flowers? 莫隱宮中玉樹花

II

I walk west to Long Mountain,
weep at the Tartar skies; 西行隴上泣胡天
To the south I face the clouds
that show the road to the Wei. 南向雲中指渭川
In felt tents when night comes
I sometimes toss and turn: 毳幕夜來時宛轉
How could this ever resemble
sleeping at the side of Han's prince? 何由得似漢王邊

III

The Tartar king knew well
of my unendurable grief; 胡王知妾不勝悲
So his musicians only played for me
the songs of the land of Han. 樂府皆傳漢國辭
This morning, on horseback,
a melody on the harp 朝來馬上箜篌引
Is a little like those in the Palace
played on an idle night. 稍似宮中閒夜時

IV

At dawn the sand startles up
and wild snow blows about. 日暮驚沙亂雪飛
My attendants still urge me
to change my thin gauze robes. 傍人相勸易羅衣
I force myself to the front hall
to see the songs and dances. 強來前殿看歌舞
Together we await the khan's
returning at night from the hunt.[40] 共待單于夜獵歸

Notes

Title: Mingfei (The Illustrious Consort) was a title given to Wang Zhaojun in later ages and a common name for her in Tang times.

I.3: The Wei River Bridge was constructed during the Han dynasty over the Wei River from the capital at Chang'an 長安; Wang would have to pass by this bridge to head north into Xiongnu territory.

I.4: Li Shan 李善 in his commentary in *The Literary Anthology* describes a magical tree of coral and jade erected by Emperor Wu of the Han in one of his palaces. If Chu is attempting to be

historically accurate, this may be the allusion. However, "Flowers from the Jade Tree in the Back Garden" was also a *yuefu* piece composed by the last Chen 陳 emperor and is often alluded to by Tang poets as an image of the luxury of court palace ladies.
II.1–2: Mount Long is in Gansu, northwest of the capital; the Wei River flows from Gansu past the capital.
III.3: "A melody on the harp" is more specifically the name of a *yuefu* tune, also known as "Sir, Don't Cross the River" 公 無 渡 河, that supposedly describes the tragedy of a suicidal drowning. Chu may merely be implying here that the Xiongnu musicians are playing a Chinese melody to soothe Wang; or the tune may have more ominous connotations.

Since quatrains like this were usually sung rather than recited, and since they often took the frontier as their subject matter (perhaps because of the popularity of central Asian music during this time), it is not surprising that they would focus so much on music itself.[41] As we have seen, Wang was already associated with the *pipa* by this time. Here, Chu cleverly plays on the legend, first by having non-Chinese musicians accompany her and then by having them play Chinese music for her to relieve her anxiety (rather than have her or her Chinese attendants play as she departs for the border, as Shi Chong would have it). In general, Chu portrays the "barbarians" as sympathetic to Wang's grief and eager to help. Nonetheless, the verses assert a great difference between "here" and "there." Instead of portraying Wang as the Chinese body assaulted by the desert sands, Chu sees her as a temperament unwilling to accept her new accommodations and sensitive to differences. "In felt tents when night comes I sometimes toss and turn." To toss and turn is *wanzhuan* 宛 轉, an insomnia often produced by anxiety or by sexual longing. Here, it suggests a sexual loyalty on Wang's part, who finds the experience of sleeping with the khan negligible compared to the nights she anticipated having with her original master. Although she grants that the khan's musicians can produce a music somewhat similar to that of the Han court, she is later summoned to a musical performance in the middle of a blizzard, while she waits for the khan's return from the hunt. Chu here refers ironically to the rusticity of the Xiongnu's accommodations: in Chang'an, Wang might attend a similar performance while waiting to see if the emperor would come and choose her as his partner for the night. Everywhere Wang looks, she sees a parody of Han palace routine. Life with the Xiongnu is not primarily one of suffering (in spite of the snow and the cold); rather, it is a life she refuses to adopt just as she refuses to trade her gauze robes for more suitable furs. Chu no doubt wishes to make Wang a figure of pity, as previous poets have done, but he also suggests a strong willfulness in refusing to surrender her old role as imperial consort. As in the account in *A Manual for the Harp*, Wang manages to temper her pathos as victim by reasserting her Chinese identity.

Another subgenre that poets used to discuss Wang was the epigrammatic quatrain on history. These often ironic poems made a distanced judgment on an issue from the past, eschewing empathic connections with historical figures for more "objective" analysis. Here, Wang becomes merely a point of argument in the debate over foreign policy. Three examples:

Su Yu 蘇郁 (fl. ca. 810): On Marriage Alliances 詠和親

Moon over the pass at night hangs down: a mirror over Verdant Mound.	關月夜懸青冢鏡
Chill clouds, thin, in the autumn: the gauze of Han palaces.	寒雲秋薄漢宮羅
My lord, you shouldn't put your faith in marriage alliance plans:	君王莫信和親策
It'll only give birth to Tartar fledglings— the savages increase their numbers.[42]	生得胡鶵虜更多

Wang Rui 王叡 (fl. 820s):
Relieving Wang Zhaojun's Resentment 解昭君怨

Don't resent the craftsman who disfigured your painted form.	莫怨工人醜畫身
Don't doubt your enlightened lord who sent you off for an alliance.	莫嫌明主遣和親
If then you hadn't been married to a savage,	當時若不嫁胡虜
You'd only be one more dancer in the palace.[43]	祇是宮中一舞人

Wang Zun 汪遵 (*jinshi* 866): Zhaojun 昭君

The emperor of the House of Han was lord of the Ring of Lands;	漢家天子鎮寰瀛
But north of the frontier, the nomad men had yet to surrender arms.	塞北羌胡未罷兵
Fierce generals, scheming ministers, count their worth in vain:	猛將謀臣徒自貴
A single smile from mothlike brows clears the frontier dust.[44]	蛾眉一笑塞塵清

Imagery is not important in these poems, although they sometimes give a peremptory nod toward the conventions that had already developed around

Wang (as in Su's comparison of northern clouds to the gauze of a Han lady's gown). Much more important is the use of poetic structure to drive home a (sometimes controversial) position on marriage alliances. Most of these poets inhabited the fragmented world of the later Tang, when the interpretability of the past had come under doubt.[45] What may also be of note here is the way the Xiongnu continue to be treated as the savage enemies of the Han; in the post–An Lushan world, few poets could accept the cosmopolitanism of the seventh and early eighth centuries.

We must turn elsewhere, however, for more complex treatment of Wang and her dilemma. Perhaps the most famous Tang version of her is Du Fu's 杜甫 (712–70) poem written at her home town, the third of his "Meditations on Ancient Sites" (*Yonghuai guji* 詠懷古跡):

A herd of hills, myriad valleys— I reach Jingmen;	群山萬壑赴荆門
Here still stands the village where Zhaojun was born and grew up.	生長明妃尚有村
Once she left the Violet Terrace, northern sands stretched on;	一去紫臺連朔漠
Alone there remains Verdant Mound facing the twilight.	獨流青冢向黃昏
In painting once she could recognize a face of spring breezes;	畫圖省識春風面
Her rings and pendants bring back in vain her soul on a moonlit night.	環珮空歸月夜魂
For a thousand years the lute speaks a Tartar tongue—	千載琵琶作胡語
We make out grief and hatred expressed within the tune.[46]	分明怨恨曲中論

Although this is, like the quatrains, an examination of history rather than an empathetic evocation of Wang's frontier experiences, Du Fu makes his meditation more intense through a characteristic projection of his imagination over distance—an ability to recapture the remote in time and place that increasingly characterizes his poetry in his last years.

Other poets had fragmented Wang into a series of components that could be described in parallel structures: her sorrow, her music (materialized in the lute), her thin gauze robes, her tears. Du Fu picks up on this technique for his own purposes, and the result is more disturbing. The princess's movements are evoked as a passage from one elevated structure to another:

from Violet Terrace (a Han palace) to Verdant Mound. Yet, in spite of this fatal movement and the claim that only the tomb remains, the third couplet resuscitates the princess through a series of broken images: a face in a mirror described as a spring breeze (a cliché that threatens to make her features literally dematerialize, leaving nothing) and a revenant that manifests itself only through the tinkling of jade pendants. If Wang cannot return in a familiar form and her mound is left isolated in desert sands, what *can* return? Only the music of the lute, which is possessed by her spirit: yet now it has become incomprehensible, transformed into a barbarian language. The return of Wang's song of sorrow as a Central Asian music tradition becomes yet another kind of miscegenation, and a disturbing one: a trapped Chinese soul attempts to make itself heard and understood through a mongrel art.[47] Du Fu is too intense and extraordinary a poet not to make this vision a compelling one, although his view of the desert as a land of imprisoned ghosts is even darker than previous versions of Wang's dilemma. Does this imply a greater xenophobia on Du Fu's part? Probably not; Du's late verse is filled with images of physical and psychological disintegration and of the tragedy of the good man who goes unheard and unconsoled.

After Du Fu, the most extraordinary verses on the topic of Wang are by Bai Juyi 白居易 (772–846). In two moderately long poems, the first on the subject of Verdant Mound, the second on the subject of Wang's natal village, Bai discusses the moral significance of her story with a much greater originality and complexity. Moreover, the figure of the discarded man of talent emerges here once again—although it shares the poet's concern with the common people and their exploitation by the imperial court.

The first of the two, "Verdant Mound," has the open, didactic quality of the author's "new music bureau" (*xin yuefu* 新樂府) poems. Here, the story of Wang exposes an abuse:

Verdant Mound 青冢

Above, a starving goose cries out;	上有飢鷹號
Below, a withered tumbleweed rolls.	下有枯蓬走
Vague amid the border snow,	茫茫邊雪裏
A handful of sand forms an earthen pile.	一掬沙培塿
It's told that this is Zhaojun's mound;	傳是昭君墓
Where long her mothlike brows are buried.	埋閉蛾眉久
Congealed rouge has turned to mud;	凝脂化為泥
Nothing left of face powder and mascara.	鉛黛復何有
Only a dark vapor of resentment	唯有陰怨氣

At times rises from the tomb.	時生墳左右
Gloomy, like a bitter fog	鬱鬱如苦霧
It does not rot away with her bones.	不隨骨銷朽
Women have but a single talent:	婦人無他才
Glory and decline are tied to their looks.	榮枯繫妍否
No way to change Mingfei's fate	何柰明妃命
Which depended alone on a painter;	獨懸畫工手
Once he sinned in his greens and reds,	丹青一詿誤
Then black and white were thrown into confusion.	白黑相紛糾
He made the eye of his lord	遂使君眼中
Turn a Xi Shi into a crone.	西施作嫫母
While her companions tasted favor to the full,	同儕傾寵幸
She made a match with an alien race.	異類爲配偶
How may one know of good fortune or bad?	禍福安可知
A lovely face is worse than an ugly one.	美顏不如醜
Who could have said that a passing affair	何言一時事
Could serve as a warning for a hundred years?	可戒千年後
Tell the beauties who are to come,	特報後來姝
"Do not rely on your looks!	不須倚眉首
Do not refuse to don thornwood hairpins,	無辭插荊釵
Marry and become a poor man's wife.	嫁作貧家婦
Haven't you seen on Verdant Mound	不見青冢上
Travelers making libations to her soul?"[48]	行人爲澆酒

"Women have but a single talent," and yet that talent is subject to misrepresentation and others' malignity. It seems an odd assertion to make, considering the implications of the poem's end: women are advised to live rustic and simple lives and to depend on more "domestic" virtues. But Bai's conception of "talent" (*cai* 才) may be tied to the rhetoric of "use"; ever since Zhuangzi praised useless trees for their ability to escape the axe, writers have commented on the possible satisfactions of not employing their talent at all and thus escaping the dangers of a political life. This is hardly a surprising point: frustrated literati have always cited this in their fantasies of retirement. But Bai's view here is more disturbing, more unyielding in its undermining of the traditional conception of *cai*. Wang's one "talent"—her physical beauty—is hedged about by contradictory images that destroy its meaning. Her body is nothing more than brows and cosmetics—so are we to see the cosmetics as a constitutive element of her "talent" or an assistance to it? No matter; when they have passed away, the only thing that remains is a dark, resentful vapor that haunts Verdant Mound. Even when she was still alive, her fate depended on a painter, and here Bai twists the old confusion of painted surface into a clever rhetorical paradox:

> Once he sinned in his greens and reds,
> Then black and white were thrown into confusion.

"Greens and reds" is a basic term for the art of painting; "to confuse black and white" means either to be unaware of basic differences (especially moral distinctions) or deliberately to muddy the situation so that others cannot determine what is right. Talent that is subject to this inversion is worthless.

If, as we have seen in the past, physical beauty was routinely regarded as the female equivalent of male talent, then Bai's narrator shows a cynical disinclination to believe that worthy men can ever be employed successfully. More pointedly, "talent" for the Tang literatus was skill in literary composition and mastery of culture, the ornament or *wen* 文 of the court. The educated men of Bai's generation (that is, the figures we associate with the "mid-Tang") were increasingly troubled over the role of literature as a service for the ruler. Should not literature have some value independent of imperial selection? And does not the long-established comparison of female beauty and male talent suggest some disturbing consequences for the earnest young official? Perhaps in such a case the plain ornament of the thornwood hairpin suits a sincere style better.

Of course, Bai speaks here in the voice of the *yuefu* reformer, who embraces simplicity, honesty, and a rigid sense of right. It does not necessarily reflect his constant vision of the world or even his most typical vision of it. But it does problematize certain basic literati perspectives on their own abilities. Once these perspectives are questioned, they begin to crumble—and with them, the basic structures that guided female representation since the Han dynasty. For many Tang literati, neither women nor themselves can be described in the same way any longer.

Bai's poem on Zhao's origins is even more intriguing:

I Visit Zhaojun's Village 過昭君村

Original note: The village lies forty li northeast of Guizhou 歸州.

Magic pearls are produced without cultivation;	靈珠產無種
Bright clouds emerge without roots.	彩雲出無根
So it was also with this girl,	亦如彼姝子
Born in this remote and impoverished village.	生此遐陋村
Perfect beauty is impossible to conceal,	至麗物難掩
So she was swiftly selected to enter her lord's gate.	遽選入君門

Solitary beauty meets the crowd's jealousy,	獨美衆所嫉
So in the end she was discarded by the frontier walls.	終棄出塞垣
How could this beauty alone, rare in the age,	唯此希代色
Not receive the regard of her lord?	豈無一顧恩
When affairs are set, circumstances must fade,	事排勢須去
It was not because of the emperor's will.	不得由至尊
Since black and white could be altered,	白黑旣可變
Useless to talk of "reds and greens"!	丹青何足論
In the end they buried her bones north of Daizhou,	竟埋代北骨
And her soul never returned to the east of Ba.	不返巴東魂
Sad and pale, the water and clouds at evening;	慘澹晚雲水
Faint and vague, her old village gardens.	依稀舊鄉園
Her seductive beauty has long been transformed,	妍姿化已久
Leaving only the name of her village behind.	但有村名存
In the village there are left some old men,	村中有遺老
Who point and say to me:	指點爲我言
"If you don't heed the warnings of the past,	不取往者戒
Then probably you'll buy the regrets of the future.	恐貽來者冤
Now the faces of our village girls	至今村女面
Are burnt with candles and scarred."[49]	燒灼成瘢痕

Here Bai speaks more as the curious ethnographer and historian than as the crusading moralist. Already he reverses himself from the previous poem: here, Wang's recruitment into the harem was only to be expected, because beauty will inevitably find an appreciative audience. He repeats the argument of his famous ballad on Yang Guifei 楊貴妃, "The Song of Lasting Pain" (*Chang hen ge* 長恨歌):

Lovely substance born of Heaven cannot be cast aside;	天生麗質難自棄
So one morning she was chosen to stay by her prince's side. (ll. 5–6)[50]	一朝選在君王側

Bai introduces this point in order to account for something unmentioned by other poets: Wang supposedly grew up in a wild and isolated village in the heart of the mountainous Three Gorges region. Although Du Fu had mentioned crossing myriad hills and valleys, he had not made much of it; only the author of *A Manual for the Lute* had mysteriously commented that "because she came from a distant area, she was not favored by the ruler." Bai finds Wang's origins extraordinary and must find parallels in nature to account for it. Yet if Wang was fated to be discovered, she was also fated to be cast aside. The image of "black and white" is repeated again, but here with a more fatalistic emphasis; Wang is here more the conventional pathetic heroine than the figure of abused *cai*.

More strikingly, however, he creates a new perspective by evoking her natal village. Now she no longer moves from civilization to barbarity; instead, she moves from periphery to center to periphery again. Bai sees Wang's career as one of rise and fall, rather than a stunning decline. Moreover, he uses her story as an origination myth to explain the scarification of the maidens of the village. Bai's elders explain it as a way of avoiding imperial conscription. Women actively choose to make themselves ugly as a form of protective coloring and thus escape the inevitable judgment that made them vulnerable in Bai's first poem:

> He made the eye of his lord
> Turn a Xi Shi into a crone. . . .
> How may one know of good fortune or bad?
> A lovely face is worse than an ugly one.

Scarification is more drastic than donning thornwood hairpins: permanent altering of the visible surface fundamentally alters one's essence. Tattooing was one of the original "five punishments" that marked the criminal, along with the amputation of various body parts; moreover, a passage in Sima Qian's account of the Xiongnu suggests that tattooing meant barbarism to the Chinese. He mentions that the khan required Han emissaries to surrender their insignia and have their faces tattooed with ink before they could enter his tent. At first, the Han makes use of a borderman sympathetic to Xiongnu custom who is willing to do so, but he is later replaced by a Confucian scholar who is outraged by this demand and refuses. For him, tattooing would be associated not only with criminality but also with an irreversible abandonment of what it meant to be Chinese (*Shi ji* 110.56–57).[51]

However, traditional history provides one famous example of a tattooing that had positive results. In chapter 31 of the *Historical Records*, Sima Qian describes the founding of the Wu state, which was originally well beyond the cultural boundaries of China. Wu Taibo 吳泰伯 and his brother, both uncles of King Wen 文, the renowned founder of the Zhou, realized that their younger brother Jili 季歷 (the future father of King Wen) deserved the throne more than they did. However, primogeniture did not allow him to become ruler. In an act of altruism, they cut their hair and tattoo themselves, knowing that this act of disfigurement would disqualify themselves from rule. They then flee south and found the Wu state. Sima Qian is here giving a rational, Chinese motive to account for a barbarian custom: he attempts to make Wu seem less strange and ultimately part of the Chinese ecumene.[52] Yet Wu Taibo's gesture is ultimately the Confucian virtue of *rang* 讓, "demurring": the refusal to serve for the good of the state. Bai Juyi, too, creates a virtue out of what was probably a tribal custom of the Three Gorges area, as Arthur Waley has suggested.[53] If the girls of Wang's village allow themselves to be scarred, they, too, are demurring from performing a public service so that they can live out a more virtuous private life. Ironically, the mark of the barbarian Other becomes instead a defense that allows one to withdraw from the public world and to avoid the obligation of service to an ungrateful lord.

CHAPTER SIX

From Ritual to Romance

Women become more interesting in the Tang. By asserting this, I merely mean that they begin to be represented in texts to a much greater extent and that the roles they play in those texts become more complex: literati themselves seem to find them more interesting. However, this complexity does not imply a more accurate representation of some previously ignored reality; rather, it reveals new motivations for writing on the part of a male elite.

Those motivations have their roots in the social changes confronting the literati class. Already by the early Tang, the literati had begun relating to imperial authority in a new way. They saw themselves increasingly as participating in a large bureaucratic government rather than in the circles of decentralized aristocrats and nobility; this in turn affected their attitudes toward writing. But after the An Lushan 安祿山 rebellion of 755, what literati represented in texts expanded tremendously. The later Tang world seems to us more detailed than earlier periods because writers chose to describe more things, to express a more complex range of reactions to phenomena, and to find new ways to use language to convey ideas. Just as women are increasingly represented and take on a greater (simulated) autonomy within the text, men, too, seem to achieve a greater individuality and complexity. But there are still important motivations for writing that affect these representations and control the material. Most prominently, men continue to represent their own male communities in writing—communities now much larger and more diverse than Six Dynasties aristocratic culture but still very much defined by issues of alliance, competition, and prestige. Of course, great clans continued to assert their primacy and continued to make marriage alliances a key to high status. On the other hand, the examination culture attempted to make literary talent (in poetry and prose) a vital tool for defining status, and as the dynasty progressed, networks of patronage grew up around examiners, examinees, and other prominent men who owed their position to the system. While seeing themselves as loyal servants of the emperor and the dynasty, they had their own self-interests, just as the great clans had—as well as their own hierarchies and forms of prestige. We have hints of other worlds: the warlords who controlled increasingly large parts of the empire after 755 and who often set up competing courts, bureaucracies, and hierarchies; Taoist and Buddhist clergy; the increasingly sophisticated and wealthy merchant communities; and the "underworld" of marginal men and women: bravos, courtesans, outlaws, and the vast array of questionable occupations most evident in an urban environment but also seen in the countryside. It is obviously the world of the literati (particularly "examination" literati) that we know best—because they were the writers, but also because their status depended in part on literary talent. They could *create* their own identity by writing about it. But a man could play many roles in this society at different times, and the different worlds of power could intersect and influence one another.

As we shall see, this complexity is expressed in the Tang narrative, which speaks in all sorts of complementary and contradictory languages. One language that emerges is the discourse of sexual exchange, of special relationships that can exist between men and women. As the Tang progressed, erotic exchange began to move out of the realm of the formalized salon and entered the more complex and sophisticated world of narrative.[1] However, women were more than objects of desire. They were pawns in marriage alliances, a cement for social bonds of friendship or political faction, and prizes that allowed for competition and male display. The vast quantity of Tang informal narrative provides a rich source for social history (women's history in particular), but it also continues to illustrate male concerns and provides not so much a mirror for Tang society as a key to male writers' dreams and obsessions.

Rituals of Seduction

Although Taizong 太宗 (r. 627–50) saw the encouragement of literary endeavors, including literary circles and salons, as part of the responsibility of the ruler and his administration, poets no longer found their area of operations circumscribed by the insular world of the Southern Dynasties aristocracy. They could roam the capital cities of Chang'an 長安 and Luoyang 洛陽, as well as the empire itself, which stretched from the borders of Korea to the Tarim basin, from the Gobi to Vietnam. The imperial court was now a microcosm of the empire itself and directed policies that shaped the world, and a literatus serving in the bureaucracy might find himself carrying out policy thousands of miles from the centers that dictated his cultural identity. The phenomenon of isolation (whether from exile or from duty) became a much greater part of poetic experience.

Of course, isolation from the center (as I suggested in Chapter 3) was also a prominent factor in the development of poetry during the Six Dynasties. Xie Lingyun wrote from his great estates while undergoing banishment (voluntary or mandatory); Xie Tiao 謝朓 (464–99) composed skillful poems about his official journeys; Yu Xin did his best work, most believe, while detained by the Northern court after a failed embassy. But for most poets during the fifth and sixth centuries, poetry remained the expression of a social urbanity, whether that urbanity was incarnated in the imperial court or in the salon of an imperial prince or nobleman stationed at one of the strategic posts of the crumbling dynasty. Quite different would be the experiences of seventh-century figures who were compelled by circumstances to find forms of expression away from the really important imperial center: Lu Zhaolin 盧照鄰 (ca. 634–ca. 684), Shen Quanqi 沈佺期 (ca. 650–713), Song Zhiwen 宋之問 (d. 712). As Stephen Owen has suggested, it was the experience of exile in the seventh century that helped forge the sense of poetry as an unofficial pastime, as a form of communication between literati rather than as a competition played before the prince.[2]

If poets began to detach their poetic selves from the courtier self, and if poetry groped its way toward some new interpretation of the old adage that it express "what is intently on the mind," then we might suspect that the forms of erotic representation would shift as well. We saw in Chapter 4 that the erotic vignettes of "palace poetry" were inextricably caught up in contradictory impulses: the lingering allegory of courtier as palace lady as well as the social competition that invited the poet to evoke erotic imagery through mannered manipulation of poetic language. The palace poem was caught between the vertiginous pull of these two needs and often could not resolve itself into an unambiguous text. But there was room in the Tang empire for many things; perhaps sex, too, could find a realm outside the court, where the male poet could detach the female from previous habits of representation and make her more identifiably "other." Women would no longer just be an emblem for certain aspects of literati self-identity; they would be presented in texts next to the literatus, helping him, hindering him, and confirming or disputing his portrayal of the self.

The earliest known erotic text in the Tang is a mysterious narrative most likely authored by Zhang Zhuo 張鷟 (660–732), the *You xianku* 遊仙窟 (Dalliance in the immortals' den; for a translation of the complete text, see the Appendix, pp. 313–54). Probably composed in the 690s, it found its way to Japan, where it circulated among literary men and exerted a considerable influence on later Japanese literature.[3] In China it disappeared completely until nineteenth-century Chinese travelers to Japan rediscovered and published it. The plot is basic: the first-person narrator, on his way to a distant post in the far west, happens upon an isolated valley where two widows dwell. They exchange a great number of poems, most of them erotic or mildly obscene, and he sleeps with one of them; he then takes his leave the following morning. Zhang uses the game of seduction and flirtation as a plot device: although literary exchange eventually leads to sexual exchange, in the end sexual performance and desire are merely the explicit aspects of literary display and a "lust" for talent. Not only do we have dozens of poems in which the author must simulate both his own voice and those of several women, but we also have elaborate parallel-prose descriptions on set topics such as banquets, gardens, home furnishings, and women's apparel. Throughout, Zhang engages in a sort of literary and sexual boasting in front of his male audience: he caps the women's poems, jokes with them, sometimes trades light-hearted insults—all with the inevitable result that the women are overcome by his literary talents and his sexual desirability. At the same time, by granting the women themselves the ability to compose poetry, to perform on musical instruments, and so forth, Zhang gives himself not only worthy sexual partners, but worthy literary ones as well.

The tale begins as the hero encounters a mysterious, isolated valley when wandering through a remote countryside (Section 1).[4] At first, we might suspect the conventions of *zhiguai* 志怪 (accounts of the strange), short anecdotes of the bizarre and supernatural whose collection had been popular since the third century A.D. Such anecdotes, although not usually related in the first person, do tend to begin with a factual place-setting.[5] There are some important differences, however. Zhang identifies his location and then proceeds to identify its importance: the ancient sage-emperor Yu 禹 engaged in flood control here. By moving west to the end of the world, Zhang travels into the primitive, to where things begin (notably, the Yellow River). It is a place of impressive landscapes, but landscapes that have been shaped by the work of the ancients. In this sense, Zhang inhabits a textual landscape. This is confirmed further when he reaches the fictional Immortals' Den: he cites elders to attest to its existence. These signs tell him that he is passing into an extraordinary world.

The move westward recalls the travels of others besides Yu. Zhang refers explicitly to Zhang Qian 張騫, the Han emissary who supposedly sought the source of the Yellow River. Less explicitly, Zhang imitates King Mu 穆 of the Zhou, whose legendary stallions brought him to distant Kunlun 崑崙 Mountain, where he was banqueted and romanced by the Xi Wang Mu 西王母 (Queen Mother of the West). King Mu's travels had long been a part of Chinese consciousness, and they form a part of one early text, the *Mu tianzi zhuan* 穆天子傳 (An account of Mu, Son of Heaven). His dalliance with the Queen Mother allowed for her portrayal as a mistress fit for emperors. Later, legends had her visit Emperor Wu of the Han.[6]

Zhang is not a ruler, however. He is a literatus—and a literatus under compulsion, engaged in a toilsome official mission. This, too, distinguishes him from typical *zhiguai* figures. Take, for example, the following story of Liu Chen 劉晨 and Ruan Zhao 阮肇; this and similar *zhiguai* were a likely influence on Zhang when he composed his own tale:

Liu Chen and Ruan Zhao had entered the Tiantai 天台 Mountains to pick medicinal herbs. They traveled far and were unable to find their way back. After thirteen days they were in a state of starvation. From afar they saw a hill on which grew peach trees with ripe fruit. Clambering up steep places and pulling themselves up by vines they reached the spot, where they devoured several peaches and assuaged their hunger. After they ate their fill, they set out down the mountain. As they took some water in their cups they saw some turnip leaves floating down the stream, fresh and lovely. Then a cup came floating down with some sesame meal inside.

"There must be some people nearby," they said to one another.

Crossing over a hill, they found a large stream and by the side were two women, extraordinarily beautiful. When they saw the two holding the cup, they laughed.

"So Mr. Liu and Mr. Ruan, you have brought our cup back!"

Liu and Ruan were startled. The two women then spoke pleasantly, as if they were old acquaintances: "Why have you been so late in coming?"

They then invited them back to their home. The south and east walls were hung with crimson gauze curtains, and from the corners hung gold and silver rafter bells.

Each had several maidservants in attendance to serve her. There was sesame meal, dried mutton, and beef, all of superior quality. After they had finished eating, they started to pass around the ale. Suddenly a crowd of girls came in, peaches in their hands: "We welcome the arrival of the bridegrooms!" They grew tipsy while music was performed; after nightfall, each retreated to the bed curtains for the night.

The ladies' seductive charms were unsurpassed. After ten days the men sought to return but were earnestly entreated to stay. After half a year the weather and the plants remained as if it were spring, and all the birds sang. But the men still wished even more to return to their village, and their longing grew keen. The ladies then saw them off, showing them the road back home. When they arrived, their village was in ruins. Ten generations had already passed. (*Taiping guangji* [hereafter, *TPGJ*] 1: 310)[7]

The Tiantai Mountains were a sacred place for Taoist believers; they were immortalized in a rhapsody by the early Taoist poet Sun Chuo 孫綽 (314–71). Liu and Ruan enter the scene engaged on a suitably Taoist enterprise—herb gathering. After they discover the goddesses' refuge, they remain as long as they like. Nothing compels them to leave except their own homesickness. For Zhang, however, circumstances are different. He has not lost his way when he comes upon the Immortals' Den, nor is he unaware of its magical qualities. Whereas Ruan and Liu are startled to see a mysterious cup drifting past, he recalls the words of local elders that tell of similar floating phenomena. He thus becomes a well-informed traveler interested in exploring the source of things. Perhaps as a reader of *zhiguai*, he is aware that otherworldly sexual delights await him as well. But since there are goddesses in the offing, he must first purify himself. "I then put myself in a serious frame of mind and observed three days of ritual fast"—unlike Ruan and Liu, who were starving quite unintentionally. In both cases, however, the body is purging itself of earthly food and is preparing itself for the spiritual feasts to come. The adventure is thus not forced upon Zhang. Rather, it is an intentional detour made by a bureaucrat in need of rest before resuming the task assigned him. Zhang is under compulsion and yet not under it: he will enjoy the same privileges as King Mu in his own journey west, but employment in the emperor's service will force him to leave his mistress soon enough. And perhaps this is a good thing: duty will prevent him from falling victim to the unearthly (as Ruan and Liu did) and losing his own world to eternity.

Although Zhang recognizes the magical nature of the meeting, he also perceives it as a social encounter. When he happens upon the goddess's habitation, he engages in an enquiry into her identity. As it turns out, the "goddess" is a certain Lady Cui 崔, from a gentry family (Section 1). Here, a maid (later identified as Guixin 桂心, "Cassia Heart") reveals Lady Cui's family background: she is of very high lineage (the Cuis of Boling 博陵).[8] Cassia Heart also compares her favorably to beauties of the past. Notably, a number of these beauties are men who combined literary talent with physical attraction: Song Yu, Cui Yan 崔琰, Pan Yue. Her appeal and talent are in some aspects independent of gender: she can participate in the same categories of appreciation and behavior as a man. Of course, these comparisons also go far to socialize the encounter. Lady Cui is not an anonymous goddess, and the deliberateness of these social courtesies (which will become even more evident as the story proceeds) confirms for us the degree to which "Immortals' Den" is itself a sophisticated narrative within which suspense is relatively unimportant. Once we recognize the tropes with which the author plays, we are in little doubt as to the outcome. Although he must make gestures of seduction toward the woman he meets, they are merely stages in a courtship ritual: the pleasure comes in carrying out the gestures as deftly as possible.

I will not attempt a theoretical discussion of "ritual" here, although a number of scholars have begun to handle the subject with the sophistication that it deserves.[9] The formality of exchange in "Immortals' Den" is not ritual in any societally recognized way; rather, it is a performance that has been ritualized. There is an implicit code governing the response speech and action should elicit. Again, as with earlier poetry composition, ritualization is here a process that channels competition: it determines what forms of behavior are unsanctioned and unruly, and how sanctioned behavior should be tested and evaluated. Of course, if the composition of palace poetry was a ritual as well, then what advantages lay in transposing eroticism from ritual male composition to a fictional ritual narrative involving women as participants? The question almost answers itself. Narrative here gives the illusion of real experience; the author is simply relating what occurred. Zhang can thus create an ideal environment in which to exercise his wit and talent, unburdened by the pressures of extempore composition. Moreover, he writes for an audience of his peers.[10] As literati, any one of them might have experienced what he experienced; any one of them might have had the opportunity to woo Lady Cui. Zhang thus presents a challenge to the male readers: Could you have performed this impressively? How long could you have kept up the poetry?

At the same time, however, "Immortals' Den" does provide a new venue for the presentation of palace-poetry eroticism. As we have seen, palace poetry was not addressed to the *object* of desire. It was a poetry shared by the male community: desire was the currency exchanged in mutual male evaluation. But "Immortals' Den" presents palace poetry as a form of exchange between the sexes. This exchange commences almost immediately: after the servant invites Zhang into Lady Cui's house, he overhears a musical performance by the mistress herself, who is more popularly known as Shiniang 十娘. There begins an elaborate series of exchanges that sets the mood of the text (Section 2): the man gives a flattering, but plaintive description of the woman and her charms, and the woman responds humorously on the exaggerations of his language.[11] It is her ritual task to resist, even if she resists wittily.

This dynamic of invitation and resistance should hardly surprise us; it is perhaps part of the universal language of literary courtship. But Zhang resorts to the tools of the trade as he knows them: his poems are not so much a direct address to Shiniang as they are little moments of palace-poetry lubricity ("Often she puts forth a slender hand, / Toys for a time with the slender strings"). Both the sensuality of the woman and her talent for music making are praised, as if this objectification of his addressee will produce results. Shiniang's response is to deny his flattery and to turn her attention to the facts: "Why do you, engaged in your lord's service, / come here in vain pursuit?" This abrupt and seemingly rude inquiry also involves a subtle probe into his public office and possibly his social status. Her second question ("Why must you so waste your time in teasing?"), although a rhetorical rejection, invites him to explain himself further in literary terms. Zhang already knows something about *her*—Cassia Heart has told him. But she still knows little of him. He proceeds to rectify the matter with a love letter, one of the great set-pieces in the text (Section 3). Here, Zhang refrains from giving his family background and instead portrays himself as a wandering sexual adventurer—the sort we have already met in *Master Dengtu the Lecher*. However, he subtly shifts his self-portrayal from that of wandering roué to that of government emissary: "Now, on this second mission bid by the same Lord of Heaven, from afar I caught scent of your fragrance." On the one hand, this answers her earlier inquiry: her fragrance has acted as a beacon to the passing traveler—that is the reason he has come. But his linking of this visit with his official duties reinforces the idea that he is taking "time out" from compulsory tasks.

Above, I suggested that Zhang's ties to his duty serve as a defense against the dangers of a promiscuous goddess: his posting is a guarantee that he will have to leave her bed eventually. But since this is Zhang's own fantasy, one might ask why he must set up this opposition between work and leisure. Quite simply, he needs it to define who he is. It is part of a new conception of the literatus to see himself as someone who has imperially appointed tasks to carry out, who has a job he must complete. Obviously, literati had long been familiar with conceptions of duty and leisure, employment and retirement. But Zhang foregrounds it in a place where he does not need to. Writing for other literati men participating in the same world, he thus redefines the sexual adventurer / goddess-romancer as a government bureaucrat. The letter does much more than this, of course: by making his plea in a form that enables him to rehearse a substantial amount of love lore from the Chinese tradition, he continues to answer Shiniang's inquiries about him through his own demonstrations. Rhetorically it conveys more about himself and his abilities than about his love for Shiniang.

Shiniang still responds in the prescribed way: she rejects his advances as excessive and insulting. She also refuses to respond at this point: evidently she needs something even more extraordinary from her guest to soften her heart. He entreats her in a more persuasive tone, ending with an injunction to "gather her rosebuds":

> Do not say you'll always have
>   a face worth a thousand in gold—
> In the end you'll become no more
>   than a handful of dust.
> While alive and in the sun
>   only take your pleasure!
> After death no spring season returns
>   to call you back again.
> If only you make dalliance
>   your goal throughout this life,
> No need to wastefully betray
>   your threescore years and ten! (Section 3)

These verses are written in a language radically different from the letter, and in a diction fairly different from most Tang poetry as well. Composed in a flexible, vernacular meter, they likely reflect a more "vulgar" level of erotic verse than the refined lines canonized in *Jade Terrace*. "Immortals' Den" is striking for the way it incorporates these different stylistic levels; once again, the informal environment of circulation likely encouraged Zhang to experiment.

Shiniang shows no interest in this second attempt at persuasion, and so Zhang goes to bed. Inevitably, he dreams of her. The powerful emotion stirred by this experience results in a further entreaty that finally breaks down Shiniang's reserve. He writes her a poem on the dream; then, after she throws his new verses in the fire, he reacts bitterly in yet another poem. Visibly moved, she goes out finally to greet him (Section 4). Why in the economy of the tale does this exchange work when the more sophisticated ones have failed? Within the formal ritual of seduction, there still must be a way of signifying sincerity and "true" passion. As the author of the letter and the earlier poems, Zhang proves himself *talented* enough for Shiniang's bed—but the old distrust of rhetoric plays a part here as well; there must be an allowance for the authentic. The very colloquial line "Is that wretched woman taunting me?" breaks down any remaining decorum. Zhang is "authentically" suffering from lovesickness. This draws a passionate response of rejection from Shiniang, but the emotion betrays her weakening. It requires only one more simple quatrain for Zhang to complete his conquest.

Now, as the two finally meet face to face, there is a long, elaborate passage of self-introduction and exchanged compliments (Section 4). If we do not keep the necessity of courtship ceremony in mind, this passage may seem unnecessary or even ludicrous. But there is a striking resemblance here to Western and Japanese epic conventions, in which enemies about to engage in individual combat proclaim to each other their background and acknowledge each other's worth. Although Chinese literature did not pursue the "love as combat" trope to the extent the Petrarchan tradition in the West did, the resonance of this passage is clear: in combat, one needs to recognize the foe as an equal in order to confirm one's own value. The exchange begins with rhetorically balanced compliments: Shiniang matches Zhang's speech phrase for phrase and even teases him by suggesting that his looks do not live up to the marvels of his composition. Then the two exchange family data: Shiniang's identity is confirmed from her own lips, and she explains how she comes to be living alone with her sister-in-law in this remote place. Zhang makes his background clear for the first time as well: he tells first of his family and then of his rank in the bureaucracy. Although he couches this information in decorous and self-effacing language, his self-satisfaction is quite evident.

With this exchange over, Shiniang makes the significant gesture: she invites him into her residence proper (Section 5). This is partly a recognition that he ranks highly as a guest and as a talented man: the language here echoes the famous lines of Confucius describing a disciple: "It may be that [my disciple] You 由 has yet to enter the inner chamber, but at least he has reached the guest hall."[12] But Shiniang is also defining the field of battle. Zhang's earlier "conquest" of her through his poems is only a partial victory: all he has won from her so far is an agreement to a formal engagement.

However, it is also clear that this contest will be decorous and, above all, playful. Unlike palace poetry, in which the concept of exchange was limited mostly to the male circulation of texts, "Immortals' Den" dramatizes dialogue and with it courteous exchange, jokes, and witty banter. Eroticism and sexual combat take on the form of leisured amusement and indulgence in the social graces. In order to facilitate this approach, the narrative requires a third person: in this case, Shiniang's sister, Wusao 五嫂 (Fifth Wife). If the male readership of "Immortals' Den" is the silent third party to the text, Wusao is its visible observer and facilitator. She socializes and makes public the private processes of seduction and resistance passing between Zhang and Shiniang.[13] Although her sexual desirability is made clear, it never becomes possible for Zhang to bed both of them; her role as commentator is too vital. She also reinforces the importance of ritualized sex: multiple partners lead to a promiscuity in which sexuality loses its sense as a reciprocal motion of exchange. She is the guardian needed to guarantee that there will only be two. This is demonstrated superbly by their first literary game: Wusao suggests that they quote poems from the *Odes* to express their feelings (Section 7). As we saw in Chapter 1, apt citation from the *Odes* was a literati art that dates back to Zhou dynasty days. Here, the poems are reinterpreted as courtship verses (thus providing further proof that readers often saw them that way, in spite of the significance of Mao's commentaries). Yet here the game ritualizes courtship further by making Wusao's role as facilitator more evident. She begins by citing the famous first Ode, the epithalamium that exemplifies the ritual power of "proper" courtship.[14] She thus reaffirms her role as go-between. Zhang replies with a quotation that suggests sexual longing and tension; in his poem, Shiniang continues to be unapproachable. Wusao replies with yet another poem asserting her own role and the importance of propriety; then, perhaps anxious that she is selecting too many discouraging verses, she adds another poem that suggests the joy with which Shiniang will eventually greet her lover. Shiniang adds to this a testy quotation anticipating fickleness on Zhang's part, thus continuing her rhetorical resistance to his advances. He caps this finally with verses suggesting his undying fidelity to her. Although he has the last word, the game has finally elicited from him another ritual—an oath of loyalty.

However, there are stresses and problems caused by Wusao's role as ritual manager. Although polygamy in the traditional family structure allowed for multiple female partners, it was not so happy about multiple partners at the same time. The inevitable result of such strictures is an obsession with favor or the lack of it, with jealousy and envy. This is illustrated slightly later in the tale, when Zhang shows an unwarranted interest in one of Shiniang's maidservants, Zither Heart 琴心 (Section 10). When Shiniang shows jealousy, Wusao jokingly takes both lovers to task: Zhang for not knowing his proper limitations, and Shiniang for not showing enough interest in Zhang, and so forcing him to seek female company elsewhere. She then leaves open the possibility (not seriously) that Zhang may transfer his affections to *her*. This is expressed rhetorically through a poem that presents the lover with a choice of two blossoming trees: from which will he pick the flower? When Zhang attempts to reinterpret this image by demanding the right to pick flowers from both (and makes the imagery even more sexual: "Playful butterflies lean on red stamens; / sporting bees enter their violet buds"), Wusao accuses him of being greedy and then finds a more appropriate image to deny his improper desires: "You really *are* too greedy; you shoot at two targets with one arrow." The male organ can only aim at one place at a time.[15] By elevating her position as outsider within the discourse of courtship, Wusao guarantees the duality of standard heterosexual ritual.

This subtle canalization of desire toward its proper culmination is played against an elaborate rehearsal of all the various forms of human aesthetic pleasure: one by one, the lovers and Wusao discuss and write poems about singing, dancing, musical instruments, archery board-games, food and drink, and the beauties of the landscape. This is in part to reinforce the "testing" aspects of the text: in a simulation of examination, Shiniang and Wusao give Zhang an opportunity to test himself in a further display of male literati ornament. As Confucius said, "If you remove the hair, the pelt of a tiger or leopard is just the same as a dog's or sheep's."[16] In courtship, presentation matters. Earlier in the text, we have a small drama on talent and recognition, when one of the maids performs on a lute (Section 6). First Wusao comments on the extraordinary appearance of their guest; then Zhang composes a poem on the maid's performance, which results in a humorous poetic response from Shiniang. This allows Zhang and Shiniang to engage in a round of complimentary exchanges. This is a mutual admiration society. Nonetheless, within a very short space the author has managed to engage in several forms of cultural display. Wusao's recognition of the hero's superior qualities plays on conventions of character evaluation or *pin* 品, of the sort we have already noted in the *Worldly Tales* anecdotes. The hero replies by citing two famous examples of talented women who were able to intuit quickly the value of the men they spied upon (thus neatly complimenting both them and himself), not to mention introducing the associations of sexual passion suggested by the mention of Zhuo Wenjun 卓文君 (who eloped with the talented poet Sima Xiangru 司馬相如 after hearing him play the zither). This display of cultural literacy leads to yet a further display of poetic talent, on the subject of the lute. He claims for himself the talent a sensitive listener possesses of discerning the emotional state of the player. Shiniang then shows *her* wit in another characteristic gesture of resistance: she teases him for making a claim for deep understanding before the music has been performed—and thus for having the audacity of praising her sexual talents before he has had the opportunity to observe them. The hero in turn diverts this attack by praising Shiniang's poetic abilities extravagantly—which results in an exchange of carefully balanced compliments. Again, cultural abilities are tested.

And so the exchanges continue; there are far too many such gestures to deal with here, especially because their structure and purpose are generally the same: the sisters provide Zhang an opportunity to shine, and his response provokes witty responses on their part, combined with continuing admiration, teasing (usually on Wusao's part), and decorous resistance on Shiniang's. When Shiniang finally concedes to his wishes (mostly through Wusao's urging), we have a more explicit drama of assault and resistance: the scene is Shiniang's bedroom, to which all three have repaired (Section 16). Now, as Confucius might have said, Zhang has earned his way into the "inner chamber." As he attempts to bring Shiniang forcibly to the bed, he continues to compose poems of persuasion. She replies with resistance poems, and Wusao continues with mocking poems addressed at both. As long as Wusao, the third element, is in the room, sex continues to be a public performance, even while it borders into the realm of assault. However, we are moving into an area where ritual no longer governs courtship and sex: nothing but the act can bring the narrative to a satisfactory conclusion. Here Shiniang's resistance in words turns into a physical resistance; the very long passage here is meant to cater to more basic instincts among the male readership. That this passage provides the inevitable climax to the narrative suggests connections once more to the purposes of pornography, whose narrative structure tends to mimic the male experience of sex. However, this moment is relatively brief; Zhang's one explicit descriptive passage is awkward and somewhat ludicrous in comparison to the more elegant parallel-prose passages that run through the rest of the text. This is, I suspect, because writers had not yet developed a sophisticated pornographic vocabulary (and perhaps would not do so until the late Ming). Zhang is much happier sublimating sexuality through ritual and game.

As if to compensate for this coarse moment, the narrative explodes into a flurry of parting poems as dawn breaks and Zhang must go on his way. Shiniang and he exchange verses several times; he and Wusao exchange verses; he and the maids exchange verses. We then have the departure and the closing of the piece (Section 19). Inevitably, sex is followed by melancholia and pain over parting. But this passage also resumes the ritual dance. Although earlier poetry exchanges had played on images of marriage as a way of concealing the casualness of the liaison, here Zhang must situate himself within a new conventional situation. Returning to the original frame for the story (the encounter with goddesses), he composes a few short verses in the meter characteristic of the shaman songs of the *Songs of Chu*.

My reading of "Immortals' Den" is not intended to be comprehensive; nor perhaps is such a discussion of the text possible, granted its anomalous position within the Chinese tradition. No other texts remotely like it have survived. Nonetheless, we do have a hint of some interesting developments in the literati portrayal of women and of erotic relationships. Zhang is obviously avoiding the pressures of *spontaneous* performance and competition by narratizing a situation in which he can take control of all the competing voices and write *for* them. This act of narrative control is reinforced by its repetition of an old palace-poetry dynamic: the controlled representation of the sexually arousing scene produced for *other* male readers. He thus becomes a provider of sexual pleasure and can control the pace and intensity of that experience. At the same time, the narrative couches itself in terms comprehensible to the seventh-century literatus: the hero is portrayed as a bureaucrat on a mission who stumbles on the goddess's lair, rather than as a ruler visiting the deity or as a Taoist herb-picker who has lost his way. His sexual adventure then becomes compartmentalized within literati experience: it is a moment of leisure stolen from the pressures of official duty. Although this creates a figure who acts under compulsion—who must leave the bed the following day rather than continue the dalliance—it also gives him purpose and a public life that properly opposes private desires and lusts: it gives him the strength to leave that bed in the end and to resume the tasks that create his own identity. Perhaps most interestingly, it partially separates sex from the political life of the literatus; the self-image of the literatus as the sexual subordinate of his ruler is replaced by the conception of sex as something that takes place away from the tensions and desires of an official career.

That compartmentalization is also part of the ritualization of the sexual encounter; it makes a space for sexuality but also gives it boundaries. Within that ritual space, courtship is played out as a series of exchanges, at times serious, at times playful. Women are visualized as playing an active part in this game playing, even if they end by surrendering their bodies to a literal sexual congress. The sexual act itself may matter little, since it is really only a figure for the more significant textual exchange that had preceded it. Women become the testers and judges of men—although in a text controlled by a male author, their primary role is to confirm male aspirations. Zhang's authorship here may suggest another act of controlling: an assurance that male performance with the woman will be satisfactory and that there will be no loss of literary or physical power during the courtship. Yet a door is opened up here that will be exploited in texts to come: that men might fail not only in relations with other men but also in their relations with women.

This competition is further ritualized through the sanctions of a third presence who acts as mistress of ceremonies, as go-between, and as procuress. Wusao is the "inside outsider" who guarantees that the courtship proceeds in a comprehensible ritual manner and that the laws of monogamous heterosexual alliance are (at least for this time) obeyed. Even if Zhang has attempted to escape from the pressures of court competition by narratizing the sexual encounter, he still ends by ritualizing it in a formal manner. With this movement back to ritualization and courtly exchange, he simply overwrites a heterosexual encounter onto the template of male-male competition. As the character Zhang seeks to impress Shiniang in order to gain access to her bedroom, the author Zhang seeks to impress his readers and gain their approbation and admiration. As long as textual exchange continues to act as a figure for sexual exchange, then literary competition produces corresponding consequences of desire and jealousy. This makes "Immortals' Den" quite different from our modern conception of pornography as fundamentally autoerotic. Rather, it is a social text in which men share their own desires within channels prescribed by a formal and mandated art of seduction.

Romancing Talent

Although later Tang narratives were no longer quite so ritualized, they continued to be social texts for the literati. We turn to a disarmingly simple story, "Liu shi zhuan" 柳氏傳 (The story of Miss Liu; *TPGJ* 4: 541–42). Although we now read it as a love story, it deals just as prominently with distinctively male concerns. The ways in which it parts from our own conventional conception of a love story gives important clues as to what mattered to the male readership.

In the Tianbao 天寶 reign, Han Hong[17] 韓翃 of Changli 昌黎 had a reputation as a poet. He was rather reckless in nature, and he was very poor and relied on others for support.

A certain Mr. Li 李 was a close friend of Han's; Li's household was possessed of a thousand in gold; he was independent-minded and fond of talent. His favorite concubine, Miss Liu, was an outstanding beauty of the age; she enjoyed conversation and joking and was good at chanting poetry. Mr. Li established her in a secondary household and held banquets with Han there, housing Han next to her.

Han had always enjoyed a wide reputation, and those who paid calls on him were all accomplished men of the time. Miss Liu would spy on them from the doorway and would say to her servants, "How could a man like Master Han always be poor and obscure?" From that time she took a fancy to him. Li had always valued Han, and he refused him nothing. Afterward he learned of Han's inclinations and held a banquet for him. After they were in their cups, Mr. Li said, "Madame Liu's beauty is extraordinary, and Master Han's literary abilities are exceptional. I wish to present Madame Liu to you to serve as your bedmate. What do you think?"

Han was astonished and withdrew from his seat. "You have already graced me with your own food and the clothes from your back. How would it be proper for me to take what you cherish?"

But Li resolutely insisted. Miss Liu knew that he was sincere, and so she bowed twice to him, pulled him over to her by his robes, and joined their mats together. Li sat Han in the guest's position, and they emptied their cups with great pleasure. Li also supplied thirty thousand cash to contribute to Han's expenses. Han looked up to Liu's beauty, and she in turn admired his talent; thus having attained each other's affections their delight might well be guessed at. (541)

The poor but talented literatus is the focus of many Tang narratives, and the attraction for Tang male readers of tracing such a figure's success hardly needs to be explained. In the Tang world, however, one does not pull oneself up through sheer talent: friends are needed. This does not mean that the Tang author here is cynical or believes that one gets everything done through "connections"; rather, it is Han's talent that attracts male admirers to him and makes them willing to help him. Mr. Li assists him first financially and then with a place to live. This pied-à-terre allows Han to make connections and establish friendships with important men. Then, in a classic homosocial maneuver, Mr. Li cements his affection with Han by giving him Miss Liu, his own most cherished concubine. Mr. Li here very likely acknowledges the sexual needs of his client. But the text goes beyond this motive by making Miss Liu participate in her own presentation. We have already seen talented women in earlier texts portrayed as competent judges of character—most famously, the wife of Shan Tao, who spied upon Ruan Ji from behind a screen, and the poet Xie Daoyun, who was so skeptical of the abilities of the great Wang family.[18] However, in those cases female judgment was divested of desire and keyed instead to codes of male self-evaluation and snobbism: the women were to some extent placing a value on themselves by evaluating the worth of their husbands and brothers. Miss Liu takes pleasure in Han in the same way that Mr. Li does—they are both attracted to his aura of brilliance and so wish to assist him. Unlike Mr. Li, however, she can express this attraction through a different, more intimate relation. Whereas Mr. Li consummates his relation with Han by proxy, Miss Liu comes to *transcend* her role of proxy by actively and explicitly linking appreciation for talent with sexual desire. Moreover, by possessing great beauty herself, she externalizes and enacts the hermaphroditic metaphor of the talented man as "beauty" (*meiren* 美人), the representation of internal male talent as external female desirability. They now become an idealized couple, the *caizi jiaren* 才子佳人 (talented scholar and lovely lady) so characteristic of later romantic literature in China. Miss Liu is a constant reminder to Han of his cultural worth (a mirror for the self) as well as a constantly accessible reward for that talent, sexual pleasure as a foretaste of the pleasures of a successful career. She is the gift that keeps on giving.

Narrative interest develops out of the separation of this pair. At first, Han is so happy with Miss Liu that she herself must convince him that his talents require him to give her up for a time so as to obtain the greater prize. She is only a temporary symbol of what he deserves:

The following year, Yang Du 楊度, the vice minister of the Board of Rites, recommended Han for the examinations. He lived in reclusion for a year. Miss Liu said to him, "A glorious reputation extends to one's relations; this is a fact respected by the men of the past. How is it right for a 'humble washerwoman' like myself to obstruct your chance to pluck the orchid? Besides, I have enough utensils and goods to last me until your return."

Han then went off to stay with his family in Qingchi 清池. After a year, she ran out of food and had to sell her cosmetics and toiletries to meet her needs. At the end of the Tianbao era, rebels overthrew both capitals, and the populace fled in fear. Because Miss Liu stood out in her beauty, she feared she would not escape; so she shaved her head and disfigured herself and took refuge at the Faling 法靈 temple.

At this time, Hou Xiyi 侯希逸 moved from Pinglu 平盧 to become military governor of Ziqing 淄青. He had heard rumors of Han's reputation and so asked him to be his secretary. After Suzong 肅宗 succeeded in returning the state to right by dint of his divine martial prowess, Han sent a messenger privately to look for Miss Liu. He filled a silken bag with pieces of gold and wrote on it the following verse:

Zhang Terrace willow,	章臺柳
Oh Zhang Terrace willow!	章臺柳
So green in bygone days, Are you still there?	昔日青青今在否
Even if your tender withes sweep down as of old,	縱使長條似舊垂
No doubt they have been pulled and plucked by another's hand.	亦應攀折他人手

Miss Liu sobbingly accepted the gold, and while those around her grew sad, she responded:

Willow branch	楊柳枝
In the season of fragrance	芳菲節
Resents that, year upon year, she is given to those who part.	所恨年年贈離別

Once one leaf goes with the wind
in answer to the autumn, 一葉隨風忽報秋
Even if her lord should come,
what would be left to pluck? 縱使君來豈堪折

Shortly afterward, the border general Shazha Li 沙吒利, who had recently achieved merit, learned secretly of Miss Liu's beauty and stole her away to his mansion, where she occupied all his favor. (541–42)

Miss Liu's act of unselfishness here occurs in other tales as well and will become a stock motif in later fiction and drama—left behind by the ambitious lover/husband, she maintains her chastity and must sell her possessions to do so. In other tales, such abandonment suggests selfishness or disloyalty on the part of the man;[19] here, it merely introduces the more serious intervention of outside circumstances: before Han can return to her, his career and her chastity are derailed for a time by war and rebellion. She attempts to defend herself by becoming (at least temporarily) a Buddhist nun and thus signifies to others her unavailability. But in time of war this makes little difference, and she falls victim to raw force. Before this happens, however, Han attempts to make contact with her: once he is established in a semi-official position (note that here it is not a true "bureaucratic" post but, rather, a place on the staff of a general), he sends her money and a poem inquiring whether she has remained loyal to him. A true love story becomes possible: this passage makes clear that Han in fact does *not* see Miss Liu as a momentary substitution for success but as an essential element in his own conception of happiness. Of course, his gesture here is literary and passive, but the exchange of poems reasserts the ties of mutual admiration and sexual desire that had defined their relationship earlier. At the same time there is a disquieting note: the act of composition is a way of demonstrating literary talent and mutual understanding, but the language in which the poems are couched alludes to the world of sexual anxiety and competition. Miss Liu's surname means "willow," and Zhang Terrace was a place-name associated with brothels or at least places of sexual pleasure. Just as willow withes were broken off and presented to departing friends in the hopes that they might be "detained" (*liu* 留), the woman/tree is used by men who inevitably abandon her, leaving her diminished by the encounters. A famous Dunhuang 敦煌 song makes the connection explicit:

Don't break me off! 莫攀我
If you break me, my heart will incline too far. 攀我太心偏

I am a willow by Serpentine Pool—	我是曲江臨池柳
This man snaps me, that man breaks me,	者人折了那人攀
Their love is only a moment.[20]	恩愛一時間

At first, the content of Han's and Liu's poems might seem strange: they sound more like an exchange between client and prostitute than they do like verses exchanged between two friends and lovers who truly seem to understand each other, and we realize that she is socially not much superior to a prostitute. Granted the vagaries of war and the difficulties Han is having in his own career, it would have been reasonable for him to assume that she would find another patron, at least temporarily. A close examination of Miss Liu's poem, however, reveals a more complex response. The opening couplet suggests the expected answer, not too different from the Dunhuang song: she has grown weary servicing men's needs. I doubt if we are meant to take this as a literal account of what Miss Liu has done. Rather, it is a generically appropriate verse meant to harmonize correctly with Han's inquiry: it is indeed true that the willow wearies of being torn by the hands of transient males.[21] But the closing couplet shifts the direction of the imagery: autumn is evoked in the inevitable loss of leaves the willow undergoes with the change of the seasons. Convention reads it as a lament over growing old and losing one's beauty. It is not men's hands that are destroying me but the natural cycles of time: I am growing old. In this way Miss Liu both declares her loyalty and asserts her independence from modes of behavior expected of women in her class and in her situation.

Of course, the poems are conventional enough, and it is not impossible that the author of the text inserted generic verses as he thought fit to flesh out the tale.[22] However, even if that were true, we still have a vivid realization and contextualization for love poetry in plain anecdotal form. In *You xianku* Zhang Zhuo created a narrative frame within which men and women exchanged poetry within a ritualized male fantasy. The essential nature of "Miss Liu" as a sort of historical anecdote demonstrates that later Tang audiences could easily conceive of such an exchange actually occurring between the sexes. They are reading as fairly commonplace the ability of women to react culturally to male actions and to participate in the male world of literary exchange.

However, the tale does not end with the exchange but instead introduces new issues and characters:

When Hou Xiyi was appointed vice-director of the left, he came to the capital for an imperial audience, and Han was able to accompany him. When he arrived in the capital, he had already lost track of where Miss Liu had gone; he sighed and thought of her unceasingly. By chance at Dragonhead Hill he encountered a gray-haired man driving a bridled ox that pulled a carriage with two maidservants accompanying. Han was following along behind them when he heard from within the carriage, "Aren't you Supernumerary Han? I'm Miss Liu!" She sent a maidservant to inform him secretly that she had been given to the General Shazha Li. Since they were prevented by other travelers in the carriage, she could only ask him to wait for her the next day at the gate to Daozheng 道政 Ward [of the capital]. When he went to this appointment, she had tied a jade box with a piece of light silk and, filling it with scented oil, gave it to him from the carriage. "Although we are now parted forever, let me show you my sincere affection." She then had the carriage turn, but she waved to him. Her light sleeves fluttered as her scented carriage rumbled away. Soon his sight of her ended, and he was lost in thought as she vanished in a cloud of dust.

Han could not overcome his passion for her. It so happened that Hou's subordinates were meeting at a house for a drinking bout and sent someone to invite Han. Han forced himself to go, but his expression was depressed, and his voice was choked with sobs. There was a certain military inspector, Xu Jun 許俊, a self-assured and independent man. Grasping his sword, he said, "You must have cause. I beg to be of use." Han had no choice but to tell him. Jun said, "Write a short note. I will fetch her straightaway." He then donned his armor and strapped two quivers at his waist. Taking along a horseman, he chose a side road to Shazha Li's mansion. He waited until the general had gone out for a half-mile or so; then, pulling his robe close around him and grasping his bridle, he shoved his way through the gates. He rushed in, shouting "The general has fallen ill! He has sent me to bring his wife!" The servants all fell back in fright, and none dared raise his head. After he ascended the main hall, he showed Miss Liu the letter from Han, tucked her underneath his arm, and leapt astride his horse. He rode hell-bent for leather, returning immediately to the party. He pulled her forward by the folds of her robe. "Fortunately I have not failed in my mission." All the guests sighed in surprise and admiration. Han and Miss Liu grasped hands and wept. All gave up their drinking. (542)

Han's failure to take immediate action does not reflect cowardice or indifference to his mistress. Rather, the literati author accurately reflects the difficulties of the situation: it is only by chance that Han meets Miss Liu again, amid the chaos of rebellion. Once he has found her, however, their bonds are reasserted. Although Miss Liu is to some extent a pawn that is circulated among men, she shows enough initiative and emotion to make her own preferences evident. Not surprisingly, she demonstrates this through the bestowing of a gift—thus repaying in a minor way his gift of the gold earlier.

However, desire on both sides goes for naught until Han earns the attention of a violent military bravo-figure, the impetuous Xu Jun. We are not told why in particular Xu decides to help Han: respect for his talent? loyalty to a colleague? pity for an unhappy lover? rage over Shazha Li's injustice? In any event, the narrative's attention now adopts Xu as its central figure. We are told with obvious enthusiasm the ruse Xu employs to gain access to Miss Liu; we experience amusement and admiration as we see him present Miss Liu a letter with one hand and tuck her under his arm with the other. However, the story still does not end here: impetuous actions by bravos can have their consequences and must be defended by more socially sanctioned elements:

At this time Shazha Li was enjoying exceptional favor, and Han and Xu Jun both feared repercussions. They went to visit Hou Xiyi about it. Xiyi was surprised. "So Xu himself has managed to carry out the sort of action which I have done all my life!" He then presented a memorial:

> The censor Han Hong, presently acting as a secretary and supernumerary of the Board of Revenue, has long been ranked among my assistants. He has repeatedly displayed his achievements, and he has recently sat for the examinations. He had a concubine, a Miss Liu. She was separated from him by the recent rebellions and was compelled to remain where she was, where she took refuge with an eminent nun. Now Your Majesty's civilizing influence has been restored to its governance, and those far and wide have been transformed. But General Shazha Li has flouted the law at will, and relying on his insignificant merit, he has abducted a woman set on defending her virtue and has thus violated your sagely governance. Since my subordinate commander Xu Jun, a palace aide to the censor-in-chief, hails from [the rough border territories of] Youzhou 幽州 and Jizhou 薊州, he possesses a valiant heart of considerable courage. He took Miss Liu back and has returned her to Han Hong. He harbored a sense of justice and has manifested the sincerity of his feelings; but since I had not heard of the affair beforehand, I have certainly been remiss in obeying your commands.

The emperor responded immediately: Miss Liu should indeed return to Han Hong, and Shazha Li should pay them two million cash in reparation. After Miss Liu returned to Han, Han held several offices in turn, culminating with [the position of] document drafter in the Secretariat. (542)

This is the section of the story that would puzzle a Western reader most: why the seemingly anticlimactic intervention of Hou Xiyi at this point? The image of Han and Xu thinking somewhat better of their impetuosity and seeking help of a more mundane kind seems to take away from the nobility of the action. Even more so, why include the text of Hou's memorial, written as it is in a highly ornamental style? There are two possible practical reasons that immediately suggest themselves: the author includes the memorial either as a sign of authenticity (this story really occurred, and here is official proof) or as a demonstration of his ability to draft government documents. However, there are good intrinsic reasons for this passage: although Han had the help of a strong-arming man at first (a man who exemplifies the positive qualities of *wu* 武, or the martial), he cannot hope to get away with what he has done without the assistance of a representative of the civil order (*wen* 文). That Hou is a general is not as important as the fact that he holds a position in the imperial bureaucracy and has the ear of the emperor. Thus Han has won assistance from "men of respect" in every realm—just as he had received assistance from Mr. Li earlier. This is the reward of talent and sincerity; it is an act of wish-fulfillment for the literati reader: raw talent will win one a career and an ideal lover, as well as enough assistance from the powerful to keep both.

However, a very odd development occurs here. Although Hou no doubt had respect for Han, the factor that affects him directly is the spontaneous sincerity and heroism of Xu Jun. The focus of the narrative continues to follow Xu. It is as if his abduction of Miss Liu has given him some figurative right to her, as if some new system was being worked out in determining narrative balance. This is only suggested by the juxtapositions of structure and not by narrated events, of course: Han gets Miss Liu back, and we have the obligatory mention of his later official career. However, the story closes with a "judgment" that is addressed to Miss Liu and Xu Jun alone, as if they were the main characters:

Miss Liu was a woman whose will was set on preserving her chastity, and yet she could not; Xu Jun was a man who admired the ideal of righteousness but was unable to carry it out fully. If before this Miss Liu had been selected for imperial service because of her beauty, then she could have continued the tradition of forthrightness typical of previous court ladies like Lady Feng 馮 (who defended her lord from a bear) or Lady Ban 班 (who declined a seat in the imperial carriage). If Xu Jun for his part had been promoted in imperial service because of talent, he would have established the same merit as Cao Mo 曹沫, who heroically defended the interests of the King of Qi 齊 at Ke 柯 City, or Lin Xiangru 藺相如, who prevented the humiliation of the King of Zhao 趙 at Min Lake 澠池. Such famous occasions as these become known only if they are written down; on the other hand, great merit must depend on such occasions in order to establish itself. If those who possess talent remain obscure and do not meet with occasions for success, then courage and a sense of justice will expend themselves in vain and will not enter into the sphere of the proper. Their behavior could hardly be seen as acting properly in accordance with changing circumstances. But this was due to what they themselves happened to encounter. (542)

Han has dropped out of the picture entirely—or rather, his ghost is present as a phantom replacement of the emperor, who has failed to employ two such capable people as Miss Liu and Xu Jun.[23] In any event, Liu and Xu are now introduced as two classic examples of "gentlemen who do not meet with their times" (*junzi bu yu* 君子不遇). What was originally a romance is now refigured as a narrative of talent frivolously expended: a virtuous woman whose vows of chastity came to naught when her powerless lover was separated from her, and a brave hero whose strength was employed to abduct a concubine. Although such ethical endings should not be taken as the last word of the author, they do represent the complexities of Tang narrative in general, the refusal to commit to the lessons or expectations of any one genre or story type. It also reflects the complex ways that literati readers can figure themselves within such narratives: at first, the reader could identify with Han, who lived the romance of having talent, the perfect woman, and powerful friends; then he could identify with those powerful friends, who could gain fame and prestige by helping others; finally, he can make an identification with noble figures whose talents are wasted. As we have seen, literati identification with virtuous court ladies was already an old tradition by the eighth century. Oddly enough, a story that had begun with asserting the relative autonomy of the female to bestow love on a literatus of genius ends with reasserting the old trope of woman as a version of the subordinate male.

Reasons for Romancing (and for Not Romancing)

Perhaps the main reason for the shifts of attention and interest in even so short a narrative as "Miss Liu" lies with the compositional attitudes of the literati author. It is not so much a coherent tale as a series of interlocking exercises in composition. There are three coherent "scenes" in the story: the banquet at Mr. Li's, Han's two brief encounters with Miss Liu in the capital, and Xu Jun's pledge of rescue and its commission. In brief anecdote form, the story of Miss Liu could have dispensed with all of them; they obviously represent an author working out ways of telling a narrative effectively and with interesting details. However, even more unnecessary to the story are the two poems exchanged, the memorial of Hou Xiyi, and the "judgment" at the end. The author's approach to narrative here is quite different from the conception of fiction writing in the West: here, storytelling is an excuse to engage in a wide variety of styles and forms; there is no need to aim for unity of effect. We do not even know how many hands contributed to this narrative: although it is attributed to a single author (Xu Yaozuo 許堯佐), it may have passed through numerous elaborations. How true was the kernel of the tale? Was the memorial real, and if not, who composed it? Were the poems imported from some other place because they fit in well here? Or if they were composed for this narrative, who wrote them? To call this sort of approach to storytelling "dialogic" in the Bakhtinian sense would be to make overly subtle a process that is explicit and cheerfully exercised. We do not have an author through whom different discourses speak; we have an author or authors who indifferently employ any discourse they feel like using. These narratives are evidently forms of literati vaudeville.

These issues cause problems only if we are exercised over the development of "fiction" in the Tang. Most of the texts examined in this chapter are ones that modern scholars conventionally label *chuanqi* 傳奇 (transmissions of the marvelous), but that does not imply that they share the qualities of a coherent, well-defined genre. The conception of *chuanqi* as something analyzable in stylistic or literary historical terms derives from the work of the Ming scholar Hu Yinglin 胡應麟 (1551–1602); he distinguished the more well-developed "tales" of the Tang from the *zhiguai* characteristic of the pre-Tang era.[24] Critics earlier in this century (particularly Lu Xun)[25] employed this division to create a history of the development of fiction; they argued that Tang *chuanqi* authors were increasingly aware of fictional narrative as an independent form and that they attempted to create stories with plot, characterization, and suspense. However, more recent discussions of short narrative in the pre-Tang and Tang eras (notably, studies by Robert Campany and Glen Dudbridge)[26] have tended to emphasize the "reportage" aspects of these narratives and their ties to religious and cosmological thinking. I am interested here not in the debate over what constitutes "fiction" in the Chinese tradition, but rather in the way in which narrative gets elaborated in the hands of literati authors. In this sense, many of these pieces do constitute a category of sorts, not because of their self-conscious "fictionality" but precisely because they are literati performances—written or compiled not for religious or philosophical reasons but for circulation within literati circles.

When literati narratives became more widespread in the decades after the An Lushan rebellion, there was audience enough for these performances. This period coincided with the rise of a more extensive and self-conscious examination culture: articulate writers of this period are increasingly men who hope to launch an official career by going to the capital, finding a prominent and powerful patron, and passing the examinations as a prelude to a lucrative or prestigious government post. Modern scholars, particularly Chen Yinke, have explicitly linked the writing of narratives to this phenomenon by suggesting that *chuanqi* were sometimes included in the scrolls that a young ambitious scholar would circulate among potential patrons in the capital to demonstrate his literary abilities.[27] Although we do not need to accept this hypothesis completely, it would not be unreasonable to assume that literati narratives *were* composed for circulation among the members of the male literati community and that (like poetry itself) they were ways of demonstrating the author's skill with language. Granted the social circulation of such texts—laboriously copied and passed from hand to hand or read aloud at gatherings—the social lives of the literati become reinscribed in them in various ways. Shared experiences lead to emphasis on certain plot elements, particularly those involving literati failure or success in the official world and the competitions (formal or informal) to which they are subject through interactions within the various male social communities of the Tang.

As we have seen with "Miss Liu," it is this world of social exchange, of mutual reading and criticism, that affects the way women emerge in the texts. In one obvious sense, it creates a unique Chinese example of the "romance," as Stephen Owen has recently hypothesized.[28] Owen defines "romance" broadly, and I take him to mean quite simply stories in which a young man and woman meet, fall in love, overcome obstacles thrown in their way, and then are either happily reunited or parted forever. From this rather flexible point of view, romance can constitute the main element of a story or enter in as a theme or as a subsidiary plot. For Owen, romance is another manifestation of a development he sees as characteristic of the early ninth century in general—an interest in personal self-expression and self-fulfillment, the creation of "private" spaces independent of the public sphere, and an accompanying conflict between the demands of the public and the private. The lovers in the romance attempt to create a small space independent of society within which they can indulge their personal desires for happiness: "I hope it will not seem too strange to suggest that not only is the culture of romance another version of the private sphere, but it is also an even more perfect counterpart of reclusion. In the public world, things change; the world of the recluse and the bliss of lovers are unchanging states of being and require commitment."[29] From Owen's point of view, this uncompromising need for changelessness puts often insurmountable pressures on the lovers. He constantly evokes the ways in which these romances go wrong—the collision of economic necessity and "true love," the barely concealed power relations that exist between courtesan and client, and the continuing requirements of the world of family and officialdom. But, by and large, he sees at least the attempt to construct an idealized romantic relationship between a man and a woman as the backbone of many of these tales.

This reading notwithstanding, I believe that many of these sexual encounters can be put in a broader context that speaks to additional factors beyond the creation of a private space and the elevation of the erotic paradigm. Their private world is also public in a different sense—it is open to the informal male communities of readers. Although Owen grants the power of the male community to evaluate the behavior of its own, he emphasizes its power to pass judgment over those who would deny the claims of the private. In creating and interpreting sexual relations in these narratives, however, the male community often does more than create the dynamic that Owen describes: it also redefines the way women and men deal with each other in terms of existing codes of requital, competition, and prestige seeking. Affection and romance between literati and their women are interpreted through structures of literati experience, and new codes of public behavior are created. This is borne out by the role of "romance" within Tang narratives in general. Although its existence is indisputable and it emerges constantly, in few stories is it the most important or significant plot element. This is quite different from other cultures, in which the emergence of the idea of romance has usually been accompanied by the development of a well-defined genre that repeats the same plot over and over with minor variants.[30]

An examination of how some narratives work through issues of love and romance will clarify this point. In one famous story, "Wushuang zhuan" 無雙傳 (The story of Wushuang; *TPGJ* 4: 546–49), the hero, Wang Xianke 王仙客, knows his beloved from childhood. Left fatherless at a young age,

Wang goes to live with his maternal uncle Liu (a powerful official). Liu has a beautiful daughter named Wushuang (Peerless). On her deathbed, Wang's mother tries to make her brother swear that he will wed Wushuang to Wang, but he cagily avoids committing himself. When Wang grows up, he becomes increasingly infatuated with his cousin and even bribes the servants of the family to ensure his access to her. But even though his aunt seems to be on his side, his uncle refuses to agree to the marriage. Then a military uprising changes things—as so often happens in literati narratives composed in the post–An Lushan world:

One day, Liu hurried to dawn court but returned again by morning. He abruptly galloped into his mansion, dripping sweat and out of breath. "Lock the main gate! Lock the main gate!" he commanded. The entire household was thrown into a panic and could not guess what his reasons were. After some time he said, "The troops in Jing 涇 and Yuan 原 have revolted, and Yao Lingyan 姚令言 has brought troops into the Hanyuan 含元 Palace. The emperor has departed by way of the north gate of the garden, and all the government officials are hurrying to accompany him. But I took thought of my wife and daughter, and so I've returned to manage things. Quickly summon Wang Xianke to take charge of the family's goods, and I will marry Wushuang to him."

When Wang heard his command, he was surprised and delighted. He bowed to Liu and thanked him. Liu then loaded up twenty animals with gold, silver, silks, and brocades. He said to Wang, "Change your clothes and take our goods out by the Kaiyuan 開遠 Gate. Find a secluded inn somewhere for refuge. I will bring your aunt and Wushuang out by way of Qixia 啓夏 Gate and will come around the city walls to meet you." (546–47)

Wang thus has an opportunity to prove himself and win the girl of his dreams. His potential intervention here is reminiscent of that of Scholar Zhang in the "Tale of Yingying" 鶯鶯傳, who relies on a friend in the military to repress a rebellion threatening the monastery where he and Yingying are staying.[31] But the uncle's motives here are worth noticing. As a maternal cousin, Wang has a certain obligation to assist the family. By marrying him into the family, however, Liu attempts to guarantee his loyalty and his goodwill in protecting the family goods (of which Wushuang will eventually be a part). He becomes the Liu family heir and responsible for the family's preservation. Wushuang in turn becomes a token of this. Yet ironically Wang is left helpless and ineffectual—perhaps because he is not a military man himself:

Wang did as Liu directed. By sundown, he had waited some time at the inn, but no one had arrived. The gates of the city had been locked since noon, and he had spent the time gazing to the south without seeing anyone. Taking a torch he rode a donkey around the city walls to Qixia Gate. The gate was already bolted. There were quite a few gatekeepers there grasping clubs, some standing, some sitting.

Wang got off his horse. "What's happened in the city?" he asked hurriedly. "Has anybody come out this gate?" "Marshal Zhu Ci 朱泚 has made himself emperor," they replied. "Some time after noon a man in a black traveling hat tried to get through the gate with a group of four or five women, but people in the street recognized him as the Special Supply Commissioner, a certain Minister Liu. So the gate guards didn't dare let him through. Toward nightfall a group of cavalry came in pursuit and took him back north into the city." Wang broke into muffled sobs and returned to the inn.

Toward midnight the city gates burst open and torches made things as bright as day. Troops rushed out grasping weapons, and guards were summoned out to search for court officials who had fled the city. Wang fled in terror, abandoning all his baggage. He returned to Xiangyang and lived at his country estate for three years. Only after he heard that the revolt had been put down and the capital and the empire were at peace did he return. (547)

Again, there seems to be no expectation that the male lead will engage in flamboyant heroics, and we cannot help but wonder if he wouldn't be as ineffectual looking after Wushuang as he is in looking after the Liu family's baggage. He does not give up, however. When he returns to the capital, he learns that Liu has been executed for collaborating with the rebels and that Wushuang has been forcibly enrolled as a concubine in the imperial establishment. This does not prevent Wang from trying to acquire her once again. His desires are romantic, but she is also the last salvageable remnant of the Liu family, and by rescuing and marrying her, he redeems his failure to look after the Liu family goods. Of course, the difficulties of his courtship have been increased tenfold. Earlier he had been forced to bribe the servants of the Liu household in order to get word of Wushuang. Now he has an even more difficult household to infiltrate.

However, Wang does not immediately have an opportunity to seek Wushuang's rescue. In fact, he shows no sign of even attempting it. Rather, he finds a minor post as magistrate of Fuping 富平 country near the capital; as a subsidiary duty, he takes charge of the Changle 長樂 post station. At this point, he fortuitously meets Wushuang again, when an entourage of thirty court ladies stop at his station on a ceremonial visit to the imperial tombs. He manages a brief meeting with his beloved on the road the following day, and Wushuang contrives to pass a letter to him:

> It was five pages on flowered notepaper, all in her own hand. In clear but aggrieved language, it related her story in detail. As Wang read, it his tears fell at the thought that they were parted forever. But at the end she said, "I often heard my father, the late censor, mention an officer named Gu 古 in Fuping County who is a man of true feeling. Perhaps he can help us?" (548)

The similarities between this scene and Han's re-encounter of Miss Liu suggest that this is a common motif in tales of separated lovers. Wushuang's letter to Wang serves the same purpose as the exchange of poems and the gift of the box in "Miss Liu," with one important difference: Wushuang has a suggestion to make concerning her own rescue. This is the only episode in the story that shows Wushuang participating in events. But as with Miss Liu, help must come not from Wang himself but from some third party. The next passage is a particularly significant one in relation to the dynamics of literati narrative:

> Wang then notified the prefecture, requesting that he be allowed to resign the post station but retain his position as magistrate of Fuping. He then sought out Officer Gu.
>
> It turned out he was living on a village farm. Wang went to visit him. He then did his utmost to fulfill Gu's every wish, giving him presents of silk, gems, and jade without number. Yet for an entire year he did not mention his business once.
>
> After his term of office was over, he went to live in retirement in the district. One day Gu came to him and said, "I am merely a soldier and am growing old and useless. You, Sir, expended all your energies seeking to please me. You must have something you want from me. I am a man of true feeling and am moved by the great kindness you have shown me. I'm willing to sacrifice my own life to help you." Wang bowed to him, weeping, and related the situation to him. Gu stared up at the sky and tapped his head a few times. "This affair certainly won't be easy, but I'll give it a try for your sake." (548)

Unlike Xu Jun, who helped Han out of pure impulse and goodwill, Gu must be courted. Wang sets up a relationship of patron and client with Gu, creating ties of obligation and repayment. Gu very likely knows that Wang's presents are not given merely out of goodwill but entail a request for services: this is part of the code of master/retainer relations. There is no disingenuousness on Wang's part. He merely demonstrates his ability to find and employ the best men to get a task done. In this way, he redeems himself from his failure to look after the Liu family goods. He may have failed in his own commission, but he is successful in commissioning someone else.

However, the elaboration of this episode and the length of time Wang needs to win Gu over suggests another type of relation as well. Wang "courts" Gu for over a year through kind treatment and presents. Their alliance acts as a substitution for Wang's courtship of Wushuang. She will be abruptly abducted through the machinations of Gu, a violent matchmaker who violates decorum. The courtesies required in proper wedding rituals are here expressed in a relation once removed from the parties involved. Wang symbolically maintains a certain propriety in his seeking of Wushuang through his show of patience and goodwill and thus earns his right to her. Perhaps this is why the story remains centered on Wang and Wushuang to the end of the tale, rather than shifting its attention to Gu, the way "Miss Liu" shifted its attention to Xu Jun.

Next we have a detailed description of Gu's rescue of Wushuang; the author enjoys relating the details, partly because the exciting and dangerous abduction of an imperial court lady would thrill readers. One main element of the abduction is Gu's ruthless execution of any minor participants in it (including Wang's loyal manservant Sai Hong 塞鴻), to prevent anyone from divulging it; the mayhem culminates in Gu's own suicide. This last gesture is necessary for a number of reasons. First, it is a stock gesture of loyalty on the part of a retainer.[32] Second, it pays for the deaths that he himself has inflicted on others—he holds his own life as cheaply as those he took to ensure the success of the stratagem. Third, Gu's removal from the narrative at this point removes the necessity for his participation later in the lives of Wushuang and Wang: he remains a tool and hence less a focus of attention. The author makes this clear in his own "judgment," which keeps Gu in an important but ultimately subsidiary role: "Wushuang met with times of disorder and so was taken into the ranks of palace women. And yet Wang would not give up his determination even when facing death. In the end he met up with Officer Gu and adopted his extraordinary plan—and more than a dozen innocents lost their lives" (549). The author speaks of loss of life here more in awe than in disapproval—but he also separates Gu and his plot (necessary as it is) from the *true* virtue demonstrated in Wang's dedication and persistence.

Without a doubt, "The Tale of Wushuang" warrants consideration as a romance: ultimately, it is Wang's love that compels him to go to such extremes. However, this should not misdirect us from much of the interest of the tale. In spite of the heroine's brief contribution to her own rescue, she is mostly the idealized and abstract goal in the tale: Wang's failure to obtain her puts the narrative in motion, and his gaining of her brings it to an end. She is a motivation for desire, for machinations, for killing, and ultimately for celebration. Her very name, which means "without peer," turns her into a prize worth possessing (the term is often applied to rare and valuable objects). This is figured most dramatically in the story by having her enter the service of the emperor himself: the hero must thus infiltrate the most difficult interior space of all, the eroticized sanctum sanctorum of the political body. Ultimately, Wushuang may be a figure for the ambitions of the young literatus seeking a career.

As a woman-to-be-obtained, Wushuang is connected to Wang through a number of comprehensible social ties: primarily through kinship, betrothal, and childhood friendship; secondarily, through the ties of "romance" that are becoming increasingly important in the Tang, as well as through the intervention of the "bravo-retainer" figure of Gu. In other tales, however, even those of a supernatural nature, the object of affection is still subject to male-defined social codes. In one of the longest Tang narratives, the "Liu Yi zhuan" 柳毅傳 (Tale of Liu Yi; *TPGJ* 4: 144–50), Liu's romance with the daughter of the dragon-king of Dongting 洞庭 takes a backseat to the author's fascination with the underwater dragon's court and the political functioning of the supernatural order. Within this context, romance is filtered through a series of social relationships humorously or surprisingly reinterpreted on the "supernatural" level.[33]

As the story commences, Liu Yi has an encounter with the daughter of the dragon of Dongting Lake. She begs him to take a message to her father: she is married to the second son of the dragon of the Jing River, who is mistreating her. However, this meeting with her is not quite as accidental as it first appears. The tale begins:

> During the Yifeng 儀鳳 reign period [676–79] of the Tang, a certain scholar by the name of Liu Yi was about to return to his home by the Xiang 湘 River after failing the examinations in the capital. He recalled that a fellow countryman was living at the time in Jingyang 涇陽; so he went there first to take his leave of him. After he had gone two miles or so, a bird flew up and startled his horse. The beast galloped out of control off to the left, and he only managed to stop it after he had gone another two miles. Then he saw a woman herding sheep by the side of the road. Liu Yi

found this strange. Upon looking more closely, he discovered her to be of extraordinary beauty, although her lovely face looked worried and her clothes were unkempt. She stood there, listening intently, as though she were waiting for someone. (144)

An attentive reading suggests that the dragon-daughter has deliberately arranged for this meeting. Having a horse bolt and get his rider lost is an established motif in folk stories; it represents a mortal's movement from the known world into the world of the supernatural or unearthly. And she looks as if she is expecting someone. It may be of equal importance, however, that he has failed the examinations: with the entrance to one realm blocked, Liu finds himself entering another.[34] This is fully realized by the end of the tale, when he obtains Taoist immortality and thus removes himself from the ordinary literati world altogether. This beginning is also a defining moment of Liu's character: as the tale progresses, it becomes clear that Liu Yi will not gain renown as a scholar or as an official; he is more a "man of justice" (*yi fu* 義夫) concerned with helping those in need. He is more Xu Jun than Han Hong. Of course, this is not to say that literati cannot be bravos, or vice versa; it merely puts the defining spin on Liu Yi's personality.

The description of the dragon-daughter is also important: "He discovered her to be of extraordinary beauty, although her lovely face looked worried and her clothes were unkempt." This is not an occasion for "love at first sight." Rather, beauty here is more a mark of social distinction and of worthiness to be rescued: she is a figure demanding Liu Yi's chivalrous attention. The shabbiness of her clothes cannot conceal her value: they are signs that she has unjustly come down in the world.

Indeed, it is Liu Yi's chivalrous instincts that make him willing to help her, in spite of the difficulties he faces in crossing the realm between human and dragon. "I am a man of justice. After hearing your story both my spirit and my blood are stirred. I resent only that I do not have wings to fly" (144). Nonetheless, he proceeds on his mission after receiving instructions from the dragon-daughter. When he arrives at the dragon-king's palace, the king extends full courtesies to him and makes plans to rescue his daughter. However, the woman's uncle, the hot-tempered and violent dragon of Qiantang 錢塘, discovers what has happened to his niece. He breaks the chains that have been restraining him (a precaution undertaken by his older brother to prevent his future excesses), terrifying Liu Yi in the process. Flying to the Jing River, he carries off his niece while inflicting widespread carnage. When he returns to Dongting, he reports on his indiscretion:

"In the morning I left Numinous Void Palace and soon arrived at Jingyang. At noon I did battle with them. Before I returned here, I hurried up to the Ninth Sky to inform the Emperor of Heaven what I had done. He knew the reasons behind our grudge and so pardoned my trespass—and pardoned what I had done earlier as well. But since I was stiff-necked and stirred up by things and did not pause a moment to take my leave of you, I shook up your palace and frightened our guest. I am ashamed and sorry that I did not realize my own fault." He then stepped back and bowed twice.

The dragon lord asked, "How many did you kill?"

"Sixty thousand."

"And you destroyed crops?"

"For about three hundred miles."

"And where is the ungrateful husband?"

"I ate him."

The dragon lord was perturbed. "It was really unbearable for that naughty boy to behave in that way. But you were surely too hasty. We have relied on the sage wisdom of the Emperor of Heaven, who has forgiven us due to the severity of our grudge. But if he had not done so, what could I have said in apology? From now on you cannot act in this manner." (146)

This is the great comic moment of the story, but it is also significant in the way it exaggerates the impulsive, if justifiable, violence of the Qiantang dragon. It also anticipates the Qiantang dragon's display of impropriety later on. Dongting has kept his brother under control in order to maintain public peace and to fulfill the promise of his good behavior that he has personally made to the Emperor of Heaven. Although Qiantang may act out of a sense of family loyalty and of justice, he in fact endangers the various contracts that maintain and preserve human relations.

Liu Yi has done the family an immense service by bearing the dragon-daughter's message from one world into another; he now becomes an honored guest and friend of the two brothers—a fact reinforced by their later exchange of poems at the banquet held in his honor (note that he does *not* exchange poems with the dragon-daughter). However, this special relation with the two male dragons is put at risk by the overbearing nature of Qiantang, who tries to browbeat Liu into a marriage with his niece:

The following day they feted Liu once more in Clear Brilliance Hall.

The Prince of Qiantang grew drunk and began to put on airs. Sprawling out arrogantly, he said to Liu, "Haven't you heard that 'a hard stone can be split but it cannot be rolled, and a man of justice can be killed but he cannot be shamed'? I have a proposition that I wish to tell you. If it meets with your assent, then we will both fly to the cloud-filled sky above. If it doesn't, then we will sink into the filth and muck. Tell me what you think."

Liu replied, "Let me hear."

Qiantang said, "Jingyang's widow is a beloved daughter of the Lord of Dongting. She is virtuous and attractive and valued by all her kin. Unfortunately she was shamed by a scoundrel, but now her connections with him are at an end. Now I would like to entrust her to your lofty sense of right and have us become kin for generations. She who has received your favor will have a home to go to, and those who love her will know to whom she has been given. Would this not be first and last the way of a gentleman?" (147)

This speech in turn inspires the guest to make an eloquent defense of propriety:

Liu grew solemn, then broke into a laugh. "I really did not know that my lord of Qiantang could be so troublesome. Earlier I heard how you spanned the entire land and embraced the five sacred mountains just to release your anger and rage. Then I saw you break your fetters and uproot the jade post in order to rush to the assistance of another. I thought that no one was as firm or decisive, as enlightened or upright as yourself. You did not avoid death in punishing the wrongdoer, nor did you cherish your own life in aiding those you cared about. This is truly the aim of a real man. Now, amid the harmonies of flutes and pipes and the friendliness of guest and host, how can you disregard the way of the true man in order to intimidate others? How could I have expected this?

"If I had encountered you amid the great waves, or on a gloomy mountainside, with scales and whiskers bristling, covered in rain clouds, driving me to my death—then I would have seen you as merely some beast and would not have resented your actions. But now your body is clothed in cap and gown, and you sit here, discussing propriety and justice with me. You have a nature that fully fathoms the Five Principles and can comprehend the subtle precepts of all human conduct. Even worthies and outstanding men in the human world do not come up to you—how much less so the spirits of the rivers? And yet with this doltish manner and harsh disposition you put on airs; under cover of your drunkenness you oppress others. How is this in keeping with the right? My own physical form is too small to fill up the space under one of your scales, and yet I dare to withstand your improper manner with my unyielding heart. I wish for you to think about this." (147)

Later, Liu Yi tells the dragon-daughter his main reason for refusing an alliance: marriage to a widow whose husband's death he caused even indirectly would be dishonorable. Yet here he chooses only to criticize the manner in which Qiantang presents the offer. There are unwritten codes on how favors should be repaid; Qiantang violates that code by concealing compulsion under the illusion of a pact that would cement their friendship. Why he does so is not altogether clear: perhaps the sense of obligation he feels to Liu Yi, a mere mortal, makes him try to force a repayment on him so as to clear the books as quickly as possible. Note, however, that Qiantang phrases this invitation in the language of the bravo: compliance becomes an issue that will decide eternal friendship or eternal enmity. Liu Yi is quick to distance himself from this bluster masquerading as the hero's code; although the dragon terms both of them "gentlemen of justice" (*yi shi* 義士), Liu complains that a truly just man does not intimidate others into a course of action, particularly when they are observing the rites of the banquet ("amid the harmonies of flutes and pipes and the friendliness of guest and host"). Liu Yi goes on to take Qiantang to task for not living up to his human form: beasts may be violent by nature, but once caps and clothes are donned, there are rules that must be followed. A true code of the hero is being formulated here, one in which the gentlemanly Liu Yi participates, but one that excludes bullying and violence. Perhaps the author attempts a defense of "heroic" behavior by contrasting Liu with the coarse dragon-bravo.

When Liu leaves shortly, he feels a brief pang of regret at having turned down marriage to such a beautiful woman. However, this separation proves to be short-lived. After two wives die in quick succession, a further marriage with a "young widow" is suggested to him. This turns out to be the dragon-daughter, although she keeps this fact hidden at first. Are we to suspect supernatural involvement in the death of his first two wives? Perhaps; but more important is the code of honor that the dragon-daughter herself pursues in her marriage to Liu. Although he tells her after their marriage of his adventures and comments on her resemblance to the maiden of Dongting, she refuses to reveal her identity. But when they have a child together, she confesses that she has loved him all along and had deeply regretted it when he turned down her uncle's invitation. When his first two wives died, she decided to approach him as a human bride:

"I didn't speak of this earlier because I knew that you had a heart that did not value physical attractions. I mention it now because I know you have some feeling for me. A woman is a worthless thing and unworthy to warrant your eternal faith. But now because you love your son, I entrust both of us to you.[35] I don't know what you think of this. Both grief and fear burden my heart—feelings I cannot overcome." (148–49)

The dragon-daughter's love for Liu is not in doubt here; her refusal to remarry her own kind out of loyalty to him suggests the same degree of commitment found in Miss Liu and Wushuang. But how she phrases this affection is striking: A repayment must be made. Her husband does not care much for "romantic adventures"; her son will guarantee his affections even if he feels nothing for her. Although we might dismiss this speech as courteous exaggeration, it in fact matches Liu's rhetoric. The dragon-daughter's conception of dignity and justice is just as demanding as Liu Yi's and requires just as much attention to social proprieties. In Tang terms, "love" can be conceived of precisely as the exchange of payments: the longing to repay can sometimes be indistinguishable from the longing for the loved one. This desire for exchange leads the dragon-daughter to engage in a mild deceit: in order to satisfy her obligation/longing, she will marry him as a human and not tell him the truth until she has turned the tables on him and can require a new payment from him: the debt of affection he now owes to their child.

Ironically, she is compelled to this deceit through her inability to read the real motives of Liu Yi. Why precisely had he turned down Qiantang's offer? Was it out of a sense of justice or out of distaste for the woman? She is reassured by another long speech from Liu Yi, who explains his reasons for refusing her. He confesses his true affection for her and laments that propriety would not allow him to repay the debt of affection *he* now finds himself owing: "In the end I was restrained by [the obligations of] human affairs and could not repay you with my thanks" (149). Of course, he could have made this explanation clear the first time, when lecturing Qiantang. But it was ultimately more tactful to take Qiantang to task for his bullying ways than to confess that he could not marry the wife of the man whom his future brother-in-law had murdered.

The love story within "The Tale of Liu Yi" becomes a sort of ritual dance in which the participants are constantly maneuvering about within the space that social propriety allows them. There is little of the conventional love story here; major decisions are accompanied by long elaborate speeches on proper and improper behavior. But perhaps romance does lie in the dragon-daughter's novel interpretation of repayment, and the stratagem she adopts in order to repay. As Owen has suggested, romance tends to make claims on individuals that transcend the ordinary demands of social codes. What is interesting here is the way those very proprieties are redefined to make allowance for romance: had the dragon-daughter not been innovative in her interpretation of propriety, the two would have been separated forever.

Romancing Replaced

In other stories, the role of the female is more complex, and the element of romance becomes inextricably tied to competing compulsions and conventions. Oddly enough, this shows itself most clearly when the object of the romance (the woman) stands fully outside those compulsions and conventions: that is, when she is either a prostitute or a supernatural being. For example, Shen Jiji's 沈既濟 (ca. 740–ca. 800) "Ren shi zhuan" 任氏傳 (The story of Miss Ren; *TPGJ* 4: 340–45), the most famous fox story from the Tang, sentimentalizes a standard *zhiguai* plot, in which a man falls into the sexual snares of a demon who deprives him of his life force. In this version, the poor and hapless scholar Zheng 鄭 spends the night with a fox but is so drawn to her that he refuses to give her up even when he learns of her true nature. In gratitude she promises him her eternal loyalty. Once again, the erotic encounter is reshaped as a legitimate private moment independent of the ordinary demands of the social. This is not the main point of the story, however. The most important character is not Zheng (whose personal name, the narrator tells us, has not even been preserved); rather, it is his wealthy and powerful cousin, Wei Yin 韋崟. When Wei learns that Zheng is keeping house with a beautiful woman, he cannot stand that his friend should have something that he himself cannot possess. He goes to visit Ren:

When he arrived, Master Zheng happened to be out. Yin entered the gate and saw a young servant boy sweeping the courtyard with a broom. There was a servant girl at the gate, but he saw no one else. When he made inquiries, the boy only smiled and said, "No one's here."

Yin looked around inside the rooms and saw a red gown visible from under a door. He rushed in to look and saw Miss Ren hiding behind it. Yin pulled her out into the light to look at her, and she surpassed what he had heard. Yin was seized with a desire to possess her, and he embraced her, but she would not yield. He attempted to subdue her by force, and when he had pressed hard upon her, she said, "I give up. Just let up for a moment." But when he loosened his grip, she resisted as before.

This continued several times. Yin then used all his force to hold her. Her strength evaporated, and her perspiration fell like rain. Knowing she could no longer avoid him, she let her body go and no longer resisted, but her expression grew sad.

Yin asked, "Why do you look so unhappy?"

Miss Ren gave a long sigh. "Zheng indeed is to be pitied!"

Yin asked, "What do you mean?"

She replied, "Although Zheng may have the stature of an adult, he cannot shelter one lone woman. How could he be considered a man! Now you are young, noble, and wealthy, and you have many beauties to choose from. Surely you must meet many women who can compare with me. And yet Master Zheng is poor and obscure, and I am the only one that he cares for. How can you bear to indulge your heart set on excess and take something from one who has so little? I feel sorry that he is so poor and hungry and cannot support himself. He wears your clothes and eats your food, and so he is tied to you. If he could but supply himself with even modest means, we would not have come to this."

Yin was a heroic sort with a sense of justice. When he heard this, he let her go. Straightening his clothes, he apologized to her. Soon after Master Zheng returned, and the two friends greeted each other cheerfully. From this time on, Yin supplied Miss Ren with all her daily needs. (342)

From the viewpoint of romance, Miss Ren defends Zheng's claims on her and supports his ties of affection; as with the dragon-daughter, she expresses this affection through the language of gratitude and repayment. However, if Ren's decision to stay with Zheng establishes the purity of their "private" relations, it comes at the expense of Zheng's character in general. Here as elsewhere he is portrayed as ineffectual—not only poor and powerless but timid.[36] Ultimately, Zheng is incapable of having even supernatural forces help him and is doomed to remain subordinate to more daring men like his cousin. Wei Yin, however, is the "man of honor" who surrenders the brutal right of the powerful to take what they want. By doing so, he becomes the patron of Zheng even in this; he essentially grants Zheng his rights to Ren, just as Li granted Miss Liu to Han Hong. "He wears your clothes and eats your food, and so he is tied to you," says Ren, echoing the very words Han used when he attempted to refuse the gift of Miss Liu.

As Wei becomes the patron of both Zheng and Ren, a new relationship develops:

Yin went out with Ren daily and was delighted to do so. They would not stop at the most intimate associations with each other with the exception of sexual relations. For this reason Yin cherished her and valued her highly and refused her nothing. He would not forget her even for the length of a meal.

Miss Ren knew of his feelings for her, and she said to him by way of apology, "I am ashamed that you are so fond of me. But I am a base creature and don't deserve your deep concern. Besides, I cannot let Master Zheng down, and so I cannot fall in with your hopes. . . . But if there should be some great beauty that you like but can't obtain for yourself, let me bring her to you. In this way I could repay your favors." (342)

The sexual tension between Wei and Ren does not disappear, but in order to disarm it (or at least keep it dormant) Ren takes up the role of the enabling mediator commonly found in many of these narratives stories, the mysterious outsider who obtains the woman for the hero—like Xu Jun and Officer Gu. As a fox, she escapes being abducted and thus suffering the fate of Miss Liu and Wushuang, but as a woman who would preserve her relationship with a man who cannot defend her, she takes on the role of abductor herself. The author enjoys relating the details of these schemes, which may have some bearing on Ren's foxlike nature—in spite of her virtue, she is still a trickster figure and seems to find pleasure in taking on the greatest challenges. Meanwhile, the tawdriness of her errands weakens her position as a romantic lover. In either role, however, the narrative portrays Miss Ren more as a male fantasy—either the dubiously moral but loyal woman that will not leave you, no matter how ineffectual you are, or the procuress, the most convenient source of an inexhaustible supply of women.

The dialectic of power that exists between Wei and Zheng is a microcosm of male literati culture at large. Wei and Zheng are the best of friends, but this does not prevent Wei from taking what he wants from Zheng. Literati friendships are usually portrayed as hierarchical, with one side as patron, the other as client. This connection prevents Zheng's relation with Ren from ever being completely private or removed from the demands of society. The only difference is that the society to which it is subject is not that of family or public service, but the complex informal relations of educated upper-class men. It may be the stresses and demands of this world that make the heroes of these tales so passive; for many literati, success in such a competitive environment must have seemed either hopeless or possible only through the intervention of more powerful and impressive patrons and colleagues. Tang literati narratives often speak to the dream of being able to achieve happiness and high office without the accompanying struggle.

Of course, the social space in which male literati interacted was the city, particularly the "underworld" of the Chang'an brothels. Bai Xingjian's (775–826) 白行簡 "Li Wa zhuan" 李娃傳 (The story of Li Wa; *TPGJ* 4: 533–38), one of the most intricate of Tang narratives, gives a fascinating portrayal of this alternative male society. Here, an unnamed scholar sent to the capital to take the examinations falls victim to a predatory madam and her courtesan, Li Wa. After he runs out of money, the madam and Li Wa drop him by abandoning their rented house while he is away. He soon falls ill and is carried by his landlord to an undertaker's establishment, where he recovers and earns a living as a professional mourner. He participates in a singing contest, and an old family retainer recognizes him. His father, rather than reconciling with him, beats him severely and leaves him for dead. Eventually the scholar recovers enough to become a beggar; one winter day he is discovered by Li Wa, who is ashamed of her previous treatment of him. She leaves her madam and sets herself up in an independent house, where she nurses him back to health and makes him resume his studies. He passes the examinations and then, on his way to take up an official post in the provinces, he is reconciled with his father, who insists that he marry Li and make her his principal wife. When he later carries out exemplary mourning for his parents, both he and Li are rewarded with noble titles.

Even as we summarize the plot in this way, we detect a conventionally moral direction to the story. Unlike "Huo Xiaoyu zhuan" 霍小玉傳 (The story of Huo Xiaoyu; *TPGJ* 4: 550–55), which condemns its hero Li Yi 李益 for denying the claims of romance, "Li Wa" creates an anti-romance in which the attractions of romance almost destroy a young man. It also exemplifies a sort of family narrative of a distinctively Chinese sort: a story not of generational conflict or of feud but of reintegration and the acceptance of responsibility. The author is also intent, however, on showing the hero's inabilities to deal with the demands of informal literati society in the capital—a society often indistinguishable from that of the demimonde. Although the text sees this courtesan world as contrary to, or threatening, the moral order that will eventually lead the hero to official success, it also portrays that courtesan world as a complex and intriguing one, guided by its own rules. In this sense, the hero is a victim largely because of his own ignorance and his failure to adapt to the pressures of the big city.

In an exhaustive study of the text, Glen Dudbridge has discussed many of these points, in particular the conflict between "good society" and the entertainment world: the scholar and his father on the one side, the courtesan Li Wa and her foster mother or madam on the other. He has also discussed at some length the themes of symbolic death and resurrection that run through the narrative.[37] What is of interest for the present discussion is the social relations behind these themes.

The scholar begins the tale vulnerable and overconfident, as he departs his family's country estate for the capital. The countryside (or at least, the provinces) is a space within which clear hierarchical relationships are maintained and everything is as it actually appears. As he departs for the city, intent on winning a name for himself, we have a sort of Chinese Rake's Progress, the story of a naïve youth unprepared for the city. When he first encounters Li Wa on the street, he easily falls victim to her, having no defenses against wicked city women. Although a chance meeting on the street does initiate the erotic affair in many Tang narratives, this is not its purpose here; here, it emphasizes the vulnerability of the scholar, the sensitivity that will result in his fall. He is not misled by his companions in this, who make the law of the courtesan quarters quite clear to him: if he wishes to win Li Wa, he must be prepared to pay:

> He investigated her by making private inquiries of those of his friends who frequented the streets of the capital. His friends told him, "That was the residence of Miss Li, a courtesan."
>
> "Is she accessible?"
>
> "She is rather expensive. Those in the past who associated with her were all from aristocratic clans and powerful families. She obtained great amounts from them. If you don't put out a million, you cannot have her will."
>
> He replied, "I only worry that the affair won't work out. How could I begrudge even a million?" (533)

It is easy sometimes to see "true love" as a victim to the economic necessities of courtesan culture or to believe that writers saw "true love" as transcending economic pressures, but generally in Tang narratives and anecdotes even men who are truly "in love with" their courtesans are well aware of the expenditures necessary to maintain them. The scholar is aware of this, at least at first; his crime, ultimately, is not to have enough money to be able to compete with the "big boys." He cannot live up to the financial code of the demimonde and ends by bankrupting himself. However, as time passes, he gradually falls victim to a romantic delusion; after he runs out of money, he hangs on, hoping against hope that the end of his resources will not matter to Li. But then again, he simply has no other place to go. His life with Li Wa has not taught him to fend for himself in the urban world. Without her and money, he is as helpless as a child—and commences the process of infantilization that will mark his development later.

In the early stages of their relationship, the scholar is well aware of the money he must spend to win Li Wa. He hopes to use that money to recreate a family for himself with Li and her foster mother. The scholar himself requests that he rent a vacant court in the madam's establishment, thus acting as the husband (or wife) moving into the family dwelling (it was a convention of the brothel quarters to call steady customers "wives" and madams "mothers"). He seems to take these conventions seriously (which were perhaps meant ironically), reading as sincere the cynical undermining of Confucian family structures that the courtesan quarters provided. But he only rents his dwelling, and like most of the inhabitants of Chang'an, he is a transient, not a property owner. This transience creates a vagueness in social relations typical of the big city, and this in turn creates a moral vagueness.

The "vagueness" of social relations is also suggested by the failure of the scholar to create a space for his relations with Li that would allow him to continue with his own studies. Most obviously, he surrenders his role as the son of his father; the madam hints at this when she says, "Great desires exist between man and woman. If by chance their feelings find each other, then even the commands of their parents cannot control them." In reply the scholar abases himself by inverting "proper" social relations: "I am quite willing to make myself your servant" (this he says to the madam, not to Li herself). In addition, however, and perhaps more seriously, he breaks off contact with his fellow literati and so dooms himself to isolation within his new pseudo-family. "From this point the scholar hid himself away and did not associate with kin or friends. Daily he would meet with singers and entertainers and the like, carrying on with them in excursions and banquets. Soon his purse was empty" (534). Oddly enough, this attempt to create a private family for himself and Li defeats the whole literati motivation behind the courtesan quarters: the courtesan quarters were, first and foremost, a place for men to meet each other, to form friendships, to show off to each other, and to make solid social connections. The scholar is unaware of this purpose, because he has bought into the belief that his relation with Li Wa is all that matters. He does not realize that one goes to the courtesan quarters to meet men, not women.

By leaving himself vulnerable through this isolation, he makes himself a non-person, undefined through any social ties; later we learn that his father assumes that he has been killed by bandits. The significant decline does not really begin until the scholar's first illness, brought on by the sorrow and rage he feels over being abandoned. After the employees of the funeral parlor nurse him back to health,

> the establishment then employed him on a daily basis to hold the muslin drapes at funerals. He supported himself with his earnings. After several months he gradually regained his health. But whenever he heard the songs of mourning, he would sigh to himself that he was not as fortunate as the departed, and he would invariably break into tears and sobs and could not control himself. When he returned from the funerals, he would imitate their songs. He was clever and quick-witted, and before long he had mastered the mysteries of this music. In all of Chang'an no one could equal him. (535)

This is a curious passage that speaks much about the central themes of the entire piece. "The Story of Li Wa" is fascinated by the details of urban living, and nothing illustrates this more than its excursion into the world of funeral parlors and mourning singers. This is also the first environment that the helpless scholar manages to master to some extent and thus learns to earn his own keep within the intricacies of capital life. There is much to ponder in this. Two explanations are given for the scholar's skill as a mourner: first, his emotional sensitivity and life experience, his ability to identify emotionally with the bereaved; and second, his intelligence, which allows him to imitate, practice, and master the songs. There are ironic undertones here: the direct expression of emotion supposedly makes for successful poetry, and perhaps the scholar's newfound suffering gives him a skill he had previously lacked. More obviously, however, the learning of mourning songs mimics the studies he has abandoned; the intelligence that should have been applied toward passing the examinations is diverted into another art that is far below him socially. Yet this skill does earn him a reputation of sorts. The contest in which he participates with the competing funeral parlor is a cruel mockery of the government examinations. It is a Tang literatus' worst nightmare.

Not that he is ineffective in the contest, of course. His own emotions—which come from real experience, so to speak—move the audience as good singing should, and this results in his rediscovery by his estranged father. But this display of emotions and skill does not end with the reconciliation we might expect. A Western reader anticipating a prodigal son parable is shocked when the enraged father leads his son to a spot near the Apricot Garden and beats him almost to death. Yet another irony occurs here: the Apricot Garden was a site for literati parties and a haunt for courtesans and their customers, as well as a common place to hold banquets for those who had successfully passed the exams. The beating is the scholar's reward for winning the singing contest.

This shocking development unmasks the scholar's regeneration as a wrong path; he has tried to integrate himself into an inappropriate community. But he requires a more drastic regeneration, one closer to a genuine resurrection, in order to put him on the true path to rebirth. When the funeral employees find him nearly beaten to death, they carry him back to the parlor and nourish him by feeding him liquids with a reed pipe. Although he survives, he cannot use his arms or legs for a month. Earlier, his vulnerability once he left the protection of Li Wa's household suggested an infantilization. This continues here; in coming back from death, he must pass through a second childhood.[38]

As we have seen, however, it is not the task of the undertakers to save him. He soon breaks out into infected sores so severe his former friends can no longer tolerate him and decide to abandon him. He can no longer be left to die even in an undertaker's parlor. At this point, he does recover enough to take up beggary; again, his ability to convey his emotions aids him in his career: "His voice as he begged for food was very painful, and all who heard him were sorely aggrieved" (536).

Li Wa now rediscovers him as he goes begging in the middle of a blizzard:

> He cried out repeatedly in a sharp tone, in the very excess of his hunger and cold. The sound was piercing and unbearable to hear. Li Wa heard him within her house. She said to her servant, "That must be him! I recognize his voice!" She hurriedly went out and found him, stiff and starving and covered in sores. He looked barely human. She was moved by his plight. (536)

Li is in this case the warm, contented householder hearing the noise of the abandoned foundling, crying out in "the very excess of his hunger and cold." And yet this adoption scene is played against a recognition scene as well—Li is not only moved by his voice, as others have been, but recognizes the specific human being behind it. Although this act of recognition brings remorse, it also reveals her role as the scholar's own peculiar *zhiji* 知己, or intimate friend. She thus in one action becomes both the scholar's mother and patron.

Li Wa has counted for very little in the story up to this point. She has had no independence of her own and simply obeyed the wishes of her madam. One enigmatic comment immediately after we learn that the scholar has squandered all his money does point to Li Wa's true feelings: "Lately the madam's treatment of him was increasingly indifferent, while Li Wa's affection (*qing* 情) grew more and more earnest" (534). Of course, this only hints that Li has a mind of her own; it does not prevent her from assisting in the scheme to abandon the scholar. When Li discovers the scholar begging in the streets, however, the situation goes beyond simple affection. In a speech to her foster mother, Li pinpoints the source of her guilt and worries about the practical repercussions of their actions:

"This is the son of a good family. Once he drove a high carriage and held a purse filled with gold. Then he came to our house, and before a year had passed he spent everything he had. Together we concocted a scheme and then threw him out and drove him away. This was hardly humane conduct! We forced him to abandon his intentions and to lose his place in human society. The way of father and son is inborn, and yet we destroyed the sentiments of a father and drove him to abandon and nearly kill his own son! And now his misery has come to this!

"Everyone in the world knows this was my fault. His relations fill the court. One day those in power will thoroughly investigate the matter, and then disaster will fall upon us. Moreover, if we cheat heaven and betray others, the spirit world will not come to our aid. Let us not bring misfortune upon ourselves!" (536–37)

There is little love (*qing*) for the scholar in this, only dread over violations of the social order and an awareness that their actions have become known to possible avengers. She recognizes her own role in a web of urban deceit and chicanery, and she fears the scholar's true family and what they might do. She also acknowledges the ultimate crime: she has destabilized the moral hierarchy of the literati clan. Her moral horror here compels her to leave her role as daughter (although she does promise to continue to provide for her old madam) and to set up a new household with the scholar. Although this action mimics the daughter's leaving home to go and live with her husband, Li is actually taking up the role of mother.[39] For example, her nourishment of him succeeds where that of the funeral employees had failed; the scholar is now in the right hands.

Li now convinces him to return to his books. Soon he is successful in the examinations, and his reputation has been recovered. At this point, Li offers to leave him so that he can marry a girl from a distinguished family; her role as foster mother is now over. However, the scholar refuses to be parted from her. She finally agrees to accompany him as far as his first post; it is on this trip that he encounters his father and is reconciled with him. The father insists that the scholar marry Li, and she is now reintegrated fully into the proper family system. The moment Li leaves her role as foster mother, she, too, like the scholar is born again into a new role: the properly accepted daughter-in-law in a ritually observant family. Immediately after this, the story tells us that the son's mourning for his parents was so exemplary that auspicious mushrooms sprang up at the mourning hut. Of course, this is meant in part to reverse as completely as possible the earlier developments in the story: from the ultimate in unfilial sons, the scholar has become the most filial. But the story suggests that this ritual observance grows out of his following the example of his model wife.

Bai Xingjian's narrative may be constructed at first glance as a certain stereotyped romantic encounter between young literatus and morally ambiguous woman. But as in the case of Miss Ren, this is a relatively minor element in the story. The narrative is much more focused on the tribulations of the scholar, and Li tends to hold a series of fluctuating and shifting female "roles" in relation to him that spell out aspects of his fate: mistress, mother, wife. Arguably, one way of defining male identity is to examine carefully the behavior or prescribed duties of the women who are closest to him; they create, so to speak, the space that he occupies. Li Wa does this for the scholar, and more: she introduces him into the incomprehensible wilderness of urban society and becomes his harsh teacher.

My discussion of six texts here is not a conclusive overview of the representation of women in Tang narrative, but it has touched on certain recurring ways women and romance are deployed in these stories. There are, quite obviously, developments in Tang society and in narrative itself that allow for the description of more complex relationships between men and women; how far this might reflect a change in women's position in society is another matter. However, we can say that the introduction of romance into Tang narrative allowed writers a new way of exploring their own self-identity in the more complex Tang world. Whereas earlier society had prescribed literary expression within a fairly narrow scope of conventional themes and genres, the Tang literatus was subject to wider and more various social, cultural, and political claims. Aristocratic circles had been replaced for the most part by a larger and more heterogeneous bureaucracy. The educated male could also move beyond these bureaucratic concerns and find himself writing about new social worlds that had previously gone unrecorded: he could stroll the new urban centers; he could dabble in Buddhist or Taoist religious practices; he could roam as a "man of justice," eager to make friends with noble men like himself; or he could experience the regional differences of the empire through travel. Women were an equally novel experience: although always part of literati life as mothers or wives, the possibilities of dangerous liaisons or of mysterious romantic encounters enabled the literatus to make expanded claims for his own spontaneity, sensitivity, and individuality. However, even these relations tended to be circumscribed by social codes as well: the demands of family, codes of free male heroism, or the competitive pressures of the literati community as a whole. Nonetheless, if eroticism had previously been ritualized, either within the palace poem of the Six Dynasties or in the sort of elaborate gaming portrayed in the *You xianku,* it now broke free of these choreographed moves and expanded into the more aleatory environment of Tang narrative itself.

CHAPTER SEVEN

Honor Among the Roués

When an older classmate of mine, Zheng Hejing 鄭合敬, passed the examination, he spent the night in the brothels, where he composed the following:

When spring comes, every place is suitable for a stroll;	春來無處不閒行
When I gaze now upon the lovely Churun, I am seized by a novel feeling:	楚潤相看別有情
How fine it is to wake at dawn, recovering still from the ale,	好是五更殘酒醒
And hear the sound of her words to me: "Mr. Valedictorian"!	時時聞喚狀元聲

The examination of *Li Wa zhuan* in the preceding chapter introduced one of the most visible erotic phenomena in the Tang: the relationship between literati and courtesans. But what do we really know of such courtesans and their lives? Our only information comes almost exclusively from one laconic source: around the year 884, the minor literatus Sun Qi 孫棨 committed to paper his memories of the Pingkang 平康 Ward (the government-sanctioned brothel neighborhood in Chang'an).[1] After having accompanied comrades and patrons there for many years in the period before the disastrous Huang Chao 黃巢 rebellion (which largely destroyed the capital in 880), he desired (or so he claimed) "to record these matters to provide a topic of conversation for later ages" (22). The result is the brief miscellany, the *Beili zhi* 北里誌 or "Account of the Northern Wards" (Northern Ward or Northern Village had been a euphemistic term for a red light district since at least the Han).[2] Sun's short, apologetic preface is followed by a description of the residences of the prostitutes who lived in Pingkang Ward, and by twelve short "biographies" of famous practitioners of the trade. The text concludes with some random descriptions of literati behavior (which may have been inserted by a hand other than Sun's) and with a cautionary afterword in which Sun condemns the hazards of the district and the immoralities of the men who frequented it.

Sun's general description of courtesan habits is fairly brief and can be summarized easily. During the Late Tang era, the brothels of Chang'an could be found for the most part in the Pingkang ward in the eastern section of the city. This thriving and fairly well-to-do area lay southeast of the bureaucratic offices and the imperial palace. Nearby, to the southeast, was the large eastern market (*dongshi* 東市). The brothels were in the northeast quadrant of the Pingkang ward; several major Buddhist temples occupied other sections, and such noted figures as the classical exegete Kong Yingda 孔穎達 (574–648) and the early Tang minister Pei Guangting 裴光庭 (676–733) had had mansions located there. In surrounding wards were other mansions, such as those that had been owned by the infamous Yang Guozhong 楊國忠, chief minister at the time of the An Lushan rebellion, as well as the specialized palaces (*di* 邸) that were reserved for visiting dignitaries from the various prefectures (civil or military). The actual area occupied by the brothels consisted of three extended "neighborhoods" (*qu* 曲) stretching east to west; this term is conventionally translated in English as "lane," although this is somewhat inaccurate. The area itself was dangerous, and many visitors appear to have brought their own guards or hired ones for the visit.

The prostitutes were licensed by the government and were apparently still liable to government service of various sorts, particularly if they were enrolled in the books of the Imperial Music Academy (*jiaofang* 教坊). However, they were most evidently under the control of their "foster mothers" (*jia mu* 假母), who "adopted" their employees and thus created a simulacrum of a family, although fathers were usually absent. The prostitutes were usually girls who had fallen into their profession through misfortune or as victims of some machination: poor daughters sold by their parents, attractive infants kidnapped, young wives abandoned by their husbands. Once sold to a house, they were instructed in the arts of singing, dancing, and the rules of the various drinking games. Their freedom of movement was severely limited; a prostitute could leave the ward, for example, only if a client accompanied her and had pledged a certain daily sum to the "mother" for her use. Clients could also "reserve" a prostitute for their own private purposes through the similar payment of a daily fee. The "mothers" themselves may have been subject to certain restraints on their freedom of movement and action, and they usually operated under the patronage of some wealthy family in the capital.

As is not surprising, this insular world quickly produced a subculture of its own, replete with conventions, customs, and expectations. Most evident in surviving texts is the relationship between the culture of courtesans and that of examination candidates. The latter seem to have placed a particular value on the artistic training of the prostitutes, often preferring the company of women who could compose verse or engage in intelligent conversation. If a particular prostitute seized a man's fancy and he was financially well off, he might buy her from her place and establish her as a concubine, although this practice was hardly limited to scholars and was even more typical of the merchants and petty officers of the city.

Although some insights into prostitute life can be gained from the poetry of both literati and courtesans and (as we have seen) from literati narrative, Sun's descriptions of lifestyles, living arrangements, and attitudes have no real antecedents. Obviously, this limits the number of reasonable conclusions the historian can draw from the text. Other problems with the historical usefulness of *Northern Wards* have long been recognized. Sun's descriptions are short and often obscure; his anecdotes about individuals are gossipy and tend to focus on lurid or extraordinary incidents, and his own limited experiences and easily detectable biases render the text a very weak foundation on which to build a history of prostitution in the Tang metropolis (let alone in the empire). What can be reasonably supposed has been carefully discussed and elaborated on by a number of scholars, especially by Ishida Mikinosuke. He and others have tended to pay most attention to Sun's opening description of courtesan life, "A General Discussion of the Lanes."[3] On the other hand, the brief biographies of the women themselves have been slighted, probably because they were thought too "literary" or dismissed as "mere fiction."

Here I focus on the rhetoric of Sun's work to see *how* it says things rather than *what* it says. In this, I am not merely making a "literary" reading of *Northern Wards*. I part ways with Victoria Cass's perspective, which implies a disciplinary separation of the uses to which the text may be put:

> The value of this brief text lies in several areas. From the sociological point of view it provides a picture of the daily life and society of entertainers of T'ang Ch'ang-an.

Sun's portrayal of the women, their relationship with their clients or lovers, with their adoptive mothers, and with one another provides a rare glimpse of a little-documented world. His references to the scholar-officials who frequented the quarters are also unusual. The profiles of the women are crafted biographical vignettes, in the convention of the abandoned beauty. In fact, the *Pei-li chih* has been traditionally viewed as a work of fiction.[4]

What this last sentence means is that *Northern Wards* has been traditionally viewed as *xiaoshuo* 小說, but whether *xiaoshuo* can ever be translated as "fiction" (i.e., something that is consciously "made up") is a dubious proposition. Cass is right, however, in seeing that Sun's method of presentation is important. In fact, it might be most useful to use this method of presentation as a clue to the mentalities of literati and upper-class Chinese males and as a guide to the role prostitutes played in this form of instituted sexuality. Again, the limitations forced upon us by depending on one writer should make us cautious about claiming that Sun's perspectives represent those of all or even most of the frequenters of Pingkang, but it would be equally fruitless to dismiss his work and to refuse to come to grips with its rhetoric—the way it fashions descriptions of prostitutes, the way it justifies or condemns male behavior during literati visits, and the way it positions cultural activity within this realm and this particular field of discourse.

The Cultural Propensities of the Client

The role of the courtesan in the lives of literati has been the subject of some speculation.[5] The most detailed of these is Van Gulik's assertion in *Sexual Life in Ancient China* that courtesan culture provided a safe space for literati who felt burdened by the strain of family life and mandatory reproduction:

> Next to social factors, as a matter of course also the satisfaction of carnal desire contributed to the continued flourishing of the institution of the courtezans, but there are strong reasons for assuming that this was a factor of secondary importance. In the first place, those who could afford cultivating relations with courtezans had to belong at least to the upper middle class, and hence had several women of their own at home. Since . . . it was their duty to give those wives and concubines complete sexual satisfaction, it is hardly to be expected of a normal man that sexual need would urge him to intercourse with outside women. . . . Glancing through the literature on this subject one receives the impression that next to the necessity of complying with an established social custom, men frequented the company of courtezans often as an escape from carnal love, a welcome relief from the often

oppressive atmosphere of their own women's quarters and the compulsory sexual relations.[6]

Literary texts are rarely perfect reflections of reality, however, and an apparent absence of concern with "sex" in accounts of courtesan life may not be a reflection of male lack of interest in sexual relations. In fact, the absence of carnality from works like *Northern Wards* probably has more to do with the general indifference of early Chinese texts toward articulating explicit sexuality unless it participates in some other discourse as well. Such explicit articulations would thus serve another purpose within another kind of disciplinary language (the ideal examples are Taoist sexual manuals, where sexual performance is represented as a series of "prescriptions" or rituals related to hygienic and religious practices).

What is more at issue for Sun is the existence of the lanes as a place to go to with other young men, a place with its own social rules and organization. The reconstruction of the brothel as an institution mimicking family relationships is particularly significant in this respect: it explicitly situates the brothel household as something that both *is* and is *not* the patriarchal family. The woman stands at its head but has no husband; here the daughters are her only offspring; and at times a deliberate reversal of sexual roles is implied. Another text, the brief "Jiaofang ji" 教坊記 (Account of the Imperial Music School), tells us that the courtesans within a house would take oaths of loyalty to each other not as *zimei* 姊妹 (sisters) but as *nüxiong* 女兄 and *nüdi* 女弟 (female brothers). Clients could form mock marriages with courtesans, during which the client would take the role of the bride. If a courtesan died, one of her "female brothers" would inherit her rights as the new "husband" in what the author suggests is a parody of Hunnish custom (and may reflect, as Ishida has suggested, levitical practices common at the time throughout Central Asia).[7] In general, we can see the Northern Lanes as a classic Bakhtinian carnival, or what Stallybrass and White have analyzed in European fairs—a site where observers can become participants, where class and gender boundaries are temporarily removed, and where "the world upside down" becomes a common trope.[8] However, the degree to which the Northern Lanes were genuinely "subversive" and put women at the top of the social order in any meaningful sense is debatable; as Stallybrass and White often show, the fair can be both subversive of accepted ideologies and supportive of them, depending on the particular social context. It is obvious, however, that the quarter played an imaginative role in literati minds as a mysterious, often dangerous place, where ordinary relations did not apply: participating in the life of this other, deliberately perverse pseudo-family would signify to them both an institution with which they were familiar and something excitingly different.

There are other factors at work as well. The courtesans' oath of "brotherhood" made possible a system of alliances for women. Such an alliance would be impossible under the terms of Confucian rhetoric, which saw female participation in social relations only when a male was present: husband/wife, mother/son. By turning themselves into a "brotherhood," the courtesans form an equivalence to a similar bond of "created" kinship, the brotherhood of the examination class (Sun Qi's most common term for a man's relations to other men is that of *tongnian* 同年, "classmate," that is, one who has passed the exam in the same year as oneself). The game of calling a client a "bride" may reflect little more than the fact that he comes into the house from the outside and (ideally) stays there; what is more important is the creation of two large "brotherhoods" that mirror each other across the division of gender. Courtesan society in *Northern Wards* is the oldest surviving example of women appropriating the male "outer" world.

These complementary brotherhoods seem particularly important to Sun Qi, who as an examination candidate puts special emphasis on the intersections between examination culture and courtesan culture. Here we may see that literati motives for visiting the quarters have much to do with a need to come to an understanding of their own social positions and to test their own self-image as individuals and as members of their social class. Sun's perceived "educative" role of the lanes should draw our attention; for first and most explicitly, he argues that the lanes enable one to acquire knowledge and to learn lessons. This essential role is emphasized from the very beginning of his work, in the preface—which begins not with a discussion of courtesans but with a description of the increasing prestige of the examinations:

From the Dazhong 大中 period [847–59] the emperor [Xuanzong 宣宗; r. 847–59] cherished Confucian studies and particularly esteemed the examinations; consequently his favorite son-in-law, Zheng Hao 鄭顥, head of the Household Administration of the Heir Apparent, repeatedly took charge of the exams. The emperor often went into the city incognito, and he would speak on familiar terms with any examination candidate he encountered. Sometimes he made inquiries of the scholars of the inner court and the imperial relatives concerning what he had heard.[9] They would be terrified, not knowing from where he had obtained his information. Consequently, "presented scholars" (*jinshi* 進士) of the examinations flourished to a degree without precedent in the past.

Generally speaking, most presented scholars were from wealthy families, whereas those who belonged to ordinary families only amounted to three or so per year. From the time [that they passed], the wealthy scions would make a display of their grooms and horses, going off on excursions or holding banquets of great luxury and expense. The youngest who passed the exams were designated "emissaries in search of flowers" and were sent out through all the streets of the capital.[10] The new degree-holders then went forth, with drums and fans dancing about them. This show increased every year. (22)

Sun Qi thus rather improbably traces the increasing popularity and prestige of examinations to the surreptitious and dangerous visits the emperor paid incognito to the streets of his capital.[11] These visits, he suggests, gave the ruler access to the advice and informal gossip of the examinees, and thus the emperor gained an advantage over his court officials, who were either ignorant of such information or were keeping it from him. Whether the emperor encountered the examinees in the courtesan quarter is unstated, although perhaps Sun thought that possibility not improbable, especially since the narrative of danger foreshadows the dangers of the lanes, where the endangered party is no longer the emperor but the examinees themselves. *Northern Wards* concludes with two cautionary tales about the violence of the area:

Wang Shi 王式 of the Guards was the son of the late Minister Wang Qi 起 of Shannan 山南. When he was young, he was rather wild. Once when he was amusing himself in the lanes, a drunk happened to stumble into his room. Wang avoided him by moving to the foot of the bed. Suddenly another man entered with sword drawn. Mistaking the drunk for Wang, he cut off his head and cast it aside. "Now let's see you announce those who come to morning court!" Then he threw himself down on the bed. Wang managed to escape, but he never went back again. The people of the house recovered the head and buried it with the victim.

Linghu Hao 令狐滈 often visited the lanes in the days when his father was in power; at the time he was still an examination candidate. There was one particular house he especially enjoyed visiting. One morning he was informed by the mother that there would be a gathering of her kin. Entreating him not to visit her establishment for the day, she then sent him away. Linghu went and spied on them from a neighboring building. There he observed the mother and one of her daughters kill a drunken man and bury him behind the house. The next night when he went to visit them, he asked the girl about it. She panicked and tried to throttle him, all the time calling for her mother to come and help her kill him. The mother persuaded her to stop. The next morning after he returned home he went to inform the prefect of the capital, who sent someone to arrest them, but the family had already disappeared. (41)

Sun's warnings of the dangers of the quarters seem incongruous with the spirit of *Northern Wards* in general, which for the most part avoids a cautionary tone. Although this postscript may merely be a screen the author uses to deflect criticism for his own involvement with improper society, Sun subtly moves from a condemnation of visits to the lanes to a consideration of the didactic value of Pingkang relationships: "I have written this not merely to provide topics for romantic conversation; I also hope it will provide precepts of warning and encouragement for later generations" (42). He then concludes with a summary of the moral values to be learned from his vignettes and stresses the virtues demonstrated by courtesans through their actions.

Perhaps significant in re-evaluating these dealings in the lanes is Sun's use of the term *xia* 狎, "to act familiarly with," but often used with overtones of "improper familiarity," especially with regard to courtesans and prostitutes.[12] As noted above, Sun states in the preface that Emperor Xuanzong often went into the streets. His fraternization with examination candidates has overtones of a violation of expected codes of behavior and of proper social distance; yet the results are positive, in that the ruler gains greater insight into how his empire should be administered and has a greater appreciation of the young examinees who will later serve him in the bureaucracy. When Sun later uses *xia* to describe literati relations with courtesans, a similar double meaning of the term may be implied: an unorthodox, unsanctioned relationship carried out in the streets results in greater knowledge and life experience. The lanes are places where something can be learned—and in particular where vices can be exposed and guarded against or eliminated.

In less explicit terms, however, the interaction of courtesan and literatus produced a testing ground for social behavior and conventions. Courtesan society could function as an image of literati society—both were educated communities with their own rules and hierarchies. The literatus (or more specifically, the examination candidate) interacted with courtesans and other classmates in a space outside the imperial court within which he could exercise those particular talents and virtues he believed to be characteristic of his own identity. In other words, the quarters provided a unique public place (and by this, I mean in contrast with the family [private space] and with the court or bureaucracy [official public space]) where social lessons could be learned and competitions played out. How should men treat each other in the quarters? How should they treat the courtesans? What qualities were attractive within such a world? These issues were being worked out and codified in a "law of the quarters," and it is this law that Sun Qi implicitly recognized when he admired or condemned the behavior of his fellow literati. The overall impression one receives of the quarters from reading *Northern Wards* is not Van Gulik's sexual refuge but rather an intensely competitive and anxiety-producing place, where one's conduct was constantly being evaluated, and where the possibility of humiliation was always present. Issues of judgment, ranking, snobbery, and the like surface as much in *Northern Wards* as they do in *Worldly Tales*.

The motives here may depend to some extent on an increasing sensibility among literati that they must acquire a certain "cultural capital" or prestige in order to define themselves as a class; this will grant them some sense of identity and a degree of autonomy independent of clan background, economic status, and recognition from the imperial throne. Why literati should desire this in the ninth century implicates some of the most frequently asked questions about Tang history: in particular, the gradual rise of an examination literati "class" as opposed to an "aristocracy."[13] How far this development had advanced by Sun Qi's time will require much greater research and exploration; all I can do here is point out the "ideological" directions these tendencies are taking in Sun's text. But the tendencies are most definitely here: economic power is not particularly admired for its own sake, kinship ties to great families are valuable but not required, and what makes examination candidates a superior group of people is not that they are actually taking the examinations but that they have acquired the education necessary for the examinations. This grants them a cultural superiority over others that shows itself immediately in the "romantic" realm of the quarters. Sun is very much aware of social class and bureaucratic status: everyone who appears in the text is carefully identified in terms of his father, the official posts he holds, and the like. But what interests Sun is whether the individual can live up to his social position through his own actions.

The courtesans provide a unified cultural group that validates and approves of literati behavior while still remaining outside the literati class. Courtesans gain the power to judge literati behavior through their own constitution as a cultural elite; like the literati, they are carefully educated from an early age; like the literati, they compete among themselves for status as most talented or the best poet; like the literati, they assert that what makes them worthy of respect are their own personal talents. Within the lanes, they can serve as judges of literati and thus can occasionally function as their superiors. However, Sun also emphasizes the limitations of that privileged position. The lanes are the only place where such topsy-turvy relations are allowed (just as the inverted sexual roles of the mock-marriage remain contained phenomena). As Sun often hints, the economic and social relations that guarantee courtesans' inferior place in Tang society are always the implicit realities behind an idealized world of cultural competition.[14]

One illustration of the "social virtues" of the courtesan/client relationship is the group of anecdotes involving Zheng Juju 鄭舉舉, a woman who achieved particular fame as a "banquet judge" (*xijiu* 席糾). In this capacity, she presided over banquets, determined the playing of drinking games, and attempted to keep things on amiable terms. Sun's account of her begins with a condemnation of male boasting (parenthetical passages marked with "C" indicate an anonymous commentary to the text that seems to date from approximately the time of composition or shortly thereafter):

Zheng Juju resided in the lanes. She excelled equally in poetry and the administration of drinking games. Sometimes she and Jiangzhen took turns as banquet judge. Since they held high standing among the courtesans, they could reprimand any boors among their clients.

Zheng was charming in conversation and was much esteemed by scholars of the court. If noted men of letters held a banquet and invited courtesans to it, then Zheng would be sure to be among them. One day the present Left Master of Remonstrance Wang Zhijun 王致君 (C: personal name Diao 調), the Secretariat Director Zheng Lichen 鄭禮臣 (C: personal name Gou 穀), the Supervising Secretary Sun Wenfu 孫文府 (C: personal name Chu 儲), and Vice President of the Ministry of Personnel Zhao Weishan 趙爲山 (C: personal name Chong 崇) were attending a banquet. At the time Lichen had just received a post in the inner court and was boasting incessantly about it. Zhijun and the others grew tired of him and failed to reply to his comments; it was putting a damper on the evening. Juju, realizing this, proposed a new round of drinks. She then pointed to Lichen and said, "You're talking too much, scholar. Though it's a grand and lovely thing to be a Hanlin 翰林 academician, the talent that makes one qualify for such a post must already be present in the person. Look at Li Zhi 李騭, Liu Yuncheng 劉允承, and Yong Zhang 雍章. All were Hanlin members, but that didn't add a whit to the reputation they already possessed!" Zhijun and the others leapt up in great joy and bowed to Juju. Lichen could only fill his cup and drink by himself without saying another word. From that moment on, the banquet livened up considerably, until they all parted at dusk. Zhijun and the others presented Juju with bright-colored silks. (28)

Pride in position is justified but not at the expense of genial relations with one's comrades; Zheng's sensitivity to the smooth operation of her own realm, the banquet, enables her to pinpoint Lichen's own vice. Her position as banquet judge empowers her to humiliate him and to console Zhijun and his comrades; restrained by social decorum, they cannot protest his arrogant behavior themselves (the reasons for this social restraint are not clear: the power of Lizhen's family? fear of beginning a feud? Lizhen's prestige as a Hanlin academician?). Their gratitude to Zheng illustrates, Sun suggests, the momentary power vested in the prestigious courtesan as guardian of male values.

The anecdote continues on a more ambiguous note, as it contrasts a positive example of wit with a negative one of stinginess:

Sun Longguang 孫龍光, who passed first on the list of candidates (C: personal name Wo 偓; brother of Wenfu 文府. He attained first place in the fifth year of Qianfu 乾符 [878]), was rather smitten with Juju. With his classmates Hou Zhangchen 侯彰臣 (C: personal name Qian 潛), Du Ningchen 杜寧臣 (C: personal name Yanlin 彥林), Cui Xunmei 崔勛美 (C: personal name Zhaoyuan 照愿), Zhao Yanji 趙延吉 (C: personal name Guangfeng 光逢), Lu Wenju 盧文舉 (C: personal name Ze 擇), Li Maoxun 李茂勳 (C: personal name Mao'ai 茂藹) and others, he often went to visit her. Other clients were not so regular—thus their classmate Lu Siye 盧嗣業 informed them that they would have to pay a greater share of the banquet expenses. He sent them a poem on the subject:

I have never met your Grand Directress,	未識都知面
But I've often paid my fees twice over.	頻輸復分錢
I'm laboring away at literary works,	苦心親筆硯
But only to help some girl pay for her hairpins.	得志助花鈿
In vain I come forward to reclaim my autumn fees,	徒步求秋賦
And hold forth a cup to collect evening gruel.	持盃給暮饘
My strength is waning, I must often plead illness—	力微多謝病
It's not that I don't wish to go along with you. (28–29)	非不奉同年

Sun takes particular pleasure throughout *Northern Wards* in the ability of clients and courtesans to compose witty poems on odd topics; this is a prominent example. As it stands, the anecdote portrays Lu in a favorable light:

he has defused a potentially unpleasant social situation (his refusal to pay further dues for a banquet he will not participate in) by couching it in humorous verse. The commentator, however, makes further observations that denigrate Lu's cheapness and lack of fellow-feeling for his "classmates": "Lu Siye was the son of Lu Jianci 簡辭. He possessed literary talent from his youth, although he had a reputation for lack of integrity. Since he had not known his classmates for very long, he pleaded waning strength as the reason he could not pay the banquet fees. Thus this poem" (29).

Thus ends Sun's account of Zheng Juju; the reader may notice how little is said about the woman. Although a fairly abstract description of her abilities begins the passage (and such descriptions are fairly typical throughout *Northern Wards*), her presence does little more than provide the context for the illustration of male follies and virtues.

One such folly/virtue may be illustrated by the word I translate as "enamored" or "smitten" (*huo* 惑), although in other senses it usually means "deceived" or "confused." Used to refer negatively to the capacity women possess to lead men astray, Sun and his commentator tend to use it in a more limited, specific way to define the emotion that develops in a client when he continues to show a strong preference for one particular courtesan at the expense of others. In Sun's world, the positive or negative quality of such heightened states of emotion depends largely on whether the object of affection deserves it and whether the forces that drive the affection on the male's part are respectable or refined. In one particular case, a courtesan earns such affection because she sets out specifically to imitate literati manners and tastes:

> [The courtesan Lai'er 萊兒] was not particularly attractive, nor was she that young; but she was clever in speech and quite amusing. She laid out her apartments in the manner of some man of affairs or distinguished scholar, and for this reason many men were quite smitten [*huo*] with her. The son of the late Shanbei 山北, Guangyuan 光遠 of Tianshui 天水,[15] was considerably younger than she, and yet he was enraptured with her at first sight and could not bear to part with her. Lai'er on her part, seeing that he was clever, young, and handsome, attached herself to him and flattered him. As they continued to treat each other with respect, their love and intimacy grew. (31)

Lai'er's own stock in trade seems to have been masquerading as a literatus, which literati clients found "stimulating." The progression of her relations with Guangyuan is a revealing account of literati virtues. Although his attraction at first seems a case of infatuation, the respect the two show for each other's talents (and Lai'er admires his youth and talent, not his economic or social status) results in a more lasting alliance. As with several other courtesans, Sun makes the point that Lai'er's talents made her popular, despite her age and lack of physical attraction. Some modern readers might prefer to use such comments (often apologetically) to emphasize that a courtesan's talent mattered more than her attractiveness. Sun's tendency to tell us this fact at every opportunity proves rather that attractiveness *did* matter—and that the client's self-conscious choice of a courtesan's ability over her appearance was to be interpreted as a display of his own sensitivity and refinement in contrast to more "typical" behavior.

Consequently, it is no surprise to see this esteem for talent over physical appearance backfire with a man who is cruder in his sensibility:

Xian'ge 仙哥 of Tianshui 天水 (polite name Jiangzhen 絳眞) lived in the southern lane. Good at conversation and jokes, as well as a capable singer and director of drinking games, she was often the stewardess of drinking parties, regulating them leniently or strictly, as circumstances demanded. Her looks were fairly ordinary, but she was superior in her manner. The worthy and elegant clients of the time respected her, and so her reputation spread. Zheng Xiufan 鄭休範, an imperial diarist of the Secretariat (C: his personal name was Renbiao 任表), once composed a poem upon her:

How did this majestic flutist descend from the sky above?	嚴吹如何下太清
Her jade-white skin cannot bear the most gossamer of silks.	玉肌無奈六銖輕
I know it's not like drinking the rosy mists of heaven,	雖知不是流霞酌
But still I'd enjoy just one song from the lutes of Mount Yunhe.[16]	願聽雲和瑟一聲

Liu Tan 劉覃 passed the examinations at the age of fifteen or sixteen; he was the favorite son of Minister of State Liu Ye 鄴 of Yongning 永寧. When he presented himself for the exams as the delegate from Guangling 廣陵, he arrived with dozens of carts loaded with baggage, not to mention numerous teams of fine horses. At that time, Zheng Cong 鄭賨, one of his classmates, was his chief companion. . . .

Liu Tan gave himself over to his inclinations and desires while in the capital. Although Jiangzhen's age was much greater than Liu's, he heard the crowd acclaiming her and so he desired to meet her without knowing whether she was attractive or ugly. Jiangzhen's associates surreptitiously made arrangements so that she could always refuse his invitation with some excuse; she thus always made it impossible for him to come and visit. Nonetheless, Liu continued to send gifts to her without admitting the slightest difficulty. One day when Jiangzhen was really ill and had refused his invitation, he refused to believe her and sent her one string of cash after another. Her associates thought to make a profit from his entreaties and continued to mislead him, even though Jiangzhen never came.

At that time there was an employee of the Department of Finance, Li Quan 李全 (C: son of Li Lian 煉 of the Department of Finance), who lived in the Pingkang Ward and who had authority over the courtesans. Liu heard of this and immediately sent a messenger to summon him; he then presented him with about two pounds' worth of artificial gold flowers and silver cups. Coveting this generous bribe, Li went off to the lanes to look for Jiangzhen. He bundled her into a sedan chair and carried her off to the banquet. She arrived with disheveled hair and dirty face, with tears streaming down her cheeks. As soon as Liu lifted the curtain of her chair and had a look at her, he immediately sent her home again. He had already wasted over a hundred gold pieces on her. (26–27)

Sun Qi carefully positions the events in this anecdote to illustrate the vulgarity of Liu Tan. First, Jiangzhen's talent is mentioned, and once again the trope of "ugly and old, but quite an entertainer" is elicited to produce a ground for refined appreciation. This is illustrated by a poem by Zheng Xiufan that makes polite, conventional gestures toward her inner beauty and mentions her artistic ability more specifically. Liu Tan is then introduced, a man of great wealth who is not afraid to make a show of it. Next come the vain and dishonorable motives Liu possesses for wanting to meet Jiangzhen: "Although Jiangzhen's age was much greater than Liu's, he heard the crowd acclaiming her and so he desired to meet her without knowing whether she was attractive or ugly." This last phrase is Sun's projection of the factors that would matter to a man like Liu and foreshadow the disaster to come. Jiangzhen's own refusal to meet Liu may have several motives: her own desire to remain free of an admirer she despises or her anticipation that her own fading looks will not meet with his approval. The anecdote ends not only with Liu's humiliation but with a wry comment on his economic losses; one suspects that such waste of money would not matter in Sun's world if the motives and/or object of attention proved worthy. Once again, we also notice how little Jiangzhen herself matters in this anecdote: the focus is on Liu's predictable and justified humiliation.

Both sides of the gender divide reinforce the discourse of appreciation that runs through so many of the anecdotes in *Northern Wards*. If men recognize good qualities in certain courtesans, "in spite of" their lack of superficial attractiveness, then women in turn recognize the good qualities in men and honor them for it. Here lies another crucial relation between courtesan culture and examination culture. In a world of competition, in which passing the examination provides young literati with a source of anxiety and stress, female recognition can act either as a positive reinforcement anticipating future victories or as a consolation when the government fails to likewise appreciate the talent. In fact, a courtesan's own sense of self-worth becomes inextricably linked with the success of a client with whom she forms an alliance. The story of Lai'er continues as follows:

Before Guangyuan had actually taken the examinations, he had become quite close to her. When his time came, he assumed his own superior talent would render him the victor after a single skirmish; Lai'er also assumed that it was a sure thing. The winter before she boasted to her clients that he would "best his companions with his opening song." When Guangyuan failed, all the young gallants of the capital made their way from South Court to visit her and tease her for her failed predictions. When they arrived, Lai'er was standing before her gate, dressed to the nines, to await news of the exams. One of the young lads recited a poem from horseback to taunt her:

Everyone says Lai'er's word can be trusted always;	盡道萊兒口可憑
All the winter she boasted to us of her lover's name.	一冬誇婿好聲名
But just now as we looked at the list at Anyuan gate,	適來安遠門前見
We find that her Guangyuan has botched his "opening song."	光遠何曾解一鳴

Lai'er didn't believe them and taunted them back in the same rhymes:

The words of milk-mouthed babes cannot be trusted always;	黃口小兒口没憑
Just go right now, see the list, check the three highest names!	逡巡看取第三名
You students are just pouring water in a vase that's way too full;	孝廉持水添瓶子
Don't run about in the street, bothering us with your song!	莫向街頭亂碗鳴

This is a good example of her cleverness. That spring it took a while before she recovered from Guangyuan's failure. (31)

Lai'er's own humiliation in backing the wrong man results in at least the temporary rejection of her lover. They later exchange two highly allusive poems, in which Guangyuan begs her to take him back and she refuses.

In another anecdote, the recognition of talent motif has a subtler relation to the examinations:

Liu Tainiang 劉泰娘 was from a poor family in the northern lane. No one of any reputation lived there, and so no one knew of her. In the spring before the rebellion came to Chang'an, I was passing by Ci'en 慈恩 Temple and saw several courtesans preparing to set out to the Serpentine together. Having reached the temple, they had dismounted and were going on foot. One particularly young one among them was rather pretty. At the time the strollers were many, and quite a few were trying to exchange words with her. When anyone tried to find out her place of residence, she would only lower her head for a long time. Finally when I asked her for details, she said that an ailanthus tree stood in front of her gate.

Shortly thereafter a spring shower broke out, and all the courtesans went their separate ways. That evening, when I was proceeding north on business, I happened to pass by her gate. It just so happened that she herself was arriving in an oxcart. I then wrote a poem on her dwelling:

Normally the ailanthus is the commonest of trees;	尋常凡木最輕樗
But today I seek an ailanthus finer than the cassia.	今日尋樗桂不如
Since the Emperor Han Gaozu was the first to capture Xianyang,	漢高新破咸陽後
Though heroes came rushing everywhere they lost their chance for fame.[17]	英俊奔波遂喫虛

The acquaintances who had been in my company that day learned of this. The next morning her visitors were so many they were forced to tie their teams outside her gate. (37)

When the poor, obscure courtesans from the northern lane show themselves publicly at the Qujiang 曲江 (translated here as "Serpentine"), the city park in the southeastern corner of Chang'an, the most refined of them attracts considerable attention from the rabble. Sun's ability to recognize her greater potential makes his action analogous to that of the ruler who can pick out talented men from the ranks of the obscure (even the image of a gate thronged with carriages is a borrowed trope from archetypal "recognition" stories). Just as Sun Qi practices this ability (a commentary not so much on Tainiang's talents as on his own ability to recognize them), he may hope that the government has equal sensitivity in acknowledging his own talents.[18]

It is clear from these excerpts that poetry is a particularly powerful component in the social relations between client and client, client and courtesan. Theoretically, of course, the writing of *shi* poetry was considered a method for communicating the nature of the self publicly: it was a guide to personality and even a tool for the inculcation of virtue. Such factors may have had some influence on the presence of poetry writing on the examinations themselves and (it has been suggested) account to a limited degree for the wider dissemination of poetic literacy in the Tang.[19] However, this perspective also resulted in generations of literati trained in poetry writing who incorporated this skill into their view of what qualities constituted the civilized elite. As such, verse writing was above all a social talent and heavily conventionalized: our own idolization of the great Tang poets often blinds us to the general social sameness of most poetry of the time.[20]

Sun thus emphasizes the abilities of courtesans to write verse. Since the demimonde is the counterpart of literati culture, poetry should hold primary place among courtesans' social skills.[21] However, the courtesan's poetic abilities are portrayed as significant only in relation to their clients. There are eight poems in *Northern Wards* composed by courtesans, and twenty-four by literati; all the poems authored by women either are created in direct response to a male-authored poem or are answered by a male. Both poems and the dialogic implications of the poetic exchange are used by Sun to illustrate not only the talent, wit, and other positive qualities possessed by women and (especially) men but also their positive and negative qualities. In two specific anecdotes, the exchange of poems is a form of witty retort, meant to silence the arrogance or frivolity of the first poet:

Wang Susu 王蘇蘇 dwelled in the southern lane. Her accommodations were quite spacious, and she carefully regulated the banquets she had there. She had a number of sisters, all of whom were talented in conversation. There was a certain Li Biao 李標, a presented scholar who claimed descent from [the early Tang general] Li Ji 勣, the Duke of Ying 英. For some time he had been secretary to Assistant Remonstrator Wang Zhijun 王致君. He went to visit Susu with a younger brother and nephew of Zhijun's. While drinking, Biao wrote the following poem on her window:

By her door as spring grows late the flowered branches fly about;	春暮花株遶戶飛
Great princes seeking superior ladies get dust upon their robes.	王孫尋勝引塵衣
The goddess within this secret cave seems truly loving indeed—	洞中仙子多情態
She keeps Lad Liu here with her, refuses to let him go.[22]	留住劉郎不放歸

At first Susu didn't understand the import of this poem; but then, irritated, she said to them: "Who's keeping you here? Don't talk such nonsense!" Then she took up a brush and matched the verses:

No wonder that the dogs are spooked, and chickens fly about:	怪得犬驚雞亂飛
Skinny boys on skinnier horses come dressed in tattered robes.	羸童瘦馬老麻衣
Oh, who was it who so rashly asked these idlers in?	阿誰亂引閒人到
I'll take their cash, But as for them—I'll let that credit go!	留住青蚨熱趕歸

Li Biao, a humorless man, turned red with shame and ordered his carriage to turn back immediately. Afterward, whenever Susu saw anyone from Wang Zhijun's household, she would always ask: "Is that fellow I drove out that one time still around?" (36)

Chu'er 楚兒 had the polite name Runniang 潤娘; she was formerly one of the outstanding courtesans of the three lanes. She was clever and quick-witted, and a number of her verses deserved praise. More recently, when she grew older, she was bought as a concubine for the Thief-Catcher Guo Duan 郭鍛, who set her up in a separate establishment. Runniang had been famous in her courtesan days for her wild and unbridled behavior, and although her husband attempted to discipline her, he could not control her willfulness. Duan was in charge of many and various affairs, and since he had a principal wife at home, he only rarely visited Runniang. Whenever an old acquaintance passed by her house, they would call to each other through the window or the visitor would send someone to ask after her. Sometimes they would exchange letters written on silk.

Duan was the scion of a family from Qinren 親仁 Ward. He was exceptionally brutal and heartless, and whenever he would learn of these adventures of Runniang's, he would flog and humiliate her. Although Runniang would emerge from these experiences hurt and angry, she never changed in the least.

One day she was accompanying Duan from a trip to the Serpentine. They were in separate sedan chairs about ten or so paces apart. A certain Zheng Guangye 鄭光業, commissioner of contracts[23] (C: personal name Changtu 昌圖), who was then acting as censor-in-chief, happened to meet with her on the road. Chu'er lifted her curtains and beckoned to him, and Guangye bid a servant send a message to her. Duan noticed this and dragged her out into the middle of the street and began to beat her with his horsewhip. Her screams were pitiful to hear, and soon she was surrounded by a wall of observers. Guangye watched from a distance, shocked and ashamed, regretting that he had provoked this intolerable scene.

The next day Guangye took the road past her house to inquire after her. Runniang was already sitting under the window facing the road playing on the lute. He stopped his horse and bid a servant send in a message to her, but she already had a letter on bright note paper ready for him. It was a poem:

I must have brought from a former life some lingering resentment;	應是前生有宿冤
Nor did I expect in these present times such an evil fate?	不期今世惡因緣
My delicate brows nearly shattered under blows from the god's massive palms;	蛾眉欲碎巨靈掌
My slender frame could barely stand mighty Zilu's fists.[24]	雞肋難勝子路拳
I can only anticipate in fear the man with the iron tallies;	衹擬嚇人傳鐵券

(C: The Prince of Fenyang 汾陽 possessed iron tallies; the possessor of one of them could be excused from capital punishment. No doubt Runniang fears Duan because she possesses no such tally.)

I don't imagine he'll ask me again to dance with my feet of lotus.	未應教我踏金蓮
When I met you at the serpentine just the other day,	曲江昨日君相遇
I suddenly met with many dozen blows from his whip.	當下遭他數十鞭

Guangye, as he sat on his horse, took up a writing brush and answered her as follows:

Open your eyes wide to the world, speak not of your resentment.	大開眼界莫言冤
To please him to the end of your days must after all be your fate.	畢世甘他也是緣

Since you have no way out, don't bother us
with your constant obstinacy— 無計不煩乾偃蹇
For if you had a way out, you'd have
fled his rain of fists. 有門須是疾連拳
Since by custom you must bear
his whipping in the road 據論當道加嚴箠
Perhaps you should rather don black robes
and chant the Teachings of the Lotus.[25] 便合披緇念法蓮
But since I see you in cheerful spirits,
not downcast at all, 如此興情殊不減
I know now that on that day
he beat you with a reed whip. 始知昨日是蒲鞭

Guangye was a devil-may-care fellow and was afraid of nothing. Nor was he restrained by petty rules and customs. This is why he dared stop his horse, compose an answer, and send it off so quickly. Those who heard of this incident were terrified, because Duan had been often put in charge arresting criminals in the capital with the result that a great number of unsavory characters had sworn loyalty to him. Everyone was afraid of him. (27–28)

Both these incidents are meant to be comic and to illustrate the capacity of verse to satirize and punish those who possess undesirable social characteristics. Wang Susu deliberately deflates romantic convention by destroying the flamboyant image Li Biao had created for himself. In the case of Chu'er the situation is somewhat more ambiguous, in that her stubbornness in refusing to give in to her husband's abusive disposition as well as her own ability to recover from the violence of his assault may possibly be meant to arouse our admiration. However, the greatest emphasis is on her frivolity. She refuses to accept her new position within the family of the man who bought her out (she continues to engage in behavior proper only to her original courtesan status). Sun also hints at her continuing interest in other men, even after her beating: "He stopped his horse and bid a servant send in a message to her, but she *already had* a letter on bright note paper ready for him." Chu'er had the verses ready for whomever might first come by to offer his sympathies, although she hopes that it will seem a completely spontaneous lament. Sun likely meant this disingenuous action to be a sign of her promiscuity; Guangye for his part is not impressed.[26]

Sentiment and Poetic Exchange

In two of Sun's longer anecdotes, many of the issues noted above are revealed in particular force and create a discourse of sentiment that grows out of the examination culture of refinement. Even more significant, perhaps, is the creation of a special role for poetry: despite its heavy use of convention and its dependence on social situations, poetry creates a perspective that attempts to represent courtesan/client relations in idealized terms, supposedly untouched by vulgarity.

Yan Lingbin 顏令賓 resided in the southern lane. She possessed a romantic and carefree manner and was refined in her tastes. She was also estimated rather highly among worthy scholars of the time. She could make use of brush and inkstone and was an author of verses. Whenever she saw an examination candidate, she responded to him with the greatest courtesy and often begged some poem or song from him as a gift; as a result, her chests were filled with bits of brightly colored stationary.

Later she became seriously ill. One day near the end of spring when the weather was clear and mild, she ordered a servant girl to help her sit out on the porch. As she looked about at the falling flowers, she sighed several times; then she took up her brush and composed the following verses:

But a few breaths left to me;	氣餘三五喘
But a few flowering branches remain.	花剩兩三株
Let us bid farewell over a cup of ale	話別一樽酒
For we shall never meet again.	相邀無後期

She then ordered a small boy, "Take this for me to the Xuanyang 宣陽 and Qinren wards. If you should encounter any who have recently passed the examinations or who are candidates, then show this poem to them and say to them: 'Miss Yan of the lanes will await your Lordship's visit in spite of her illness.'" She then had her family set out ale and fruit to wait for them. Shortly after, several men showed up. They took their pleasure and reveled in drinking until evening. Then, with tears streaming down, Lingbin said: "I'm not long for this world. I hope that each of you will compose a dirge to see me off." Earlier the members of her household had supposed that she would seek to collect funeral expenses from her guests and were delighted. But when they heard these words, they were disgusted with her.

After she had died, the madam received several letters on the day she was to be buried. She opened them and glanced over them. They were all dirges. She angrily tossed them out the window. "How can this help with our daily expenses!" (30)

The first part of the anecdote constantly asserts the significance of poetry; in fact, it becomes obvious that to write good poetry is a supremely valued ability, either as the most prominent quality among many positive ones or in contradistinction to other values that would threaten it. Yan's talent as a poet is the most prominent of her skills; yet we do not immediately receive an example of that talent. Instead we are told of her interest in collecting verse from others. Here, one may note that she sought poems from "examination candidates" (*juzi* 舉子)—not simply from "gentlemen" or "literary men" or even "officials." Sun Qi makes clear that she chooses to recognize these particular men as her own male counterpart in the literary world. The currency between them is not money but poems, and she hoards the paper they are written on, believing only these things to be precious.

Chinese poetic criticism recognizes the special relation between the poet and the person who truly understands his poetry; this bond goes beyond being merely the perceptive critic and instead establishes bonds of friendship and understanding—to understand the poem is to understand its author.[27] Although this belief may function more often as a pious ideal than as reality (especially in the conventionalized world of courtesan/patron relations), it becomes clear that Yan believes in this special relation and that her male admirers appreciate her for it. By linking the two "brotherhoods" of examination candidates and courtesans in this manner, Yan helps to conceal a more accurate assessment of her own role in the social and economic world of Chang'an (money, not poems on paper, is the true currency). By appreciating examination candidates as talented and expressive poets, she helps to conceal for them their own dependence on a system invented by higher powers who intend to make use of them. The role of writing supposedly transcends politics and becomes a factor in a purer realm of social relations that is unmediated by power.

As Yan sits on her porch and beholds the spring weather, she is moved to confront her own mortality and to issue the invitation that will eventually lead to the dirges of her admirers. Her language is both deeply affecting and highly conventional: Yen compares herself to the dying flowers of late spring and expresses this explicitly in the poem of invitation she composes. The conventionality of the situation places Yan within the realm of socialized poetic activity in China; being "moved by external things" (*ganwu* 感物)—especially the world of nature—is an essential element in traditional poetics. It is this movement toward things that produces the "stimulus" (*xing* 興)

that inspires the poem. This poem in turn, with the death of Yan herself, moves its readers to compose verses in response. The dirges are texts that mourn her passing, as well as elegies to her peculiar personality and talent.

However, these dirges are threatened in turn by the outsider, the one who does not understand these "special relations" established by the talented courtesan and her clients. It should not surprise us that opposition comes from the foster mother, who can understand Yan only in vulgar economic terms and sees her imminent death only as a financial liability.[28] The "mother" is also a deliberately disruptive force and prevents the further continuance of the poetry relationship by discarding the poems. If Yan is a figure for the literati, her mother is a figure for literati anxiety: the person who may have power over him and yet cannot appreciate him or assist those who can. The literatus who sighs over the vision of so sensitive a person as Yan falling into the clutches of so heartless a mother is sighing for himself.

However, the power of poetry cannot be contained so easily:

There was a certain Hunchback Liu 劉 who lived nearby who enjoyed playing the Tibetan flute. Clever and sharp, he was a talented song composer as well. Some said he had had an affair with Lingbin. Gathering together a number of the dirges, he taught them to the mourners before the procession began, and they sang them with great lamenting. That day the entombment took place beyond Azure Gate. An idle scholar happened to encounter the cortege; some time later he asked Liu to sing the dirges for him. Liu still remembered four:

I

In days past, I sought my goddess;	昨日尋仙子
Now I find a hearse before the gate.	輀車忽在門
Must life indeed end in this?	人生須到此
The ways of Heaven can never be divined.	天道竟難論
Her clients once arrived in droves,	客至皆連袂
But who comes now to beat the pot of mourning?[29]	誰來爲鼓盆
I can hardly bear to notice	不堪襟袖上
The fading mark of mascara on my sleeve.	猶印舊眉痕

II

In waning spring I helped her ailing form to drink;	殘春扶病飲
This eve is now most heartbreaking of all.	此夕最堪傷
The dream's illusion vanishes with the dawn,	夢幻一朝畢
Windblown petals had few days in their mad season.	風花幾日狂
A lone lovebird sees his mirrored form;	孤鸞徒照鏡
A single swallow shuns his nest in the eaves.	獨燕懶歸梁

How could we describe your generous heart? 厚意那能展
Blighted with bitterness I offer a cup in memory. 含酸奠一觴

III

Those romantic feelings so hard to recall; 浪意何堪念
Deep emotions turn early to grief. 多情亦可悲
They gallop their horses to display their feelings; 駿奔皆露膽
In crowds they gather, all raise cups to their brows. 麇至盡齊眉
Though flowers fall, days of blooming return; 花墜有開日
But when her moon sets, it will not rise again. 月沉無出期
Dare we speak of the closing of the tomb 寧言掩丘後
And the lush grass that will soon cover it? 宿草便離離

IV

How could it happen so suddenly? 奄忽那如此
Early blighted peach blossom once at spring's height. 夭桃色正春
Hands clasped to heart, still moved by this; 捧心還動我
Tears wiped away—I grieve for her alone. 掩面復何人
Who will tell me the way to Mount Tai? 岱岳誰爲道
No use to ask directions when the river's flowed by.[30] 逝川寧問津
Before your death you should have wedded— 臨喪應有主
Song Yu lived right to the west.[31] 宋玉在西鄰

These songs circulated throughout Chang'an and were frequently sung by mourners.

Someone asked Liu: "'Song Yu lived to the west.' That's you, right?" Liu snorted. "There were quite a few Song Yus." Her clients all knew she was having an affair with a musician, but her neighbors thought this so shameful they hushed it up. Jiangzhen was once with Lingbin at a party of several gentlemen. They were teasing each other in turn, trading playful insults. Jiangzhen let slip, "Considering whom you're keeping at home, you shouldn't be so pushy." She immediately regretted what she had said, and later if anyone who knew Jiangzhen or the gentlemen who had been at the party asked about her blunder, all maintained their silence. (30–31)

Liu enters the anecdote because he is the medium through which the examination candidates' dirges are preserved. Sun brings up the issue of his low social status when he describes the shame many felt when they became aware of Yan's liaison with the common musician. However, the increasing involvement of poetry with music during the Tang makes the figure of the musician a particularly attractive one in many texts: poems dedicated to musicians or describing musical performances are common, and (as has been observed in particular by Ren Bantang), a large percentage of poetry written in the Tang was sung rather than chanted, particularly five- and seven-character quatrains written on erotic subjects.[32] Tang literati participated in this world, and even the most respected poets wrote poems for musicians. Even if Liu is a somewhat disreputable figure, he is a fitting mediator between the dead courtesan and the poets. He recovers some of the poems, sets them to music, and teaches them to the mourners. However, before they can become truly memorable, they must be recognized, appreciated, and communicated: and only a literatus can do that: "An idle scholar" happens by and asks Liu to sing them; presumably that is the medium by which they entered the text of *Northern Wards* itself and thus made Yan's tragedy a fitting subject for preservation.

The poems are fairly conventional dirges (literally "coffin-pulling songs"). Their descriptions of grief and of the departed draw some inspiration from the "dead wife" tradition that begins with the "mourning poems" (*daowang shi* 悼亡詩) of the Western Jin poet Pan Yue 潘岳.[33] However, they are more significant here for their role than for their content. Although we have only the single poem of invitation by Yan, we have no fewer than four mourning poems in reply. The world of poetic exchange, of one sensitive person listening to and responding to another person, is replaced once more by the world of male competition: Sun uses Yan's story as a forum to display and set off several prime examples of the funereal art, one against another. Death becomes an opportunity for a contest.[34] In this flood of male voices, Yan's own poetic talent is forgotten, and her death becomes a convenience. Her coffin becomes another receptacle for verses, like the chest she had once owned, crammed with the currency of red note paper. She ends by serving literati interest first as a sentimental model for literati anxiety, embodying as she does the literatus who perishes with none to appreciate him sufficiently and with his posthumous legacy threatened by the ignorant, and second as an opportunity for male literary production and competition. What remains of her is the only capital she generated, the texts of the poems.

The longest anecdote in *Northern Wards* concerns Sun Qi himself and his long-term relationship with the courtesan Yizhi 宜之. It is without parallel among the other anecdotes. It also bears an uncanny resemblance in some respects to Yuan Zhen's 元稹 much more famous *Tale of Yingying* 鶯鶯傳, particularly in the way it contrasts female and male perceptions of the same relationship.

The opening introduces the various "daughters" residing with the "mother" Wang Tuan'er 王團兒:

Wang Tuan'er lived in the front lane, first house from the west. (C: When the emperor returned to the capital following the rebellion [885], many court officials resided here.) When she became a foster mother, she acquired several girls. The eldest was Xiaorun 小潤 (polite name Zimei 子美); she was renowned from her youth. Vice-President of the Board of Personnel Cui Chuixiu 崔垂休 (C: personal name Yin 胤, original polite name Sizhi 似之, he passed the exams at twenty) was smitten with her about the time the revolt broke out and spent quite a lot on her. He once wrote a poem on her thigh. It was seen by Weishan 爲山 (C: personal name Jiu 就; he now possesses the polite name Gunqiu 袞求; his friends called him "Little Qiu." He was governor of Linjin 臨晉), who replied to it as follows:[35]

Below the pagodas of Mercy Temple, close by, a plastered wall;[36]	慈恩塔下親泥壁
Smooth and supple, gleaming white, quite surpassing jade.	滑膩光華玉不如
But why should this Mister Cui, fortieth of his rank in Boling,	何事博陵崔四十
Engage in Ouyang's brushwork on this golden-hillock thigh?[37]	金陵腿上逞歐書

(C: Chuixiu ranked fortieth of his surname in his generation, although this was later changed to forty-one. This is the man who became Minister Cui.)

The next girl of Wang's was named Funiang 福娘, polite name Yizhi 宜之. She was very bright; she matched the ideal in plumpness and was breezy and elegant in conversation. She was also talented in letters. The late Cui Zhizhi 崔知之, vice president of the Board of Office, presented her a poem during a visit (C: personal name Dan 澹; at the time he was serving in the inner court).

Why should not this clear breeze send forth distinctive scent?	怪得清風送異香
Gracefully swaying, this immortal child lets sweep her rainbow skirts.	娉婷仙子曳霓裳
You're mistaken if you take me for he who stole Heaven's peaches;	惟應錯認偷桃客
Though like Dongfang Shuo, I too serve as vice minister![38]	曼倩曾爲漢侍郎

(C: At the time, Cui Dan was vice president of the Board of Civil Office in the Inner Court.)

The next girl of Wang's was named Xiaofu 小福, polite name Nengzhi 能之. Although she was lacking in grace, she was quick-witted. (32–33)

Sun begins with one of the most striking incidents in the entire text: the inscription of a poem on Xiaorun's thigh. Here Sun provides us with a figuration of two forms of male competition: first the examinations, through the explicit comparison of the thigh to the wall of the Buddhist temple where examination names are inscribed (Xiaorun's body serves as a site for male triumph); second, male social competition in verse writing, through the representation not of the original poem on the body but of the male-authored poem written in response to it—thus, Xiaorun, who at first was the site of the text, now merely becomes a representation in another poem echoing that earlier one: the eroticism of the initial gesture becomes more complex through the incorporation of the male social contest. One should also pay attention to the way in which this text in turn "eroticizes" the examinations themselves, making the smooth, white wall of the Buddhist temple a site for desire and the fulfillment of that desire.[39]

The poem Cui Dan presented to Yizhi works through many of the same issues. Although he fails to mention whether Xiaorun possesses literary talents, Sun asserts these qualities in Yizhi's case, which may provoke a more subtle statement of male triumph from the second poet. Here, the peach tree becomes a symbol of conquest, just as the blank wall of the temple did earlier; immortality is daringly snatched from a forbidden garden. Yizhi's representation as the peach tree is less important than the role she plays in the allusion—that is, as the tree violated by Dongfang Shuo. This role naturally allows Cui Dan to draw attention to the more important comparison, that of himself to the noted writer of the past. In both poems, wall and tree present opportunities for males to "make their mark."

When I was in the capital pursuing my studies with other young men, whenever we were fatigued or depressed we would go to visit Wang's house and sit with Funiang and Xiaofu, engaging in refined conversation and elegant drinking. I regarded their manner highly. One day I presented Yizhi with a poem:

Goddess robed in kingfisher splendor, flesh so rosy and white;	綵翠仙衣紅玉膚
Light and graceful, her years just approach fifteen.	輕盈年在破瓜初
Drunk on cups of rosy vapor, she urges Lad Liu to drink;[40]	霞盃醉勸劉郎飲
With cloud-like tresses she lazily asks her mama to comb her.	雲髻慵邀阿母梳
Unafraid that the chill invades the jewels on her knotted sash,	不怕寒侵緣帶寶

She worries that the breeze will lift 每憂風舉倩持裙
    the skirts she wears so charmingly.
In vain you'd picture Xi Shi 謾圖西子晨妝樣
    at her dressing table at dawn;
For Xi Shi never could 西子元來未得如
    equal her in the end.[41]

Although she had received many poems before, she considered this one to be the most satisfactory. Holding the text against the red wall of her bedroom to the left of the window, she asked me to inscribe it there. After I had finished, since the wall was far from filled, she asked me to compose one or two other verses written as if in her own voice. I then composed three quatrains:

I

I wander by the walls, turn back at the window— 移壁回窗費幾朝
    thus squander several mornings.
Then fingering the door pull I secretly go out, 指環偷解薄蘭椒
    still near my orchid and pepper chambers.
There I frolic with a neighbor girl, 無端鬥草輸鄰女
    winning mock battles of straw,[42]
Claiming as reward from her 更被拈將玉步搖
    one of her jade hairpins.

II

In the chill, a red embroidered gown 寒繡紅衣餉阿嬌
    is offered to Ajiao.
Newly coiling, incense cakes 新團香獸不禁燒
    smoke away without restraint.
The eastern neighbor sets the fashion, 東鄰起樣裙腰闊
    her skirt billows about her waist;
Decorated here and there about 刺蹙黃金線幾條
    with threads of yellow gold.

III

I try to throw my darling 試共卿卿戲語麤
    some ribald jest,
But serving boys without end 畫堂連遣侍兒呼
    are called through the painted hall.
My winter dry skin cannot bear 寒肌不柰金如意
    the gold back-scratcher's touch:
Sweetie, do you have any of that 白獺爲膏郎有無
    White Otter Skin Lotion?[43]

I saw afterward that a space for several lines still remained. I visited her the following day and found that she had added a poem of her own following mine:

You struggled in your poems to encourage others' talent;	苦把文章邀勸人
And when I chant them, I do think your diction can be fresh.	吟看好箇語言新
Though they don't come up at all to Sima Xiangru's works,	雖然不及相如賦
You could say they're still worth a pound or two of gold.[44] (33)	也直黃金一二斤

Sun Qi now introduces himself into the household, as the third talented poetry-writing male. Perhaps the reason for presenting the poems of the two Cui now becomes obvious: Sun wishes to demonstrate his own abilities by placing his poems in juxtaposition to theirs. However, an important difference exists here: Sun, as a participant in the idealized discourse of poetry between intimate friends, now desires to show that he does not view Yizhi merely as a site but as a respondent and partner in verse writing. He does not write his poem on her body, nor does he use his visit as an excuse to figure forth his own status. He instead writes a longer poem centered on her own attractions. He then waits a response, which gives him the figurative victory over the earlier poets: "Although she had received many poems before, she considered this one to be the most satisfactory." Yizhi has validated his worth through her reply.

Yet here again the element of inscription recurs, as she asks him to write the text on the walls of her room. Not only that—but when he finishes, there is much more space to be covered, more areas to be inscribed. The quatrains he continues to write are all little vignettes of client/courtesan relationships, written with attention to sentimental moments of daily life. Perhaps we may see this as Sun attempting to occupy all of Yizhi's life, to prevent any other males from arriving and writing their own verses on her walls; in this way, he reserves and protects for his own experience the cozy epiphanies portrayed in the verses. The only space that he allows for further articulation becomes the space for her own response, in which she thanks him for his poetic instruction. Ironically, the poems act as an advertisement for Yizhi—not only describing her beauty but bearing witness that she has the sensitivity to appreciate poetry from a talented scholar. Underlying the sentimental exchange of verse are manipulations on both sides, a continuing reference to a less idealized situation. Nonetheless, both sides obtain what they need from this exchange.

When Yizhi initiates her own poetic dialogue, however, the results are not so satisfactory:

> Whenever Yizhi was serving at a banquet, she was always downcast and depressed, as if she could not stand what she was doing, and those around her would be affected by her mood. Finally, unable to contain myself any longer, I inquired quietly after her problem.
>
> She answered: "How could I be blind to this way of life and not wish to escape it? But what means do I have? Whenever I think of it, it is impossible for me not to get depressed." She then cried for a long time.
>
> Another day she suddenly presented me with a piece of red paper. Weeping, she bowed to me. I saw it was a poem:
>
> | | |
> |---|---|
> | Day after day downcast and sad,
to escape I have no plans. | 日日悲傷未有圖 |
> | Yet I hesitate to tell my heart
to an ordinary man. | 懶將心事話凡夫 |
> | Unlike water spilled from a jug
I can be gathered up again; | 非同覆水應收得 |
> | I ask you, my fine sir,
what you hope to do? | 只問仙郎有意無 |
>
> I thanked her for her poem and said, "I understand full well your implied meaning, but this is not suitable for an examination candidate. What should we do?" Weeping again, she replied, "Luckily I'm not yet on the registers of the Imperial Music School; so if you so desire, only one or two hundred pieces of silver should be enough."[45] Before I could reply, she handed me a writing brush and asked me to match her poem. I wrote under her own verses:
>
> | | |
> |---|---|
> | How comes it, lovely beauty,
you have these distant plans? | 韶妙如何有遠圖 |
> | I cannot yet assist you,
I truly am not the man. | 未能相爲信非夫 |
> | Though the lotus remains pure
with its roots in mud, | 泥中蓮子雖無染 |
> | To transplant it to my home plot
I cannot hope to do. | 移入家園未得無 |
>
> When she saw this, she broke into tears and said nothing further about it. Our feelings for each other immediately faded. (33–34)

Like Yan Lingbin, Yizhi sees poetic exchange as a potential avenue for self-expression and not merely as social exchange. Although she had earlier told Sun about her own sense of depression and hopelessness, her hope that he will buy her out and set her up as a concubine can be expressed only in verse. However, Sun is nonplussed by this request; his examination candidature, normally a status that allows him to engage in contests and courtesan adventures with his peers, now allows him an excuse to deny her request. Although the examination class may have been more lenient than the aristocracy in the acquisition of concubines, there is some evidence that taking on a prostitute was still a risky social move to make.[46]

Two images are placed against each other here. Yizhi claims she is not "spilled water" and can be reclaimed. Sun chooses a more loaded image: that of the lotus flower, which grows white and pure, although its roots usually go down into the muck of a swamp. At first this seems a dubious compliment: but the lotus had already become an image of the integrity of the literatus, who maintains the purity of his virtue in spite of the moral filth of his surroundings. On the one hand, Sun Qi oddly recognizes the affinity of the courtesan "brotherhood" with that of the examinees, and so produces an image that softens his rejection; on the other, he plays on the plant image in order to refuse the responsibility of transplantation—perhaps because his own "garden" does not possess the muck that the lotus requires to grow so beautifully.[47]

That summer I went east to Luoyang. I would hold parties at home from time to time, and when she grew tipsy she would say, "I don't know if the happy times I've had with you will last much longer." Then she would cry. That winter when we returned to the capital, it turned out that some magnate bought her, and I could no longer see her. (C: the courtesans of the lanes were often reserved by rich men from their mother at the rate of a string of cash a day. This was known as a "contract of purchase." Although such courtesans could be exempted from government service, they could no longer meet with clients.)[48]

That spring, on the third day of the third month, I went with some friends to perform the Purification Rites at the Serpentine. I heard the sound of strings and flutes from the neighboring tent. I went to see and saw someone sitting at the west side in violet robes, two on the east side dressed in sackcloth, and the one sitting at the north entirely dressed in hempen mourning clothes. Before this last one there were sacrificial balls of rice. To the south were two courtesans—Yizhi and her foster mother. I went behind the tent to wait for servants to come out so that I could question them. One told me that a certain Zhang Yan 張言, a silk merchant from the Xuanyang Ward, had prepared this meal for the Street Inspector. This Zhang was Yizhi's master. Jingxuan 敬瑄, the Street Inspector's deceased wife, was the mother of the two dressed in sackcloth.[49] Before I left the tent, I asked a female servant if I could come to the lanes on a visit the following day.

The next morning when I visited the quarter, I saw Nengzhi at the gate. She invited me to dismount, but I refused, saying I had other business to attend to. I continued to talk to her from horseback. Nengzhi tossed me a red kerchief rolled into a ball. "This is a poem from Yizhi." I opened it:

Long ago I showed my gratitude, 久賦恩情欲託身
wished to entrust myself to you.
And I had told you of my heart 已將心事再三陳
over and again.
The lotus already sunk in mire 泥蓮既没移栽分
has been now transplanted,
If I remain parted from you, 今日分離莫恨人
don't resent me for it.

I read this over and galloped home, disappointed. I never returned there again. (34)

Sun Qi moves rapidly on to the decline of the relationship, followed by his nostalgic encounter with her following their separation. Oddly enough, this passage echoes the last pages of Yuan Zhen's *Tale of Yingying*, where we read of Zhang's disillusion with Yingying and their eventual separation, followed by a later, aborted meeting:

About a year afterward, Miss Cui was married to someone else, and Zhang himself took a wife. Once when he happened to pass where she lived, he asked her husband if he might speak with her, using his kinship to her as an excuse. Her husband went in to ask her, but she refused to come out. The resentment and longing on Zhang's face were evident to everyone. When Miss Cui heard of it, she secretly sent a poem to him:

Since then grown pale and thin, 自從消瘦減容光
my beauty faded,
I toss and turn a thousand times, 萬轉千迴懶下床
too weary to leave my bed.
It isn't because of my husband 不爲旁人羞不起
that I am ashamed to rise—
Grown haggard rather for your sake, 爲郎憔悴卻羞郎
I am ashamed to appear to you.

In the end he never got to see her. Several days later, when Zhang was about to leave the area, he also sent her a poem to take his leave:

Forget all about it, what remains to be said?	棄置今何道
But we were so close at one time!	當時且自親
Take your memories of past encounters	還將舊時意
And use them to love the one before your eyes.	憐取眼前人

After this he knew no more about her.[50]

This similarity in conclusion may be coincidental, or it may represent a common pattern in the male conception of what constitutes a melancholy and transient affair of an "illicit" nature (I will leave undiscussed Chen Yinke's contention that Yingying was herself a courtesan).[51] What is common here, I believe, is Zhang's and Sun's mutual desire to portray themselves as sensitive males particularly vulnerable to the "romantic" side of sexual relations. In both texts, the sentimental and emotional attachments that fuse woman and man are put into place through the arts of poetry and music. If we are allowed, somewhat cautiously, to read the reasons for Sun's coolness toward Yizhi by examining Zhang's motives in "Yingying," we may note that this emotional "overcommitment" on the part of the male brings with it a fear of excess, of allowing oneself to get "too involved." This fear is represented as arising from a vice that is really (implicitly) a virtue: oversensitivity. As Yuan describes Zhang at the beginning of *Yingying*:

Scholar Zhang was by nature warmhearted and passionate, handsome to look on and pleasant in manner; however, he held to high moral standards and would not tolerate any behavior considered indecorous. Sometimes his companions would entertain themselves at banquets, disporting themselves wildly, rushing about in their revels as though afraid they would fall behind the others. Then Zhang would only look on tolerantly—he would never practice such behavior himself. And so until the time he was twenty-two, he had never been with a woman. An acquaintance asked him about this. He apologized, saying, "Master Dengtu had no appreciation of beauty—he engaged in immoral practices.[52] I on the other hand do have a true appreciation of beauty, but I have had no opportunity to express it. Why do I say this? Generally speaking every lovely object inevitably leaves its mark on my feelings; this is how I know that I haven't given up love completely."[53]

This odd combination of moral strictness and romanticism plays a particular role in Zhang's dubious apology for his conduct at the end:

All of Zhang's friends who had heard of the affair thought her an extraordinary person, but Zhang had already made up his mind to break with her. Yuan Zhen was particularly close to Zhang, and so sought the reason from him. Zhang said, "Generally speaking, those creatures that heaven has ordained to be seductive either bring early death to themselves or, failing that, an early death to others. If that Cui person had happened to meet someone who was wealthy and of high position, she would have taken advantage of the favor granted by her charms. Then if she had not manifested herself as rain, she would have done so as some evil dragon or monster—I don't know what all of her transformations would have been. In the past King Xin of Shang and King You of Zhou controlled states populated by millions, and their power was indeed great. Yet a single woman defeated each of them; their armies were destroyed, they themselves were killed, and they have become laughingstocks up until the present. Since my own virtue was insufficient to overcome such evil and bewitching beauty, I hardened my heart to her." At the time those present all sighed deeply with admiration.[54]

If Sun, like Zhang, sees himself as a man vulnerable to romantic attachments due to his own poetic sensitivity, he may be introducing the account of his affair with Yizhi to illustrate his own superiority over cruder men—men like Cui Yin, who writes a poem on a prostitute's body. This would also account for his efforts to see Yizhi again, although Yizhi obviously wishes to avoid him (his discomfort is subtly indicated by his refusal to dismount when it appears that only the servant Nengzhi has been sent out to talk with him). On the other hand, Yizhi's social status allows Sun Qi to avoid the accusation of coldness; although he has compassion for her situation, he (and presumably his readership) would recognize her status as an inevitability—tragic perhaps, but a part of life that must be accepted.

Sun's vignette closes with one of the most moving passages in the text, Yizhi's description of her early life.

I often think of her lively intelligence with delight. She once told me: "I'm originally from Xieliang 解梁, and we lived next door to musicians. When I was very young, I stayed at home learning needlework and how to chant poems. Even before I was sixteen, I was deceived by a man: I was married to this passing stranger who claimed he was going to the capital to take the examinations. But when we reached the capital, he put me in a brothel and abandoned me. First everyone in the household treated me very kindly, but after several months they forced me to study singing and the rules for drinking games. Soon they made me see clients, and I was then deflowered by a policeman named Mr. Ji 計. I was then given to the son of Grand Minister Wei Zhou 韋宙 as well as to the son of Wei Zeng 衛增 the Administrator. They must have spent over a thousand pieces of gold in this house. Two of my brothers came looking for me. They talked about carrying me off, but since I thought that my older brother was too weak to carry out such a scheme, I told them 'I'm already lost and I fear your efforts are useless.' I did manage to obtain several hundred gold pieces from my foster mother, which I gave to my older brother. Then, weeping bitterly, we parted forever." Whenever she told clients this story, she would cry a long time at this point. (34–35)

One might ask what prompted Sun to place this account here, at the end, rather than earlier, when Yizhi first tells Sun of her depression. I think by now the answer is obvious: from the time that Yizhi became the property of Zhang Yan, the attention of the anecdote has been on Sun Qi and his emotional response (mostly melancholy) to his loss of Yizhi. The relation of her misfortunes at this final juncture thus subtly emphasizes Sun's own sensitivity, since he finds these events worth recording as he thinks back on his memories of her ("I often think of her lively intelligence with delight. She once told me . . ."). We are not meant to see this as merely an example of his callousness—that he knew her life was like this, yet refused to buy her out; rather, we are to see this as a demonstration of Sun's ability to see sides of courtesan life that others do not see. In this, he returns to his self-portrayal as unconventional moralist—discovering lessons for life in odd and unorthodox places. Sun's account of his affair with the courtesan thus provides the ideal pedigree for his own right to compose a work like *Northern Wards*: for he can through his descriptions confirm the right of the examination class to visit the lanes and thereby perform culturally significant acts that help to construct their own status, even as he distances himself from their excesses.

AFTERWORD

Lost in a Sea of Coral

In grief take up the iron net— in which hang the coral;	愁將鐵網罥珊瑚
Yet the sea is broad, the sky is vast, and your place is lost.	海闊天翻迷處所

Chinese society began to change in the latter half of the Tang; by the middle of the Song, these changes had altered China irrevocably from its "medieval" stage to what most historians term the beginning of the "early modern" period. Major intellectual changes also took place among the literati; although they had been formulating their own identity as a group since the Eastern Han, the process was accelerated by the examination system and an increasing autonomy from government-approved orthodoxy. Independence of thought was accompanied by an increasing consciousness of the private world: poets wrote about refuges, safe spaces, leisure.[1] Most of these changes are beyond the concerns of this book. Yet at the end of my own narrative of gender, I can at least gesture toward changes in this area. Gendered power and dynamics took on new forms and ultimately became more complex and (arguably) more modern. The work of Li Shangyin 李商隱, greatest of the Late Tang poets, seems an inevitable place to finish—not just because of the experimental nature of his work (with its self-consciousness and high degree of "craft") but also because of the way his work dominates modern discussions of Chinese poetry and eroticism.

Controversies surround the subject-matter of Li's verse. Image after image in the poems foregrounds the act of misunderstanding, of being lost—

and each of them suggests in turn the reflexive act of reading poetry itself and the confusion it induces:

In grief take up the iron net—
  in which hang the coral;
Yet the sea is broad, the sky is vast,
  and your place is lost.

So reads a couplet from the first of Li's four "Yan Terrace" 燕臺 poems.[2] Commentators explain that the best way to procure coral is to sink an iron net into the water. When the coral has grown through its meshes, one can simply retrieve the net. But what if you forget where you had cast your net? If Chinese poetry is a process of discovering and reading the world around one, then not knowing where to lift the net means losing one's bearings in the text: lines are lost between the vast heavens and broad seas.

This sense of being at sea can be exhilarating, of course, and in our era much of poetry education grows out of analyzing difficult modern verse. Searching for the lost net is a skill that we have been taught through the techniques of New Criticism and close reading, and we continue to exercise it even as we have come to reject the implications of New Criticism methodology. As a consequence, if the academic communities of China and the West express a particularly "modern" interest in the poetry of Li Shangyin and if Li has inspired more literary scholarship than most other Chinese poets, who can be terribly surprised?

Yet there are other reasons why Li's poetry has become a particularly lively site for critical polemic and discussion. In particular, a group of his poems (I mean the ones simply designated "Untitled" and others that share those poems' imagery and rhetorical structure) have produced critical curiosity and unease because they are simply unlike anything that came before them in the Chinese tradition. Chinese poetry is generally expressed in a laconic, often vague version of the classical language, although such vagueness is more often than not resolved through the recognition of conventions and through readers' expectations of what the poem will say. Consequently, verse like Li's that gives no such guideposts is sui generis. Eschewing the conventions of such genres as banquet poem, parting poem, or *yuefu*, Li's untitled poems strike out into unknown seas; we do not know beforehand what they are about, and they do not seem to follow a clear, articulate path of vision or thought from beginning to end. The inevitable critical response, even before our century, has been to superimpose some sort of recognizable, conventional expectation on the poems in order to make them comprehensible.

The different perspectives of these expectations are well known to the student of Li Shangyin scholarship. Qing commentators, for example, attempted a comprehensive political analysis of all of Li's work. Of course, Li was a serious literatus player in the political system of the ninth century, and he would have produced (almost inevitably) a body of serious, public, politically oriented poems that would discuss both his own career and disappointments, as well as his opinions and disapproval of court policies. That some of Li's poems can be read within this tradition is rather likely. The more complex and questionable gesture occurs when such commentators interpret the "untitled" and similar poems as part of a tradition of misdirection and enigma, in which political concerns are masked by deliberately ambiguous language that conceals their import from the casual reader. In this respect, the "untitled" poems were thought to be derived from such earlier texts as Ruan Ji's *Yonghuai* 詠懷 (Singing my feelings) cycle, or from Chen Zi'ang's 陳子昂 (661–702) *Ganyu* 感遇 (Moved by what I encounter) cycle. Such political readings were by no means the only ones possible for the traditional reader, but they do represent a canonizing move on the part of commentators who wish to place these problematic texts within a high-minded literati tradition. As a result, although we can all recognize the Ruan Ji–Chen Zi'ang tradition as one that Li would have known well and could have drawn on in his own work, the peculiar anxiety the Qing commentators show in enforcing this reading expresses certain historical contingencies characteristic of the Qing era—in particular, the need to reinforce a sense of the literati tradition, to validate literati culture, and to arrive at a comprehensive canon through the standardization of commentary.

Of course, far and away the most popular move of the modern reader is to see the "Untitled" poems as "love poems." This reading was championed at considerable length by Su Xuelin in the 1920s and 1930s;[3] and even if few of us would fall in with her detailed attempts to identify the objects of Li's hidden and adulterous passions, most of us generally find ourselves reading these verses as love poems naturally and almost instinctually. The question that should make us pause, however, is not whether this reading is valid; rather, we might ask why so many of us accept this interpretation so easily. The answer is that the Western lyric tradition has pervaded not just the educated classes of China and the West but also certain cultural attitudes toward "love" reflected in popular culture everywhere. Almost all readers of Li's poetry, whatever their nationality, are aware of the Western "courtly love" tradition that gave birth to the love poem, the impassioned verse addressed to one's mistress, and (most especially) to the rhetoric of adulterous, unsanctioned, or guilty passion. For May Fourth intellectuals, such a tradition might seem to have a healthy iconoclastic quality: the use of a literary genre to express emotions that traditional Chinese culture seemed to go out of its way to repress. Of course, those same intellectuals recanonized their own tradition by trying to trace the rhetoric of love within Chinese texts that had been ignored, displaced, or misinterpreted: the rereading of Li's poetry is paralleled by similar rereadings of drama, vernacular fiction, and other previously "marginal" genres. There is nothing wrong with this interpretation, and many early Chinese readers from at least the Song dynasty saw Li's poems in such a way; I merely wish to point out the degree to which our own cultural and political interests shape our interpretation.

But if our present reading of Li Shangyin's poetry as love poetry is based to some extent on our own historical situation, we need to ask what the consequences of such a love poetry would be in Li's own time and within his own tradition. It is unlikely that Western-style love poetry arose spontaneously in ninth-century China without any cultural predecessors. I propose instead that Li's work gropes toward a different function within Chinese culture and society—a function that is made possible by new cultural and intellectual trends. How could Li as a literatus of his own time produce verse that looks so much like love poetry to us? What were the consequences to the tradition that he wrote such poetry? And why does it still seem to us a largely unique phenomenon?

Tang Public Erotic

To articulate these questions and so to answer them, one must emphasize once again how little Chinese "erotic" verse up until Li Shangyin's time resembles the "Untitled" poems. By the late Tang, the expression of sexual desire in poetry follows three broad traditions, based on their differing social uses. First, in poetry authored by men in which a female persona stands for the author himself and in which his desire is public and political, the lover or husband is the ruler, who refuses to employ the poet or listen to his advice. As we have seen, this tradition dates at least to the *Chu ci* and dominates

Chinese habits of reading poets such as Cao Zhi. By the sixth century, this reading tradition became inextricably linked with more "objectifying" tendencies, in which court poetry circles demonstrate their control of language as well as their social unity through the elaborate description of sexually desirable women—this is the so-called palace poetry anthologized in the *Jade Terrace*. Both of these kinds of erotic verse tended to use the same themes repeatedly: the most common situations are a woman abandoned by her husband, who is philandering or is serving in the army, or a palace lady ignored by the emperor. The theme is rendered less tedious through the exploration of these women's activities as they pine away: weaving, picking mulberry leaves, picking lotus blossoms. Palace poetry, with its less evident emphasis on unrequited loneliness, also contains courtesans, dancing girls, singers, and the like. One of the most common rhetorical structures of these poems is that of inside versus outside: often the poet and his audience stand outside the woman's private chambers, watching her through the window or listening to her private songs or speeches. This voyeurism maintains a predominantly male participation in poetry sustained by the idea of the female as object of attention, not as recipient of the poem herself. Although the private chambers of the woman are penetrated in a figurative sexual violation, privacy is rendered public through the explicit and open exchange of verses that occur among the male authors. Palace poetry is above all a social genre in which the seemingly intimate moments of the bedroom are turned into *jeux d'esprit* and a manipulation of common expectations. Although subjective female emotion is represented, it is rendered in conventional ways.

Although these erotic themes did not change in the Tang, a third use of them emerged through popular song. The rhetorically elaborate and medium-length palace poems contracted to the four-line erotic vignette, in which lonely and desirable females were evoked through some striking gesture or pithy comment. Some have seen this withdrawal from voyeuristic description as a sign that the political allegorical reading of eroticism reemerged in the Tang, provoked in part by the increasing psychological insecurity resulting from the examination system as a way of evaluating literati. This is true in part, but it should not keep us from acknowledging that such erotic vignettes were also treasured as popular songs and were sung by entertainers at parties and other social gatherings. As Ronald Miao has noted, without concrete extrinsic proof, it is impossible to determine if an erotic quatrain represents the poet's serious, political self-expression or whether it was written as an effective and not-too-deep song lyric.[4] Perhaps the same poem could have served both purposes, depending on the circumstances of its application.

The following apocryphal anecdote, dating from a ninth-century collection, suggests the dynamics of Tang popular song and verse:

During the Kaiyuan 開元 Period [712–42], the poets Wang Changling 王昌齡, Gao Shi 高適, and Wang Zhihuan 王之渙[5] were equal in fame. In those days, when we had yet to encounter the tempestuous chaos of rebellion,[6] the three of them often went out together. One day when the weather was chill and a light snow had fallen, the three poets visited a wine shop, where they ordered a round and sat down for a short drinking bout. Suddenly a dozen or so musicians from the Pear Garden[7] came upstairs for a banquet. The poets withdrew to a corner of the mat, and huddling around a brazier for warmth, they awaited the affair. Four lovely singing girls arrived in turn after the musicians. Charming and seductive, they were at the height of their beauty. The musicians then began to play, and all the pieces were famous songs of the age. Changling and his companions made a bet. "The three of us have gained fame as poets, and yet we haven't settled who is greatest among us. Let's secretly listen to what songs these musicians sing, and whoever has the most of his poems performed will be the most talented of us all." One of the performers then took up the music and sang the following:

A chill rain comes over the river and enters Wu by night.	寒雨連江夜入吳
At dawn I bid traveler farewell in the solitude of Chu hills.	平明送客楚山孤
If kin and friends of Luoyang should happen to ask after me:	洛陽親友如相問
A single piece of my heart in a vase of jade.	一片冰心在玉壺

Changling reached out and marked the wall. "One quatrain." Then another performer sang:

I open my trunk, tears soak my breast:	開篋淚沾臆
For I see your letters of former days.	見君前日書
At night how lonely the terrace grows!	夜臺何寂寞
Where still this hermit dwells.[8]	猶是子雲居

Gao Shi marked the wall in turn. "One quatrain." Another performer then sang:

At dawn with broom in hand when the golden palace opens:	奉帚平明金殿開

She forces herself to take up her fan 強將團扇共徘徊
as she paces back and forth.
Her pale jade face cannot match 玉顏不及寒鴉色
the black of winter crows
That come bearing the sunlight 猶帶昭陽日影來
from the Palace of Shining Glow.[9]

Changling again made a mark. "Two quatrains." Zhihuan had had renown as a poet for some time, so he now addressed his companions: "These are mediocre and careless performers. Their songs are but the vulgar village airs of Ba 巴 and cannot approach great lyrics like *Spring Light* or *White Snow*."[10] Then he pointed to the most beautiful of the singing girls. "Let's see what she sings. If it isn't one of mine, then I'll never dare cross swords with you again. But if in fact it is one of mine, you will all have to do obeisance and acknowledge me as your master." They all laughed and awaited the outcome. In a while the chosen woman, charming with her splendid hair arranged in side buns, broke into song:

Yellow sands ascend afar 黃沙遠上白雲間
into the white clouds.
A single expanse of lonely city 一片孤城萬仞山
on a myriad-foot-high peak.
Why should barbarian flutes 羌笛何須怨楊柳
resent the willow branches,
When the spring breeze cannot cross 春風不度玉門關
Jade Gate pass?[11]

Zhihuan danced about the other two. "Bumpkins! Wasn't I right?" They all broke into loud laughter. The performers were unaware of the reason and came over to see them. "What so amuses you gentlemen?" When the poets had explained, the musicians all vied in bowing to them. "Our vulgar sight did not recognize the gods. We beg you to grace our party with your presence." The three accompanied them back and drank until the end of day.[12]

The story has already been discussed in some detail by Stephen Owen; he points out the degree to which individually authored verses (some of which are even occasional in origin and thus became unmoored from their original, biographical circumstances) find a new place among the world of generalized popular song.[13] (One may thus note how the two Wang Changling lyrics reflect two different traditions: the first is an occasional parting poem general enough to apply to any circumstance, and the second is a "palace lament" on the subject of Lady Ban.) The three poets in this anecdote do not participate in the popular music world but are well aware that their own work could be appropriated for song lyrics. Rather than seeing this as a violation of their own authorial autonomy, they use it as a gauge of their own popularity. There is a mild refiguring of the competitive community here, as the poets wager whose work will be most used by the singers. There is also a refiguring of competitive anxiety, since Wang Zhihuan is obviously annoyed by the failure of the first three singers to sing his lyrics, but he reroutes his potential humiliation by turning it into a mark of his own distinction: rather than admit defeat in the category of quantity, he claims status through a link to a "discriminating" performer, the most beautiful of the singing girls. By doing so, he reinterprets the concept of intimate friend, or *zhiyin* 知音: previously a *zhiyin* might imply a person who understands a poet's work through an intimate intuition into the poet's feelings and character, but here the *zhiyin* is seen as the connoisseur, one who can judge the classiest verse of the finest quality. Wang Zhihuan's identification that the most beautiful of the singing girls will be the one to perform his own verses, and hence will be his own *zhiyin*, continues the implicit identification in the erotic tradition between physical surface attraction in the female and poetic talent in the male. The performative realignment of eroticism is so important here that the female performer need not even sing a song about the female world; her female role as singer transcends the subject matter in importance.

In such a performance context, male/female relations are somewhat altered: a woman's role is no longer simply to be represented as the object of desire, as in palace poetry; rather, she becomes the throat through which certain male verses are articulated to the public world. These verses do not themselves have to be erotic in content, as the choice here of a parting song and a frontier song indicate; instead, women singers play a part in a sort of public, ritualized form of sexual play. Although a woman singing as a prelude to sex can be traced in poetry back to the Nineteen Old Poems and becomes more evident in later verse, the choice of song is now articulated as expressing a figurative choice of sexual partner on the part of the woman: her selection confirms the poet's sense of self-worth, a sense of literary self-worth that stands for or substitutes for a sense of one's own sexual desirability. We have already seen how Sun Qi's courtesans make similar choices with similar results; here the game of sexual choice is sublimated even further into a form of popular art.

The difficulty of understanding another's state of mind is often foregrounded in the short lyric:

Li Bai 李白 (701–62): Feelings of Resentment 怨情

The beauty rolls up her pearl curtains, 美人捲珠簾
Sits withdrawn, knits fair brows. 深坐顰蛾眉
I only see the shine of tears' traces— 但見淚恨濕
I cannot know whom her heart despises.[14] 不知心恨誰

Here, the characteristic analysis of female emotions found in palace poetry has given away to the epistemological problem itself—we can detect signs only on the surface; we cannot know the "true state" within. Another famous poem of Li Bai's restates this problem more subtly:

Lament of the Jade Stairs 玉階怨

White dew springs up on jade stairs. 玉階生白露
The night is long; it soaks her gauze stockings. 夜久侵羅襪
Then she lets down her bead curtains, 卻下水晶簾
And idly gazes at the autumn moon.[15] 玲瓏望秋月

Here, again, emotion is represented only through gesture, which we may observe but not ultimately fathom. The ambiguity of emotion's representation is augmented by the poetic term *linglong* 玲瓏 in the last line—some take it as descriptive of her actions (idle, lackadaisical); others interpret it as the effect of fragmented moonlight in her chamber.

I suspect that later readers have reacted more positively to these Tang quatrains than to their palace-poetry antecedents precisely because of this aura of mystery. It is as if we are now granting greater autonomy to the female figure by suggesting that her subjectivity is unreadable. Yet this stance is heavily conventionalized as well, so that even this aura of mystery becomes foregrounded and expected and enters into a sort of public, social repertoire of imagery. It may also have its origins in certain characteristics of eighth-century poetic closure: for example, this nature quatrain by Wang Wei 王維 (699–761), "Lake Yi" 欹湖:

We play flutes as we cross to the far shore. 吹簫凌極浦
As dusk nears I part with you. 日暮送夫君
Back on the lake, I turn my head once: 湖上一迴首
Mountains green, white clouds curl up.[16] 山青卷白雲

Here, as has been noted by Shan Chou, the final image introduces an air of mystery or unreadability to the landscape.[17] Although Wang may be influenced by Buddhist metaphysics and its assumptions concerning the illusory nature of reality, the poem itself functions as a way of observing the world around one that was much imitated throughout the Tang. Read in this way, Li Bai's quatrains "conventionalize" the possibility of an independent female consciousness.

Li Shangyin's Private Erotic

Li Shangyin was familiar with these various erotic traditions and employed them throughout his work. As a well-educated literatus, he would have known the ways in which erotic relations could express political relations; in addition, he (like many poets of his generation) had rediscovered sixth-century verse and even wrote imitations of the previously discredited palace poems.[18] As for the erotic quatrain, although he was not as fond of it as his contemporaries Du Mu 杜牧 (803–52) and Wen Tingyun 溫庭筠 (ca. 812–ca. 866), he knew its rhetoric and could employ it when he cared to. The "Yan Terrace" poems themselves were exercises in Tang erotic and romantic rhetoric—often vague and incomprehensible in terms of "plot" but essentially coherent as exercises in sentiment.[19] However, the later "Untitled" poems (and their related verses) stand somewhat outside this tradition and deny the validity of reading them in conventionally sanctioned ways. This does not mean that they do not employ the same erotic language; it would have been impossible for Li to create a new eroticism, even if he had wanted to. I suggest that Li was interested not so much in reinforcing or continuing the erotic tradition as in using it as a means to create a discourse of private subjectivity, and that in his poetry, probably for the first time in the Chinese tradition, there is an attempt to create some impossible, non-social space for poetry, as if poetry could function as a means of communication directly between poet and addressee, without relying on mediating social conventions and expectations.

Here is one of the more famous poems in the series:

Untitled 無題

Stars of last night and wind of last night;	昨夜星辰昨夜風
West of painted houses, east of cassia halls.	畫樓西畔桂堂東
Our bodies do not share the paired pinions of the bright phoenix in flight,	身無綵鳳雙飛翼
Yet our hearts possess the single spot of numinous rhino horn.	心有靈犀一點通

In separate seats we pass the hook, warm the springtime wine;	隔坐送鉤春酒暖
Divided into teams, we guess at what lies hid, red the waxen lanterns.	分曹射覆蠟燈紅
Alas. I heard the watch drum and answered to my office:	嗟余聽鼓應官去
I canter my steed to Orchid Terrace, myself an uprooted weed.[20]	走馬蘭臺類斷蓬

Notes

ll. 5–6: Generally held to describe two forms of party game. In the first, a hook was passed from hand to hand, and members of the opposing team had to guess who possessed it. In the second, objects were concealed under bowls or similar vessels.

ll. 7–8: At dawn, drums sounded at the imperial palace to announce the end of the night's curfew and to summon officials to morning audience and to their own bureaucratic offices. The speaker is employed at the Censorate (Yushi tai 御史臺), poetically known as the "Orchid Terrace."

Most modern readers see in this a description of a party the poet has attended, some time in the past; most read the opening line as suggesting "last night," although there is nothing preventing the reader from taking it as "Stars of past nights, wind of past nights." Kang Zhengguo feels we should go no further than such a reading and denies the erotic or romantic connotations of the verses.[21] However, such a reading ignores the force of the hidden and taboo in the text—something deliberately *made* unclear by the failure to be explicit. After all, this poem is not called, "Written from Work, After Attending a Banquet at a Friend's House." Moreover, the image of line 3 (the shared wings of the phoenix) is generally an image associated with heterosexual erotic verse, not friendship. Above all, however, this poem is obsessed with images of blockage and frustration, just as the "Yan Terrace" poems are. The second couplet points to a liaison that dare not be acknowledged: a failure to flee together is compensated by a hidden, surreptitious connection, one that is obscure and difficult to fathom (although rendered even more poignant if we agree with A. C. Graham in thinking that the rhino horn should be interpreted as an aphrodisiac).[22] The poet's separation from the object of his affection is powerfully concretized through their mutual participation in party games, in which they are compelled to sit on different teams and participate on different sides. These games involve guessing at hidden things. Throughout Li's untitled poems, he speaks of the need to conceal and the embarrassment of exposure:

Her fan cut to the moon's shape
cannot conceal her shyness.[23] 扇裁月魄羞難掩

These party games hint at the game of romance and the game of sex, of teasing, of revealing what should be covered. They are ironic reminders of the games the two wish to play in private. The games continue through the night, as wine is drunk and candles burn—and the night is not employed for the sort of sport the poet would prefer. One might here think also of game playing at other moments in Chinese romantic culture—of the word games played by the lovers in "Immortals' Den" or of the drinking games carried out between singing-girls and clients before they retreat to the privacy of the bedroom. The games do not progress to this point at the party: rather, the dawn call to public office is heard; it is the only thing that reaches him, the only thing that traverses the distance of his isolation. The poet must ride off to his position at the Censorate. He here compares himself to the *peng* 蓬, or artemisia weed, which is uprooted when it dies and blows about in the wind, much like the American tumbleweed. It is a common image in Chinese poetry but is almost always applied to the despair of the traveler who is forced to leave home and family for distant public office, usually on the frontier. For Li, the exile from the beloved's bed forces him to the role of rootless wanderer, although he travels but a few blocks through the streets of the city. The projection of the space separating lovers as a profound geographical distance is a common one in Li's poetry: he is reinterpreting the common political theme of exile in romantic terms. One imagines Li the next day, in his office, writing down this poem, perhaps seeking some way to pass it into the hands of the woman he has left.

We might also consider for a moment the ultimate role of the party in this poem. As I have already stressed, ordinary erotic verse was a social phenomenon: although it described the private world of the lonely woman, the conventions were common coin passed back and forth between men. One could imagine that at this very party men are composing such verses, one of the many games they are playing, or perhaps singing-girls at the party are performing erotic songs. It is this sociable eroticism that Li's poetry undermines by suggesting that it is too public a setting for the erotic scenes he longs for and anticipates. For Li, privacy is not merely distance from the duties of public office—it is also privacy from the very world that produces most poetry, that produces informal social relations. The public rituals of the party block such privacy as surely as the public duties of office: indeed, the term used for "divided into teams," *fen cao* 分曹, may be used also to describe the dispersal of court officials to their various duties following the morning imperial court sessions.[24] It is ritual, social behavior in general that this poem dreams of dispensing with, although it acknowledges its own failure to do so: for the desire here transcends poetic form itself, which continues to organize events into the balanced, antithetical couplets that constitute the very act of ritualized poetic composition.

If the poet's desires are frustrated by the social world, he can at least gesture toward this privacy by evoking the party as a focus for his frustration. What would it mean to create an eroticism that was not suitable for a party and the exchange of social verses? What about a poetry that would be for hidden eyes, eyes that barely dared to read the words, that were the only eyes to understand the meaning of the words?

At this point, one might argue that romantically or erotically autobiographical readings of these poems are merely a matter of the reader's choice; after all, if there is no precedent for such poetry in the Chinese tradition, how do we know Li is even writing in this vein? Ultimately, such a reading cannot be proved, but these poems use images of blockage, failure, and impotence that draw their power from the earlier, more explicit erotic tradition. And if one were to grant that but then suppose that such a reading would compel a "political" reading of such poems (that is, to see the female figure as an allegorical trope for the ruler, and Li as the frustrated minister), then I would reply that there is already a pattern for such poems at Li's disposal (namely, the erotic song), but he refuses to use it. There is another reason for doubting that the "Untitled" poems are political. Most political erotic texts are rooted in certain common, socially perceived relations between men and women: most prominently, the emperor is the man, and the minister is the woman. The hierarchy of male/female relations is projected on political relations, and it is essential to some extent that these relations remain unchanging. Roles are adopted that must remain consistent. For example, if the poet/minister is the wife or court lady, he must more or less remain passive and immobile, pining in his isolated chambers.[25] But in Li's poems, the roles of poet and the object of his desire are shifting and complex. Li alternates passive and active roles, as does his beloved; and, more interestingly, the references are often maddeningly unclear as to who is doing what throughout the poem.[26]

In another of the more famous "Untitled" poems, Li draws this ambiguity further by projecting emotions and meanings onto things—the poem itself is strangely empty of inhabitants, the only actors being historical allusions and not "real people":

Untitled 無題

Whistling, the eastern breeze— a fine drizzle arrives.	颯颯東風細雨來
Beyond the banks of the lotus pool there is light thunder.	芙蓉塘外有輕雷
A metal toad gnaws the lock— burning incense enters.	金蟾齧鎖燒香入
A jade tiger pulls the cord— turns as it draws from the well.	玉虎牽絲汲井迴
Lady Jia peeped through curtains at the youth of Han the clerk;	賈氏窺簾韓掾少
Fufei left a pillow to the talent of the Prince of Wei.	宓妃留枕魏王才
Your spring heart mustn't vie in blooming with the flowers—	春心莫共花爭發
One inch of longing turns to one inch of ashes.[27]	一寸相思一寸灰

The atmospheric disturbance of the beginning creates a certain erotic unease, with the echoes of "cloud and rain" as well as the "eastern breeze," the seasonal bearer of spring longings, but the mention of thunder beyond the lotus pool introduces the theme of sexual frustration and of the longing of the woman abandoned by her lover. Commentators are agreed that this line alludes to a verse from Sima Xiangru's "Tall Gate Rhapsody," a poem meant to describe the anguish of Empress Chen 陳 after she was dismissed from Emperor Wu's 武 favor.

Thunder rumbles and echoes rise—	雷殷殷而響起兮
A sound like the noise of your carriage.[28]	聲象君之車音

Someone is waiting for a lover. Normally we might read this as a fictional female persona or a voyeur's description of a court lady. This is followed, however, by a very strange couplet, one of the strangest in Li's corpus. Again, commentators tell us that Tang dynasty folk wisdom held the toad to be a secretive animal, and thus it was used to decorate door locks, and that the pulleys on wells were sometimes carved in the shape of tigers. But even if we know this, we can only read the couplet subjectively or symbolically: things passing through holes or gaps, the holes often too small to admit a human presence, and secrecy. But does someone wish to turn into incense, so that she or he can pass through a locked door? Or does she or he really gain such surreptitious admittance? The well may have profound sexual overtones—liquids drawn from some unfathomable and gloomy depth. But I can think of other interesting associations with wells. In a famous poem on the creative act, the poet Jia Dao 賈島 (779–843), whose work Li surely knew, compared poetic effort to drawing water:

If I go a day without making a verse, 一日不作詩
The well waters of my mind dry up. 心源如廢井
My brush and inkstone? The well pulley. 筆硯爲轆轤
My chanting? The length of rope.[29] 吟詠作縻綆

If Li knew Jia's poem, a curious resonance is projected between the toils of sexual desire and the toils of poetic writing and perhaps the way in which writing can compensate or replace sexual activity, with brush and bucket images for male potency. There is also the story of the last emperor of the Chen 陳 dynasty, who, along with his favorite court ladies, hid in a well from the invading Sui 隋 troops.[30] There the well is a dark place of safety, a strange womb within which the male hides with his sexual partners.

These subjective visions are interrupted in the third couplet by human and divine presences, a pair of historical allusions. In the first, the daughter of Jia Chong 賈充, a minister of state, falls in love with her father's secretary, Han Shou 韓壽. They form an illicit liaison, which the father discovers when he detects the distinctive fragrance of the man on his daughter's pillow (the knowledge of this anecdote allows the reader to feel an echo of the incense that wafted through the lock in line 3).[31] The second allusion refers to the meeting between the poet Cao Zhi and the Goddess of the Luo River, described in Cao's famous rhapsody. As I noted in Chapter 2, most Tang readers of the poem assumed that it was an allegory of his love affair with his brother's wife, Empress Zhen. Thus, both couplets of the poem say the same thing: a woman takes the initiative in establishing an illicit relationship with a young and talented man. Li hints, but makes nothing clear, because the possibility of an illicit relationship compels that the poem that speaks of it speak in hints, lest the taboo be violated. Note also how the voyeurism of palace poetry is inverted in line 5, as Lady Jia, the active member of the partnership, initiates action by spying through a curtain.

The final couplet, of course, is an assertion, a comment on what has gone before: Is the strain of illicit passion ultimately worth it? No, says the poet, it merely causes early destruction. Again, are we to take this assertion in earnest? Read against the images of "Stars of last night," I believe we are meant to take it as the disappointed voice of a male figure who has found attempts to meet with his lover constantly frustrated. In such a reading, it is he who awaits with anticipation in the opening couplet and thus inverts the sexual role usually assigned to such a position; the second couplet comments on the difficulties of consummating such a relationship, while suggesting (possibly) creative compensation through writing; and the third is a retrospective glance to how the affair began. Li's worldly-wise gesture at the end is entirely romantic and sentimental, because it comes from the voice of someone committed to the "path of destruction." It is meant merely to underline the frustration of the main theme, not deny it or represent a movement away from the destiny it prescribes.

Something else is important about this closing, however. It suggests a sort of retreat from subjectivity on the poet's part—an attempt to frame and examine his subjectivity. It is a frequent gesture: note, for example, the end of "Ornamental Harp" ("Jin se" 錦瑟):

One might wait for these emotions 此情可待成追憶
to transform into memory:
Only at that very time 只是當時已惘然
we were already lost.[32]

Or, the closing of another "Untitled" poem:

It's frankly said that longing 直道相思了無益
has no use at all—
But no one can prevent its grief— 未妨惆悵是清狂
a sort of lucid madness.[33]

Just as the party in the first poem serves as a *point d'appui* from which to project oneself into the site of impossible desire (the site of the unspeakably subjective and personal), these closing couplets attempt to stand outside the incoherent subjectivity of earlier images, as if the mad narrator had a brief moment of sanity that allowed him to comment on himself. Again, the unobtainable is articulated from outside—boundaries are drawn around it. And they are drawn literally as well, because the act of writing the poem is conceived here as an act of withdrawal, of distancing, of examining sensations from a distance. I think here of Yu-kung Kao and Tsu-lin Mei's division of regulated verse into "metaphorical language" in the middle couplets and "analytical language" in the openings and closings;[34] for Li, this difference of languages within the poem creates subjective space by allowing for a final viewpoint *outside* subjectivity. Perhaps this is why we can also conceive these two poems as retrospective compositions written after the emotion: the first one from the office the morning following the party; the second following the failed tryst at which the woman did not show up. Writing is a sort of salvation—or at least a release of tension.

This is even more evident in another "Untitled" poem:

"I shall come" but empty words: you left, and cut off your traces.	來是空言去絕蹤
Moonlight slants atop the house, a watchman's bell at dawn.	月斜樓上五更鐘
Dreams present a distant parting, weeping cannot bring you.	夢爲遠別啼難喚
My letter completed in haste— not letting the ink grow thick.	書被催成墨未濃
Wax candle's shine half encages gold kingfisher plumes;	蠟照半籠金翡翠
Musk scent subtly pervades embroidered hibiscus bloom.	麝熏微度繡芙蓉
Master Liu long resented that Mount Peng was so far away.	劉郎已恨蓬山遠
Yet ten thousand peaks still block him beyond Mount Peng itself.[35]	更隔蓬山一萬重

Evidently, a more explicit retelling of "Whistling, the eastern breeze": the poet accuses the lover of breaking her word. After falling asleep while waiting, he is startled awake by a vision of her. He writes her a letter. Then, as James Liu puts it, "the speaker looks wistfully at the light of the candle, which half encompasses the kingfishers' feathers mixed with gold on the quilt, and smells the perfume, which has gradually penetrated the embroidered bed-curtain: all these elaborate preparations have been in vain!"[36] The final couplet is Li's typical distancing comment on his emotions: "Master Liu" is most likely an ironic name for Emperor Wu of the Han, who is noted here either because he sought immortality by reaching the Isles of the Immortals or because he sought the spirit of his dead mistress, Lady Li 李.[37] The poet argues that the distance from the beloved is even greater than the gap between mortals and immortals, living and dead—once again equating distance from the desired one in vast geographical terms. This is, then, one of the most explicit and comprehensible of the "Untitled" poems, although it is still enshrouded in mystery: Who is the woman? Why didn't she come? And why does the poet wait for her visit, rather than make the visit himself? These questions are part of the mystique of the texts.

If we return to the idea of writing as a place in which the poet can construct his own subjectivity in the form of a poet who has *not yet* begun to compose, then we may find ourselves drawn to the fourth line, the hasty writing of the letter. Since it is situated in a parallel couplet and juxtaposed with the dream, that raw, emotional cry of loss elicited from his nightmare of parting is inscribed on silk—the plea to the beloved becomes yet another cry that perhaps fails to summon. Once again the possibility of writing is occurring amid the strain of desire. Are we to take this poem that we hold in our hands as the letter that he writes? The entire poem consists of forms of sensual marking: the moonlight shining on the building, the cry of the dream, the words of a letter, the candlelight (an internal echo of the moonlight), the wafting incense, as insidious here as it is in the earlier poem, in which it slipped through locks. The only marking that truly concerns him, however, is absence: her own traces, marked through words that are actually "empty," have been severed, removed, cut off. The act of writing creates images of marking that surround the empty space where she is and where his desire lies.

A final technique used in these poems worth noting is the imaginative re-creation of the absent lover's presence in some other place. These are often marked through the use of the word *ying* 應, "ought to," to suggest that the poet is imagining what is likely, not what he can really see—as in this couplet:

In a dawn mirror, only merely grieve that cloudy tresses change to gray;	曉鏡但愁雲鬢改
At night, chant poems—ought to feel the moonlight chill.[38]	夜吟應覺月光寒

Li is vague throughout in his referent; the most likely reading is a male speaker imagining the beloved: since cloudy tresses tend to belong to women, he imagines the cycle of her day: gazing at herself in the morning, chanting poems at night. Perhaps that act of gazing in the mirror, as a way of constructing the self through the act of seeing, is an analogy to the act of writing a poem—once again, a form of expression that calls for distancing from the feeling self. Unlike the palace poet, who places the fictional woman within plain view, creates her as if she really existed, Li foregrounds the idea that this is what she "ought to" be doing or is most likely doing: from the lover's perspective, it is perhaps what he would like her to be doing. The act of self-gazing and of writing bears a close affinity to what he is doing himself in these poems.

In a similar poem, "Spring Rain 春雨," Li links the vision of the beloved with dreaming and through their mutual experience of late seasonal showers:

In sorrow I lie through the new spring season, in my white-lined robes.	悵臥新春白袷衣
At White Gate, lonely and desolate— my thoughts go mostly awry.	白門寥落意多違
Red houses through the rain: as I gaze toward them, chill.	紅樓隔雨相望冷
Beaded curtains billow by the lamp when I return alone.	珠箔飄燈獨自歸
On a distant road she ought to grieve as the spring turns to dusk;	遠路應悲春晼晚
Late at night, still I have vague dreams of her.	殘宵猶得夢依稀
Jade earrings and a silken letter— what means have I to send them?	玉璫緘札何由達
Through ten thousand miles of cloud-gauze a single goose flies.[39]	萬里雲羅一雁飛

The opening lines allow the poet to use the rain to express a typical series of blockages and semi-opacities: houses seen through rain; curtains billowing in the bedroom (where there is no female presence). Again we have letter writing, here combined with the idea of immeasurable distances. The gauze of the clouds becomes a barrier, like the rain and the curtain, another symbol of separation. It is entirely appropriate that the *gauze* clouds must be penetrated by a goose bearing a letter written on *silk*. But this letter, bearing this very poem, reaches her only in some future that he can only conceive of. It is in light of this poem that one of Li's simplest and most overquoted poems attains a certain added weight:

You asked for a return date—
but no date in sight. 君問歸期未有期
Night rain on Ba mountain
fills the autumn pools. 巴山夜雨漲秋池
When will we trim the lamp together
by the western window 何當共剪西窗燭
And speak of Ba Mountain,
and the season of night rain?[40] 卻話巴山夜雨時

Behind the sentimentality of this little verse lies a striking image of distancing, the desire to escape from pain by projecting into the future when it can be discussed from another point of view, a point at which desire has been fulfilled and is no longer maddening. What fails to be conveyed in English is the rhythm of repetition: *Ba shan ye yu* 巴山夜雨 (Ba Mountain night rain) are the first four characters of line 2, but in the last line, they are syllables three through six and thus falling syncopatedly on the ear as they straddle the caesura metrically required between the fourth and fifth syllables. In this repetition, the caesura splits the agonizing monotony of the rain into two parts, a past and a present; and the line is completed with the single word *shi* 時, "season" or "time," the way of marking this event as finally, definitively, and safely in the past. Li attempts to use writing once again to create a space from which to view, and hence objectify and separate himself from, painful subjectivity. But as with the earlier texts, the very techniques that allow him to separate himself from this private pain are the very ones that give it form and existence. We can conceive of Li's pain only through his attempt to put it behind him.[41]

Li attempts to extract the theme of desire from the common coin of social erotic poetry. Such a gesture is a problematic one, since it demands that the poet articulate things for which there is no extant language. As it turns out, he creates this privacy by refusing to articulate it; rather, he articulates the borders and boundaries *around* it. He makes his own pain and desire evident, not through any attempt to recapture it in writing but through the constant distancing of it from the perspective of writing itself: through an intellectual contemplation of one's own emotions, through the acknowledgment of sublimation through the act of poetry. As I have said before, these gestures seem at first to console the poet: poetry as a form of therapy (a not unusual notion nowadays). But in fact they gesture toward a space that can be acknowledged, if not fully comprehended, by his reader: a space in which some ideal poet's self might lie, stripped of the value with which society, politics, and family have adorned him. It is there that he floats, within the iron net that still lost at sea, we can never quite locate.

A Sea Change

Li's creation of a new space for private eroticism might have marked a new tradition, a new way of writing about the erotic and about desire. However, his influence became felt in a different way, through its socialization within the public song lyric, or *ci* 詞. To a very large extent, *ci* remained a social performance art; it borrowed his language of eroticized interior space while denying his anguished rhetoric of taboo.[42] Yet the erotic world expressed in the song lyric was radically new in some definitive and significant ways—ways that are not unrelated to Li Shangyin's own experimental representations of desire. A brief consideration here may begin to suggest how they represent a genuinely new world that came into existence after the Tang.

Most readers familiar with song lyrics and with the traditional and modern scholarship surrounding them are aware of a major critical issue in their history: Do song lyrics continue the "self-expressive" poetics of *shi* poetry, or does the fact that they grow out of social entertainment verse suggest a new type of poetic art, one more concerned with connoisseurship, craftsmanship, and self-conscious poetic "making"?[43] I think it is fairly clear that a certain strong tendency in song writing does indeed work toward such a distanced, connoisseurlike approach to composition, even though the mainstream expectations of the literary tradition attempted to force it back into the mode of autobiography and self-expression. Most assuredly, at its earliest stage, when its practitioners undoubtedly made little or no distinction between composing *ci* and composing popular musical *shi* verse (and this stage is best reflected in the work of the song lyricists Wen Tingyun 溫庭筠, Wei Zhuang 韋莊 [ca. 836–910], and their contemporaries), erotic *ci* were purely social entertainments, sophisticated and well crafted as they may be. The question of self-expressivity does not really emerge (at least from a later perspective) until the work of Li Yu 李煜 (937–78), the last emperor of the Southern Tang, who lived his final years in exile at the Song court. Most modern readers take many of his songs as autobiographical laments on the author's loss of his kingdom. Daniel Bryant has expressed skepticism about such readings, and although some of the poems do seem to fit in with an autobiographical reading, he is right to emphasize that such poems are relatively few, and that all of Li's verse continues to work within heavily conventionalized themes and language.[44] Li Yu's poems may be party songs with a difference, but they are still party songs.

Of much greater interest for my purposes is the work of Liu Yong 柳永 (987–1053), a minor official who revolutionized the genre through his widespread use of the hitherto underutilized *man ci* 慢詞 or "long song" form, as well as his heavy dependence on Song dynasty colloquialisms and "vulgar" diction. In Liu's work, we see the most radical break with the decorum of *shi* diction and rhythm, as well as the genre's approach to the type of relaxed, monologic speech that characterizes it at its best. Liu is also distinctive in his emphasis on male desire: the speaker in many of his poems is the lover of courtesans and singing-girls, who writes openly of his jealousies, longings, and even his lusts:

Phoenix pillow, simurgh drapes—	鳳枕鸞帷
for several years	二三載
I knew them as a fish knows water.	如魚似水相知
On fine days, in lovely weather,	良天好景
so much in love, cherishing her deeply,	深憐多愛
I would try to fulfill her every desire.	無非盡意依隨
How could she	奈何伊
have become so willful?	恣煞些兒
For nothing at all she'd torment me	無事孜煎
a thousand, ten thousand times—	萬回千度
yet how could I bear to leave her?	怎忍分離
And now, as I drift further away,	而今漸行漸遠
I realize (though I regret it) I can never go back.	漸覺雖悔難追
Useless to send any news to her;	漫寄消寄息
in the end, what could it do?	終久奚爲
And when I think to hook up with her again	也擬重論繾綣
I'm forced to ponder awhile:	爭奈翻覆思維
even if we met again	縱再會
I fear that our love	只恐恩情
would never come up to the past.[45]	難似當時

There is nothing really like this in the erotic tradition through the Tang (even among the vernacular Dunhuang songs); an interior monologue of male desire, in which the narrator not only admits his own feelings but obsesses about them. Even more extraordinary are the occasional songs in which the narrator explicitly becomes a song writer, and the "story" of the poem becomes the composition of the very song we hear:

By error I wandered into Pingkang Ward.	誤入平康小巷
Deep within the painted eaves	畫檐深處
someone raised a beaded curtain slightly.	珠箔微褰
In a throng of silks and satins	羅綺叢中
I recognized by chance an old flame.	偶認舊識嬋娟
Her kingfisher brows revealed	翠眉開
a charming stretch of distant hills;	嬌橫遠岫
her dark tresses hanging down	綠鬢鬤
were deeply tinted with spring mist.	濃染春煙
I remembered then her attraction:	憶情牽
by painted walls, once in the past	粉牆曾恁
she gazed for three years at Song Yu.	窺宋三年
As I lingered there	遷延
on the coral mat,	珊瑚筵上
she took up a brush of horn,	親持犀管
folded a piece of scented paper,	旋疊香箋
and begged for a new song.	要索新詞
Bewitching me, full of smiles, she stood before the cups.	殢人含笑立尊前
Performing a new tune,	按新聲
her pearly voice slowly steadied	珠喉漸穩
as she thought of our old feelings.	想舊意
Her lovely features grew more charming	波臉增妍
as she tried to make me stay.	苦留連
Phoenix coverlet, mandarin duck pillows—	鳳衾鴛枕
who could turn their back on such a moment?[46]	忍負良天

This extraordinary lyric writes explicitly the implicit circulation of desire present in the Tang anecdote about Wang Zhihuan and his fellow poets; here the beautiful artiste commissions a song from the talented poet and then turns and uses it as an instrument of seduction on its author. Such reflexivity might appear onanistic. Within the conventions of the erotic tradition, however, I would suggest this fantasy embodies instead a representation of desire that has been unarticulated up until this moment: the recognition of, and the self-conscious male participation in, the artistic construction of his fantasies. The woman's silence, her need of the male other to articulate her desires for him, is still present, as it has been since the Nineteen Old Poems. But here she compels a voice, demands a song: female acknowledgment of male creativity results in practical action. It is here the demand upon the male or, rather, his recognition of the demand, that breaks Chinese eroticism free from its attempts at the sort of aesthetic containment characteristic of palace poetry and its descendants. Desire is no longer yoked to fantasies of self-mastery.

But there are other consequences here. Poems such as this one almost inevitably produced autobiographical assumptions in later readers—the poet in the poem must be Liu Yong. The result was the creation of Liu Yong as a sort of folk hero in later literature, the quintessential roué and sympathetic frequenter of brothels, the patron saint of singing girls. Of course, Liu Yong's poems are so explicitly tied to the world of entertainment and the demimonde that even if they were autobiographical in origin, they would inevitably result not in a self-confessional bleeding from the poet but rather in the creation of a persona, the song poet in his environment. Liu Yong created the most enduring image of the male lover as type or as fictional character; he did not create a new poetry of male desire, as Li Shangyin did.

However, the two poets do have one point in common. In writing about desire, they write of interiority: of the bedroom and of the mind. Liu Yong may be the first great Chinese poet of the dramatic monologue, and it is easy to see that the direction he took the song lyric would bear great fruit in the rise of drama. But both poets are compelled to situate their verse within the realm of the internal (*nei*), the world that is often the female world. As we have seen, in Li's verse gender markings are no longer connected to the speaking subject or the object of contemplation; the ways in which we read conventional palace poetry (where women are the languishing inhabitants of solitary rooms) no longer quite apply. Up until Li's time, such gender markings acted as indications of conventionality and of reading expectations. Men *observed* women or *became* women; the choice was often left unclear, but the nature of the choices was unmistakable. Such conventions preserved as well the political aspects to gendered language, which we have already discussed at great length: the harem lady as spurned minister; the jealous wife as competitive courtier. The moment Li Shangyin walks into the bedroom *as himself*, or even as a *man*, the tradition collapses. The female world, the inner sphere, now admits the man as a man, and he finds his own place there, as active lover of the female. That he experiences an interior world of subjective emotion there—including desire, jealousy, longing—means that he no longer speaks as a woman. He has become the desiring man.

As we have seen, however, Li could not fully articulate such a male invasion; he could only articulate the space around it, concealing its interiority with images of blockage, madness, distancing, irony. Liu Yong articulates the space by opening the door wide and striding into it in the company of his audience, both male and female. He (and the erotic song lyric in general) creates the inner space as a new social space—boudoir not as imperial court or bureaucracy but simply as the boudoir.

It should hardly come as a surprise that this should occur in the Song dynasty, with the full-fledged development of a literati class independent of explicit aristocracy, as well as a literati class that was to become increasingly independent intellectually and culturally from the court. Obviously, these developments were occurring during Li Shang-yin's life and before. But their impact on eroticism and the subsequent development of a language of "private" romance really comes into its own in the Song dynasty song lyric. By "private," I mean a realm of action outside politics and sanctioned social interaction, a realm that began to develop in the courtesan quarters in Chang-an but could not really work there because of the links between courtesan culture and examination culture. This privacy comes at a cost, however: once it obtains a certain independence as its own world, it largely ceases to resonate with the political world, as it once did. Eroticism becomes increasingly depoliticized—or at least its politicization must be reformulated in the world of the social and not as a manifestation of elite narratives of failure or success.[47] The move accompanies the identity of the *ci* itself as a distinctive genre, separate from the more "serious" "self-expressive" *shi* form, and results in the withdrawal of erotic themes from *shi* and their ghettoization within the new genre.

APPENDIX

APPENDIX

A Dalliance in the Immortals' Den

The following is meant to improve on the only published English-language translation available of the You xianku, *that of Howard S. Levy (1965). Levy did not have a decent modern commentary to work with, and he was sometimes unfamiliar with Tang vernacular expressions. He also took some liberties with the poems. My rendering here is not the book-long study a scholarly translation would demand. Those concerned about textual issues, controversies of interpretation, and detailed discussion of allusions should consult Yagisawa Hajime's translation and commentary. I follow Yagisawa's text and rely heavily on his notes. For interpretation of Tang vernacular, I have benefited from Arthur Waley, "Colloquial in the* Yu-Hsien K'u"; *those interested should also consult* bianwen *scholarship.*

My notes are brief comments to explain allusions and puns (notes follow at the end of each section). In some places my translation is still tentative.

I

Stoneheap Mountain 積石山 is located southwest of Metal City 金城; the Yellow River flows by its base. This is the mountain the *Classic of Documents* has in mind when it says, "He channeled the River from Stoneheap, and then he arrived at Dragon Gate 龍門."[1]

I was traveling by way of Yanyang 汧陽 District on an imperial mission to River-Source 河源;[2] and so I lamented the contrariety of my fate and sighed for the home that was so far away. I followed in the ancient path of Zhang Qian 張騫, through a hundred thousand leagues of billowing waves; and in the traces two thousand years old left by Lord Yu 禹 over slopes and hills. Deep ravines encircled the land, boring and passing through the shapes of cliffs and bluffs; high peaks spread over the sky, as if a knife had cut them into forms of crests and ridges. The thin delicacy of the mists and sunset clouds clearly exposed the streams and rocks. This was truly a numinous marvel of Heaven, cut off mysteriously from the world of men. Never had eye beheld it, nor ear heard of it.

The day grew late, and my road was long; both horse and rider had grown fatigued. I then came upon a certain place with extraordinarily steep cliffs. Above me rose blue-green walls a myriad feet high, while at my feet lay jade pools a thousand fathoms deep. The elders here have a tradition about this place and call it the "Den of Immortals": "Human feet rarely reach it, and only the paths of birds can get through. Often fragrant fruit, branches of jade, divine garments, and monks' staffs and bowls come floating out, although no one knows whence they come."

I then put myself in a serious frame of mind and observed three days of ritual fasting. I clambered up slender vines and took a light skiff against the current. I felt as if my body was in flight, my soul lost in a dream. Before long I came upon a cliff of pines and cypresses and a spring of peach blossoms.[3] A fragrant breeze brushed the ground, and a bright radiance pervaded the sky.

I noticed a girl washing clothes at the edge of the water.

"I have heard that this is the Den of Immortals, so I have come to pay my respects. However, the hills and streams have blocked my way, and I have grown rather fatigued. I wonder if I might find lodging with you and rest for a bit. If I receive your kindness and consideration, I would count myself fortunate indeed."

"Our house is humble, and our food simple and coarse. I fear it will not be adequate for you, although we will begrudge you nothing."

"I am a traveler, and so I often must put up with humble or scant accommodations. If I only find shelter from the wind and dust, I will be fortunate enough."

She then left me at a thatched hut by the side of the main gate. After some time she emerged again.

"Who lives here?" I asked.

"This is the home of a Lady Cui 崔."

"And who is she?"

"She is the descendant of the Prince of Boling 博陵 and a member of the venerable clan of the Duke of Qinghe 清河. She has the beauty of her uncle, for she happens to be Pan Yue's 潘岳 niece; she has the gracious manner of her older brother, for she is the little sister of Cui Yan 崔琰.[4] Her fresh, blooming features are seductive, she is without peer in Heaven; her jade form is sinuous and slender, with few to match her on earth. A radiant face so slender, one would fear bruising it too easily; thin, slender waist, in danger of snapping with an embrace. When Miss Han 韓娥 and Song Yu 宋玉 saw her, they were aggrieved; when Crimson Tree 絳樹 and Emerald Harp 青琴 encountered her, they were greatly shamed.[5] One cannot hastily find comparisons for her thousand charms and hundred graces, nor can her fragile form and light body be captured fully in words."

Notes

1. A reference to the irrigation work carried out by the sage-ruler Yu. Stoneheap is in northwest Gansu.

2. The text has Yan[yang] 汧陽 and Long[zhou] 隴州. Yanyang is the district center for Longzhou, on the Yan River (a tributary of the Wei), ca. 140 km west of Chang'an. River-Source (Heyuan) is near Bohai in Xinjiang; here it evokes the semimythical travels of the Han envoy Zhang Qian in search of the Yellow River's origins.

3. Although some scholars believe that Tao Qian's 陶潛 short narrative "Peach Blossom Stream" is alluded to here, it is more likely that Zhang means to invoke the adventures of Liu Chen 劉晨 and Ruan Zhao 阮肇 and their encounter with goddesses (see discussion in Chapter 6).

4. Pan Yue and Cui Yan were paragons of male beauty for the Western Jin and the Wei, respectively.

5. Miss Han: famous entertainer of the Warring States period; when she sang, dust on the eaves did not settle again for three days (she is alluded to again in section 13). For Song Yu, see Chapter 2. Crimson Tree: talented singer and dancer of the Three Kingdoms period; Emerald Harp: a female deity.

2

Right then I heard the sound of a zither being tuned within the house. I then chanted the following verses:

Well aware of her graceful figure,	自隱多姿則
She cheats him by sleeping alone.	欺他獨自眠
Often she puts forth a slender hand,	故故將纖手
Toys for a time with the slender strings.	時時弄小絃
Ear heeds—I lose my breath;	耳聞猶氣絕
Eye beholds—how I do love!	眼見若爲憐
If you are so unwilling,	從渠痛不肯

I'll seek Heaven nowhere else. 人更別求天

Soon she sent out her maid, Cassia Heart 桂心, with the following verses in reply:

This face is no one else's;	面非他舍面
This heart is surely my own.	心是自家心
Why do you, engaged in your lord's service,	何處關天事
Come here in vain pursuit?	辛苦漫追尋

When I finished reading, I raised my head toward the doorway. I suddenly caught sight of Shiniang's 十娘 face in profile. I then chanted:

She suppresses a smile, conceals a remaining dimple;	斂笑偷殘靨
Bashfully she shows but half her lips.	含羞露半脣
I can withstand but one of her brows—	一眉猶叵耐
Both eyes would surely wound me!	雙眼定傷人

Again she sent out Cassia Heart with a reply:

It must be another's beauty—	好是他家好
I am not the one to attract your desires.	人非著意人
Why must you so waste your time in teasing?	何須漫相弄
How greatly you squander your wit!	幾許費精神

3

The night grew late. I brooded, sleepless, paced back and forth; there was no way to convey my feelings. If she truly intended to come to me, how could she not have responded? I told everything I felt in a letter to her:

> From my youth I have delighted in women and song; cherishing the fine season of my youth, I often involved myself in romantic encounters, as I rambled about the world. As I strummed a crane harp in Shu's 蜀 capital, I saw my fill of Zhuo Wenjun 卓文君; as I piped on phoenix flutes on Qin 秦 tower, I beheld Nongyu 弄玉 again and again.[1] Although women often presented me with orchids and untied their girdle-pendants for me, I never fully engaged my affections.[2] How could marriage rite and bridal bed bring me contentment?
>
> In former days when I had a partner to sleep with, I always resented how short the night was; tonight as I lie alone, I rage at the length of the

hours. Now, on this second mission bid by the same Lord of Heaven, from afar I caught scent of your fragrance and was wounded by thought of Han Shou's 韓壽 heart.[3] As I neared, I heard your harp, as if I were in the presence of Wenjun 文君 herself.

Earlier I heard Cassia Heart tell me that you were peerless in the world, alone in the human realm; that the fragile willows gently swaying were bound together to form your waist, gentle gleaming ripples flowed from the corners of your eyes. As soon as your cheeks were displayed, I would agree that the earth was barren of flowers, and that when your brows were exposed, I would soon think the sky had lost its moon. You would make Xi Shi 西施 cover her face and burn her cosmetics a hundred times over; the southern beauty to give her mirror a thousand blows in her sorrow.[4] The Luo 洛 River's goddess of gusting snow is fit only to fold your robes; the lady of Wu 巫 gorge with her divine clouds would hardly dare to look after your slippers.[5] I am angered by Qiu Hu's 秋胡 blindness, for he threw away his money;[6] and I take heed of Jiaofu's 交甫 madness, who repaid white jade in vain.

I now find myself taking my ease in some superior realm, lodged in your pavilion of leisure. I suddenly encountered an immortal and am overwhelmed with confusion. As the lotus grows in the bottom of the ravine, its seeds of love are surely deep; as the catalpa grows on the mountain top, its fruit of longing spreads farther day by day. Although I have not swallowed charcoal, I feel my vitals burning within; although I don't recall consuming a knife, my belly is pierced and sliced inside. The bright and heartless moon deliberately shines in at my window; the meddlesome spring wind continually shakes my curtains. When a melancholy person beholds these things, how can he bear it? My heart is vainly suspended, about to shatter. You must save a life on the edge of dissolution! If we had never met, then matters would have progressed as before—but since we have encountered one another by chance, it has brought my brain to a boil. I dare tell you of my innermost feelings, hoping with good fortune to enlighten you. If I can behold your lovely features even for a moment, how could I hope for any later good fortune?

After she received my letter, Shiniang grew stern and said to Cassia Heart, "Formerly we were just flirting. Now it seems he wants to force his attentions upon me."

I then wrote a poem as well. It read:

This morn I suddenly see
your graceful form; 今朝忽見渠姿首

Unawares I find you've snared
my heart for sure. 不覺慇懃著心口

I'm moved time after time
to send her my requests, 令人頻作許叮嚀

But she is far too cruel,
so hard to entreat. 渠家太劇難求守

I sit in repose, then my heart is startled; 端坐剩心驚

Grief comes, I'm even more ill at ease. 愁來益不平

When I see her, I won't need
to die at her sight; 看時未必相看死

But when deprived as I am now,
it's just too bad for me. 難時那許太難生

Brooding I sit in my lonely room 沈吟坐幽室

My longing turns to illness; 相思轉成疾

I resent that our paths are far apart 自恨往還疏

Who'd be willing to deliver our love secrets? 誰肯交遊密

Night after night I vainly know
my heart has been deceived; 夜夜空知心失眼

Dawn after dawn no way to join
into a lacquer-and-glue embrace. 朝朝無便投膠漆

In her garden, blossoms bloom;
they will not shun my path; 園裏花開不避人

But within her boudoir her features
fly from me in shame. 閨中面子翻羞出

Now a single step away,
but blocked by Heaven's River;[7] 如今寸步阻天津

How may I place my affections,
seek out my new attachment? 何處留情更覓新

Do not say you'll always have
a face worth a thousand in gold— 莫言長有千金面

In the end you'll become no more
than a handful of dust. 終歸變作一抄塵

While alive and in the sun
only take your pleasure! 生前有日但爲樂

After death no spring season returns 死後無春更著人

to call you back again.
If only you make dalliance 祇可倡佯一生意
your goal throughout this life,
No need to wastefully betray 何須負持百年身
your threescore years and ten!

Notes

1. Two of the most frequently recurring allusions in "romantic" literature; the romance between the poet Sima Xiangru 司馬相如 and the recently widowed Zhuo Wenjun and that between the flutist Xiaoshi 蕭史 and the daughter of the King of Qin.

2. Untying girdle pendants: an allusion to Zheng 鄭 Jiaofu, who was presented with pearl girdle-pendants by two goddesses of the Han River with whom he was flirting.

3. Han Shou was famous for his distinctive perfume, which ultimately exposed the affair he was having with the daughter of his superior. The daughter and he eventually married.

4. Probably a reference to the southern beauty evoked in Cao Zhi's 曹植 "Miscellaneous Poem" ("Zashi" 雜詩), "In the south there is a lovely woman."

5. The Luo goddess is immortalized in Cao Zhi's rhapsody; the Wu goddess's tryst with the King of Chu is told in the *Gaotang fu* (see Chapter 2).

6. Qiu Hu, after marrying his wife, went away for several years. After his return, he tried to seduce a woman he saw harvesting mulberry leaves, offering her money. Later he discovered that she was his wife. She killed herself from the shame.

7. An allusion to the Milky Way, which separates the Herd Boy and the Weaver Girl.

4

After a while I dropped off to sleep, and I dreamt that I saw her. When I awoke, I reached for her, but my hands were empty. I was grieved within my heart, but what more could I say? I then chanted:

In dream I thought her real; 夢中疑是實
Awake, so soon she's false. 覺後忽非眞
I truly know my heart will break: 誠知腸欲斷
Is that wretched woman taunting me? 窮鬼故調人

When Shiniang saw this poem, she wouldn't even read it, but threw it in the fire. I then chanted:

Not yet may I catch her with my verse! 未必由詩得
I made the verse to display my love 將詩故表憐
But I hear she's tossed it in the flames— 聞渠擲入火
It must be her desire's caught fire. 定是欲相燃

When Shiniang read this poem, she let out a gasp and arose. She removed her mirror from its case, selected a robe from her chest. In glittering apparel and lovely makeup, she straightened her slippers on the stairs. I made another poem:

Fragrant scent collects everywhere, 薰香四面合
Radiant beauty spreads about. 光色兩邊披
Brocade curtain suddenly pushed aside, 錦障劃然卷
Gauze drapes pulled half back. 羅帷垂半欹
Red of cheek blends with emerald kohl, 紅顏雜綠黛
Suitable in every way— 無處不相宜
Glamour hovers mid paint and powder, 艷色浮妝粉
Fragrance tangles in the rouge of her lips. 含香亂口脂
Her cicada-black hair cheats the cicada— 鬢欺蟬鬢非成鬢
  more perfect than its model;
Brows scoff at others' moth-brows, 眉笑蛾眉不是眉
  to compare, no brows at all.
I gaze on sensuous, graceful motion, 見許實娉婷
Throughout, so light and airy: 何處不輕盈
Adorable, her face amid her charms; 可憐嬌裏面
Lovable, the sound of her speech. 可愛語中聲
Sinuously graceful is her waist, 婀娜腰支細細許
  so delicate withal;
Slender and narrow her fine eyes, 䁠䀽眼子長長馨
  so long and sweet.
A skilled craftsman of long ago 巧兒舊來鐫未得
  could never fashion her aright.
Gifted painters heretofore 畫匠迎生摸不成
  could not model her form.
I see her now, thought yet unacquainted; 相看未相識
She would topple a city, topple a state. 傾城復傾國
Her cape catches the breeze, 迎風帔子鬱金香
  spreads saffron crocus scent;
Her skirt shines in the sun, 照日裙裾石榴色
  the color of pomegranate.
Coral hangs upon her mouth— 口上珊瑚耐拾取
  no way to collect it;

Hibiscus lie within her cheeks — 頰裏芙蓉堪摘得
no way to pluck it.
To hear her name, my vitals within — 聞名腸肚已猖狂
are driven to madness;
To see her face, my very spirit — 見面精神更迷惑
loses itself again.
My heart will soon break apart! — 心肝恰欲摧
I leap about, cannot grow still. — 踊躍不能裁
Slowly she comes—with each step — 徐行步步香風散
a fragrant breeze disperses.
About to speak, from time to time — 欲語時時媚子開
her hair ornaments glitter.
A dimple is the single star — 靨疑織女留星去
the Weaving Maid left behind;
Yellow-brow powder is a moon — 黃似恒娥送月來
the moon-goddess sent to earth.
Full of charm, shy and graceful — 含嬌窈窕迎前出
you come out to greet me;
Holding back a smile, still bashful, — 忍笑嫈嫇返卻迴
you turn away again.

I then stopped her. "If you have good intentions, then why do you always hide away?" After that she tenderly turned her face to me and came forward bewitchingly. She folded her hands and bowed to me several times. I bowed myself and extended full courtesies. "You were praised so highly before; one might have said that it was mere empty exaggeration. Who knew that when I came face to face with you, you'd indeed be an Immortal? This is the 'abode of Immortals'!" Shiniang replied, "When I read your verses, I still thought you might be shallow and vulgar.[1] Now I encounter your jade features, I find them superior to your writings. You are truly an 'Abode of Composition'!"

"Might I ask your family name and where your family comes from? Where is your husband?"

"I am the descendant of Duke Cui 崔 of Qinghe 清河; I was married to an elder son of the Yang 楊 family of Hongnong 弘農. After celebrating the great rite of marriage, I accompanied his father to Hexi 河西. At the time certain treacherous rebels arose in Shu 蜀 and infiltrated the border regions several times. My older brother and husband threw aside their writing brushes to pursue the barbarians but perished on the field of battle; their lonely ghosts could not return. So I was a widow at sixteen. The wife of my brother, then eighteen, swore she would never remarry. My brother was the fifth son of Duke Cui of Qinghe, and my sister-in-law is the third daughter of the head of the Taiyuan Wangs 太原王. We have lived in seclusion here for several years. Our dwelling has become overgrown and our manner of living is simple and poor.

"But I still do not know where my honored guest hails from?"

I solemnly replied, "My clan hails from Nanyang 南陽 and from Xi'e 西鄂. We acquired the mysterious art of the Yellow Stone and controlled the flood waters of Boshui 白水. In the Han 漢, seven generations of us held the ermine tail and cicada badge. We were five times ministers in the ancient state of Han 韓. When bells were rung, dishes were brought in; and generation after generation donned official caps. We stationed spearmen at our lofty gates and pursued the rites and ceremonial music. I inherited from my father but had no talent, and so I squandered the family fortune. I am the descendant of the Governor of Qingzhou 青州 the Marquis of Bowang 博望, and also of the Guangwu 廣武 General the Marquis of Julu 鉅鹿. But I could not avoid the vulgar world and so fell into a lowly position.[2] Yet I did not choose retreat but instead wandered about between the roc and the sparrow. Neither official nor commoner, I am not confined to the petty realms of right and wrong. Now I have come here on my way to my post and find myself unexpectedly imposing upon you. I truly am grateful for the honor."

"What post have you been granted?"

"Although we have the good fortune to live in a time of peace, I shamefully occupy a lowly position. Formerly, I was selected by my district to take the examinations, in which I took first place. Later I was selected for special imperial examination and again placed high. I was then appointed the supervisor of a small district within the pass; now I have been sent to fill a position in the army administration at River-Source. I have repeatedly accepted commands from on high, and my only concern is to repay my lord's favor. I hurry about in my lowly jobs, not finding any respite."

"If you had not been on an official mission, how could you ever have managed to visit us?"

"Before we knew each other, I failed to extend courtesies to you. After today I dare not be remiss."

Notes

1. Shiniang's odd rudeness here seems to have provoked a textual variant, where Shiniang expresses admiration for his poems.

2. I will not explicate these boasts in detail. Zhang is rehearsing Zhang clan history, particularly Zhang family service in the state of Han 韓 in the Warring States era and its connection to the founding Han 漢 hero Zhang Liang 張良 (who was adept in sorcery). Also mentioned are Zhang Qian and the Western Jin literatus Zhang Hua 張華.

5

Shiniang then turned and called to Cassia Heart. "Straighten the Central Hall and make the guest comfortable there."

I drew back and apologized. "I, a distant traveler, am utterly too insignificant to deserve this great a favor. I haven't the talent of a Jia Yi 賈宜;[1] how could I dare ascend your hall?"

"Just now, when I heard of you, I took you to be an ordinary guest, so I was clumsy in extending courtesies. Now I am deeply ashamed. I think the most proper thing to do is welcome you and bring you inside. This place is poor and humble and cannot shelter you from the wind and the dust. But if you didn't think it proper to refuse your small room, why should you turn aside from ascending the main hall?" She then led me inside.

There were golden terraces and silver towers that hid the sun and pierced the clouds. Some were like Copperbird 銅雀 newly opened; then like Numinous Light 靈光 just constructed.[2] Prunus rafters and cassia beams, as if a long rainbow drinking up the streams; curving roofs and carved tiles like lovely phoenixes brushing aside the sky. There were floating pillars of crystal glittering filled with stars; mica windows flashing as they filled with sunlight. Long verandas on all sides vied in displaying their beams of tortoiseshell; high chambers atop triple tiers all constructed of amber tiles. Pale silver made the walls, more shining than fish scales; green jade made the stairs, in uneven ranks like swan's teeth. As I entered the lofty, high-ceilinged chamber, my heart leapt with every step. When I saw the bright spacious gate courtyard, my eyes were dazzled with every glance.

"May I lead you up the stairs?"

"There are proper degrees between guest and host!"

"But there are also high and low degrees between men and women."

I paused, holding back. "I have offended in not making the acquaintance of your sister-in-law."

"She would have come of her own accord. But it is quite proper of you to desire her acquaintance." She then sent Cassia Heart to go and invite her sister-in-law.

Shiniang then engaged me in conversation. After a while, her sister-in-law Wusao 五嫂 entered. Her gauze and silks fluttered and fell, her reds and blues shone and gleamed. Before her skirt a musk scent pervaded; behind her hairbun a dragon coiled. Strings of pearls were sewn on her turquoise blouse, and her crimson slippers were edged with gold foil. I chanted:

Marvelous and strange, lovely and elegant, 奇異妍雅
Features outstanding, novel and fresh. 貌特驚新
A moon emerges from her brow 眉間月出疑爭夜
  to vie with the one of night;
On her cheeks the flowers bloom 頰上花開似鬥春
  contending with the spring.
Slender waist turns, so charmingly; 細腰偏愛轉
Smiling face suitable, even in frowns. 笑臉特宜嚬
Truly a marvel, a rarity 眞成物外奇希物
  beyond all worldly things;
Indeed, a person severed 實是人閒斷絕人
  from the human realm.
So naturally she comports herself— 自然能舉止
Lovable, incomparable. 可念無比方
Her talent will revive the duke's son, 能令公子百迴生
  restored to life a hundred times!
Her charm will cause the royal prince 巧使王孫千遍死
  to die a thousand deaths.
Black clouds trimmed to form side curls, 黑雲裁兩鬢
White snow divided into teeth. 白雪分雙齒
A unicorn foal is figured 織成錦袖麒麟兒
  on her brocaded sleeves;
A parrot chick is embroidered 刺繡裙腰鸚鵡子
  on the waistband of her skirt.
Whatever she touches brings forth passion, 觸處盡關情
Always at her most splendid. 何曾有不佳
Her ways too fine and marvelous, 機關太雅妙
Her gait so charming and graceful. 行步絕娃㛹

Every servant at her side displays red gauze stockings;	傍人一一丹羅襪
Maids at attendance three by three have slippers with green threads.	侍婢三三綠線鞋
A yellow dragon leaps upon a yellow golden bracelet;	黃龍透入黃金釧
A white swallow comes flying on her white jade hairpin.	白燕飛來白玉釵

After we had met, Wusao said, "Your Honor has climbed mountains and waded streams; you must be utterly fatigued from your travels. Has not your arrival here brought you to exhaustion?"

"How dare I refuse to do my utmost in my prince's service?"

Wusao turned to Shiniang and laughed. "This morning I heard the chattering magpie—and truly a fine guest *has* arrived!"

"Last night my eyelids fluttered," I said. "And this morning I see a fine person indeed."[3]

Notes

1. The youthful genius and writer of the Western Han.

2. Both excessively splendid palaces. Copperbird Terrace was constructed by Cao Cao; Numinous Light Palace, by a son of Emperor Jing 景 of the Han.

3. These are Tang folk beliefs.

6

They then led me to the upper part of the hall. There pearls and jade startled my heart, gold and silver blinded the eyes. Variegated dragon-whisker mats, brocade-embroidered green-bordered carpets, eight-foot-long ivory couches, rush-stuffed cushions of purple damask, mother-of-pearl and other treasures; everywhere shone rare udumbara flowers, red agate and pearls, and cords threaded with crystal beads. Patterned cypress-wood couches, decorated with panther-heads; lamps with orchid wicks, burning with fish-brain oil. Pipes and strings sang distant and clear; the musicians lined either side of the northern door. Cups and plates were piled up one on another, set out before seats under the southern windows.[1]

All deferred to one another; no one was willing to sit first.

I said, "Shiniang is the host, and I am the guest. May the host be seated first."

Wusao was clever with her tongue. She covered her mouth and laughed. "Since Shiniang is the 'mistress of the house,' you should be the master!"

"How could I dare to assume such a thing?" I replied.

Shiniang said, "My sister enjoys having her joke. Do not be flustered by her remarks."

I replied, "Since I can't avoid it, I'll take my place."

Wusao laughed. "I rather suspect you couldn't refuse it." Everyone laughed.

We then took our seats. They called for Little Incense to fetch the ale.

Before long, she brought in a huge bowl that could hold three pints, ornamented with gold knobs and brass rings. Gold plates and silver cups were fashioned of river-shell and sea-conch, with bamboo stems ornamented with slender-eye patterns, cups of carved tree-knots, and bottles narrow as scorpion mouths. There were curving pools of ale, tenfold ranks of drinking vessels, goblets fashioned of musk ox and rhinoceros horn, set up impressively before the seats. Ladles were goose-necked and duck-headed in shape, floating about on the surface of the ale.

Shiniang sent the maidservant Asarum 細辛 to pour, but none of us were willing to take the first draught. Wusao said to Shiniang, "Master Zhang's nothing but a lowly guest. He would never be willing to drink before us. You take up the cup first." Shiniang cast a side glance at me, then said angrily, "His Lordship has just arrived here. Don't keep on teasing him!" Wusao said, "Just take up the cup, girl, and don't be angry! I won't dare say anything more."

When the cup came to me, I didn't drain it. Wusao said, "Why didn't you finish it?" I replied, "I don't drink much as a rule. I'm afraid of getting inebriated." Wusao scolded me. "How can we overlook this? A husband is the wife's lapdog. She could beat you to death, and you'd have no right to complain. Drink up now, all of it, and don't waste our time with useless chatter!"

Shiniang said to Wusao, "You're really getting up to your old tricks again!" Wusao then rose and apologized. "I really have committed a terrible faux pas." She then turned and looked at me closely. "I've often had the chance to examine people, and none of them come up to you. You really possess a divine talent, certainly not common or vulgar." I then rose and apologized in my turn. "Once upon a time, Zhuo Wenjun recognized the true worth of Xiangru simply by hearing him play the harp. Shan Tao's

山濤 wife bored a hole in the wall and thus knew that Ruan Ji 阮籍 was a true genius.[2] If you are sincere in what you say, then I'll never forget the favor you've done me." Shiniang said, "Have Green Bamboo 綠竹 bring in the lute and play; I'll get his Lordship to drink."

Green Bamboo took up the lute, but before she even played, I chanted:

A heart that's stirred but cannot be guessed.	心虛不可測
Eyes so slender are filled with passion.[3]	眼細強關情
She turns it about, clasps it to her breast—	迴身已入抱
Charming sounds issue forth unawares.	不見有嬌聲

Shiniang responded to my verses:

Its cherished heart could easily break;	憐腸忽欲斷
Its eyes of longing are open wide.	憶眼已先開
But she hasn't yet strummed upon it,	渠未相撩撥
So where could those charms come from?	嬌從何處來

When I had looked over her verses, I was struck from within. Getting down off the dais I rose and apologized to her. "Up to now I have only beheld your features, but now I see your heart for the first time. You could surely make Lady Ban 班 pull your carriage or cause Mistress Cao 曹 to lay aside her brush.[4] How could I speak of myself as your contemporary, or even as living in your era?" I then requested a brush and inkstone, and I copied out the poem and then slipped it into the sleeve of my robe. Shiniang teased me: "Not only are you a superior poet, but you have a fine hand as well. As you write, your brush soars like a blue lovebird, and you look like a white crane." I said, "Madam is not only talented in expressing her feelings, but can truly chant as well. Who knew a visage of jade would possess a golden voice?" Shiniang replied, "I have been afflicted with a cold of late, and my voice is not clear."

"Well, I have been suffering from a sore hand, and cannot bring brush and ink under control."

Wusao laughed. "The girl couldn't even show off her talents! But you were fast enough with your compliment!"

Notes

1. In this and in other descriptive parallel-prose passages, there are long lists of items meant to impress with their luxury. I will not attempt to annotate them in detail (and in many cases we can only guess what many of the items are).

2. As told in *Worldly Tales, New Series,* Shan's wife insisted on spying on her husband's elegant friends. See Chapter 3.

3. "Empty Heart" is idiomatic for a heart moved by emotion; here it also refers to the hollow body of the lute. "Eyes" are the holes carved into the body of the instrument.

4. Court-lady paragons of virtue from the Han. For Lady Ban, see Chapter 4; Mistress Cao is the famous Ban Zhao 班昭, sister of Ban Gu and author of *Admonitions for Women* 女誡.

7

Shiniang said to Wusao, "Up to now we've merely been talking nonsense. There's been no order to it. Why don't you come up with some drinking rules?"

"How dare I not accept your request? Let's quote old poems. You take your meaning from quoting a line out of context that is in keeping with your feelings. If it doesn't match the situation, then you'll be fined in keeping with the offense. Why don't we start with our lady here?"[1]

Shiniang obeyed her request:

"Guan guan cry the fishhawks	關關睢鳩
on an islet in the stream.	在荷之洲
Charming and virtuous maid,	窈窕淑女
A fit helpmeet for our prince."	君子好仇

I then replied:

"In the south there are soaring trees,	南有喬木
But you cannot find shade below them.	不可休息
The Han has its roaming girls,	漢有遊女
But you cannot seek them."	不可求思

Wusao said:

"How do you cut an axe-handle?	折薪如之何
Without an axe you cannot do it.	匪斧不剋
How do you get a wife?	娶妻如之何
Without a matchmaker you cannot have her."	匪媒不得

Then Wusao added:

"When I did not see Fuguan,	不見復關
My tears fell in a stream.	泣涕漣漣

But after I saw Fuguan, 既見復關
Then how I laughed and talked!" 載笑載言

Shiniang responded:

"I have not been wrong in this; 女也不爽
My lord has been unfaithful. 士二其行
My lord has no principles, 士也罔極
And bestows his favors variously." 二三其德

I then replied:

"While we live, you shall live apart; 穀則異室
But when we die, we shall share a grave. 死則同穴
If you say I am untrustworthy, 謂余不信
Then I swear to be like the bright sun!" 有如曒日

Wusao laughed. "Master Zhang has great powers of concentration. He cites the *Odes* quite appropriately. There's an old proverb: 'If you can steady your thoughts, you can bore through stone.' If you think of [long for] something steadily enough, how can it remain far away?"

Notes

1. This is a game of *fu shi* 賦詩, or "quoting the Odes" to express one's opinions. Here, an age-old formal banqueting custom is turned into a party game—interestingly enough, with a heavy emphasis on "courtship" interpretations of the poems. Odes 1, 9, 158, 58 (twice), and 73 are quoted.

8

Green Bamboo then took up the zither and played. Wusao chanted a verse on the zither:

Heaven-endowed, pure-white features, 天生素面能留客
Able to make a guest linger;
Provoking longings, engaging one's love 發意關情併在渠
Wholly lie within its scope.
Don't find it strange that just now 莫怪向者頻聲戰
Flurried notes trembled.
Having found a companion, 良由得伴乍心虛
Its heart now grows restless.[1]

Shiniang then said, "If you're going to compose a poem on the zither, then I'll compose one on the flute."

Its many eyes now cause you To cherish it deeply.	眼多本自令渠愛
Only one mouth—this the cause It is conquered so.	口歩由來每被侵
When the wind blows it has no trouble Sounding in the ear;	無事風聲徹他耳
But if the player takes extra pains Then the hearer will be content.[2]	交人氣滿自塡心

I thanked them once again. "How marvelous! How lovely! Every line is excellent. In spite of my inferiority, I have had the opportunity to hear your lofty song."

Notes

1. There is constant play here between description of a woman and the musical instrument. The pun on "empty heart" occurs again in the last line.

2. There is some debate among commentators on this poem. The first line obviously refers to the stops on the flute. The second line (lit. "of few mouths," i.e., only one mouthpiece) has been thought to refer to the fact that Shiniang is quiet and is "outtalked" by Zhang. The third line refers to "certain bamboos that emit a sound of wind and rain without anyone blowing into them" (Waley, "Colloquial in the *Yu-Hsien K'u*," 562). I suspect a possible double-entendre here as well but have no proof.

9

Shortly thereafter, Cassia Heart brought in some food to accompany the ale: strips of mullet from Donghai 東海, dried phoenix meat from Xishan 西山, deer tail and tongue, various kinds of dried and broiled fish, minced goose, pickled watercress, quail stew, cassia-flavored rice and meat buns, bears' paws and the thighs of hares, pheasant rumps and fox snouts. Every conceivable flavor—I cannot begin to enumerate all of them.

Shiniang said, "Your Lordship must be hungry." She then called to Cassia Heart to lay out the dishes. I said, "My eyes have looked their fill; so now I feel no hunger in my body."

Shiniang laughed. "Don't tease. Let's bring out the backgammon board and gamble for drinks." "I've never been any good at gambling for drinks." I replied. "Let's gamble on spending the night."

"And what do you mean by that?"

"If you lose the round, then you have to sleep with me. If I lose the round, then I have to sleep with you."

Shiniang laughed. "So! 'If the Chinese rides an ass, then the Turk has to walk; but if the Turk has to walk, then the Chinese rides the ass.' Both ways you get what you want, and I lose out. You're too clever by far!'

Wusao said, "As I told you, girl, it's hopeless to gamble on anything, because tonight I know you won't escape."

Shiniang said, "You're always talking nonsense in order to curry favor with His Honor." I arose and apologized. "I knew it was a joke all along. I dare not hope for such an honor."

When the backgammon board arrived, Shiniang stretched a hand out toward it. Her eyes were clear and eloquent, and her hand was round and white. A pair of gleaming arms cut into my very vitals; ten fingers sliced into my heart and marrow. I then chanted a poem on the gameboard:

Eyes like stars newly shining,	眼似星初轉
Brows like moons about to melt.	眉如月欲銷
First one must press upon the back leg,	先須捺後腳
Then you'll force the waist in front.	然始勒前腰

She then replied:

Be clever and quick when you force the waist,	勒腰須巧快
And more insouciant when you press the leg.	捺腳更風流
If you make slender eyes close for a bit,	但令細眼合
Then my side will lose in the end.[1]	人自分輸籌

Notes

1. Here curiously enough the sexual meaning is clear, but the supposed "literal" meaning of the poem is not. Obviously allusion is being made to terms used in the game of *shuang liu* 雙六 ("double six," here translated as "backgammon"), but I have not seen a commentary that could explain them. Waley confesses himself puzzled.

10

After a while a maidservant named Zither Heart 琴心, a quite attractive girl, approached me. I secretly watched her, and Shiniang did not seem happy about it. Wusao scolded me loudly. "'Know what's enough and thus escape shame. Human life has its limits.' It seems the girl is frowning, and you shouldn't be glancing to the side."

Shiniang feigned seriousness and scolded Wusao. "What has His Lordship done to you that you're always bothering him?"

"You've been giving His Lordship the eye all along. Even if you don't have feelings for him, there's some communication going on. Otherwise, why are you instantly drawn by every love glance he has sent you since this morning?"

"You know you're interested in him yourself. When did *I* ever look at him?"

"If you're not interested, let me have him."

"You ask His Lordship himself. I don't know anything about it."

Wusao then chanted:

Fresh blossoms bloom on a pair of trees; 新花發兩樹
Each with its own scent, throughout the wood. 分香遍一林
As trees greet the wind, they shift their delicate forms; 迎風轉細影
As they face the sun, they turn their light shade. 向日動輕陰
Playful bees from time to time hide there; 戲蜂時隱見
Fluttering butterflies come searching from afar there. 飛蝶遠追尋
I have heard you wish to break a branch— 承聞欲採摘
But which tree moves your heart? 若箇動君心

"I'm greedy by nature," I said. "I'd like to pick from *both* trees."

For a time I'll ramble below paired trees, 暫遊雙樹下
From afar gaze upon both branches' scent. 遙見兩枝芳
Each faces the sun with fluttering shadows, 向日俱翻影
Each greets the breeze and scatters its fragrance. 迎風並散香
Playful butterflies lean on red stamens; 戲蝶扶丹蕚
Sporting bees enter their violet buds. 遊蜂入紫房
One ought to pick both of them now, 人今總摘取
Placing them one on each side.[1] 各著一邊箱

Wusao said, "You really are too greedy; you shoot at two targets with one arrow."

Shiniang said, "'If you seek out three, you aren't even given one. Look for two and you'll lose both.'"

Wusao said, "Don't try to find excuses for yourself, Madam. 'If a rabbit enters through the dog door, what else could happen?' "[2]

I then rose and apologized. "I asked for simple broth and was offered wine instead. That *is* an unexpected pleasure I've encountered before. But to hunt for rabbit and find a deer is beyond my hopes."

Shiniang said, "What kind of great personage are you, sister, that you can take over the management of my affairs! No doubt Master Zhang has no more interest in me than if I had already departed this world! Tomorrow he'll be saying outside that I have no more worth than a copper penny."

I replied, "Upon encountering your outstanding beauty, my spirit nearly perished. Then, when I witnessed your enlightened conversation, my heart broke in pieces. How could I dare to speak of you outside and spread lies about you? Grateful as I am for your kindness, how could I bear to act like that? But I am humbly willing to take my joy with you and express my feelings fully. Then I could die without regrets."

Notes

1. Some texts attribute this poem to Wusao or to Shiniang, but that makes little sense.

2. Some texts explain the meaning behind this proverb: "If a rabbit enters through the dog door, then he's bringing himself as a meal."

II

Food and drink then arrived. Fragrance filled the room, and dishes both white and red were brought forward—delicacies that exhausted land and water, fruits and vegetables from every stream and plain. For meat there were dragon livers and phoenix marrow; for ale, jade brew and jasper nectar. There were "sparrowsquawk" grain from Chengnan 城南 and "cicadachirp" rice from the Yangtze. Chicken broth, pheasant soup, turtle paté, and quail consommé, fat piglets raised under mulberries, slender carp bred among water reeds. Goose and duck eggs gleamed from silver plates; dried unicorn and panther foetuses were spread out in profusion on jade dishes. Grizzly tartare of pure white, crab eggs of pure yellow. Fresh minced fish vied in brightness with red thread, chilled livers mixed themselves with emerald string. Grapes, sweet sugarcane, dates and pomegranates, purple salt from Hedong 河東, crimson oranges from Lingnan 嶺南; clustered wood-pears from Dagu 大谷, Zhuzhong 朱仲 plums from Fangling 房陵; fairy cassia from the King-Father of the East, magic peaches from the Queen-Mother of the West; peppers from southern Yan 南燕 that clustered down like cow udders; dates the shape of chicken hearts from northern Zhao 北趙. A thousand different varieties of foods that cannot all be named.

I arose and apologized. "Though I was previously unacquainted with you ladies, because of my temporary government mission I met you by chance. You are extraordinarily generous with your fine and exceptional dishes. Though my body should be turned to powder and my bones to ash, I could not repay you enough."

Wusao said, "'If you're kin, you needn't be courteous; if you're courteous, you must not be kin.' I wish you wouldn't be so proper."

"Since I have received the favor of your command, I don't dare refuse."

I felt my breath go out of me and I couldn't help turning my gaze and spying on Shiniang once again.

"Your Lordship mustn't look at me," she said.

Wusao said, "Ah, you're still teasing him!"

I then recited:

Suddenly my passion within 忽然心裏愛
Unawares escapes as love from my eyes. 不覺眼中憐
It's not that my two eyes are warped— 未關雙眼曲
It's just that my inch of heart inclines! 直是寸心偏

Shiniang replied:

Eyes and heart are not of one place; 眼心非一處
Heart and eyes of old have been parted. 心眼舊分離
If you let your eyes behold us now, 直令渠眼見
How could they convey what the heart knows? 誰遣報心知

I answered:

The heart's always used the eyes as messengers; 舊來心使眼
The heart's longing the eyes convey. 心思眼即傳
Because the heart sends the eyes to see, 由心使眼見
The eyes fall in love along with the heart. 眼亦共心憐

Shiniang chanted:

Eyes and heart both love and long; 眼心俱憶念
Heart and eyes both search and seek. 心眼共追尋

But who knows how to employ the eyes 誰家解事眼
To help capture a lovable heart? 副著可憐心

Wusao then took up a wordplay game on the fruits:

"I only ask what your intentions are— 但問意如何
And why there was no *date* for our meeting." 相知不在棗

Shiniang added:

"Now my thoughts grow intimate: 兒今正意密
I do not dare to split this *pear*." 不忍即分梨

I added:

"Now we've met, your kindness is great; 忽遇深恩
From now on my life will be *peachy*." 一生有杏

Then Wusao:

"Now we've reached our pleasant bliss, 當此之時
Who could stand to be a *crab*?"[1] 誰能忍柰

Shiniang said, "Perhaps I could borrow His Lordship's knife for a moment so I can cut open this pear." I then composed a poem on the knife:

I cherish the firmness of its lacquered sheath, 自憐膠漆重
That feelings of longing never cease. 相思意不窮
A shame that so sharp a thing 可惜尖頭物
Should stay within its scabbard-hide. 終日在皮中

Shiniang in turn composed verses on a sheath:

Putting in the blade, you must be slow. 數捺皮應緩
In polishing, move often and quick. 頻磨快轉多
After you have pulled it out, 渠今拔出後
What's to become of the empty sheath? 空鞘欲如何

Wusao said, "Well, we're getting in deeper and deeper, aren't we?"

Notes

1. I have altered the original meanings of these puns in order to give their sense. Literal meanings: "Why didn't we know each other earlier (*zao* = date)?" "I cannot bear to be parted (*li* = pear)." "My entire life will be fortunate (*xing* = apricots)." "Who can bear to refuse (*nai* = crabapple)?" I am indebted to Levy for some of my translations here.

12

They then retrieved a chessboard, and we gambled for drinks on it. I won. Wusao said, "Skill in chess is proof of a cunning intelligence. You're just *too* talented."

I answered, "If a cunning man concentrates a thousand times, he will still lose at least once. And if a fool concentrates a thousand times, he's bound to win at least once. Let's stop for a while."

Wusao asked why. I replied,

All along I've been astute at the game— 向來知道徑
All my life could not bear to cheat. 生平不忍欺
But if I'm forced to defend my moves, 但令守行跡
What use to play chess [break a date] with her again? 何用數圍棋

Wusao then responded:

This lady of ours by nature
likes to play chess— 娘子爲性好圍棋
When she meets someone, then right away
she plays happily without a thought. 逢人剩戲不尋思
You nearly lose your breath
when she makes eyes at you— 氣欲斷絕先挑眼
You'd better stop now,
before it's just too late! 既得速罷即須遲

When Shiniang saw that Wusao continued to tease her, she feigned anger and would not smile. I then chanted,

Since I know it's worth a thousand in gold, 千金此處有
I wait for a single smile from you. 一笑待渠爲
But as I can't hope to see your flashing teeth, 不望全露齒
Even a frown for me would suffice. 請爲暫嚬眉

Shiniang replied,

Paired brows shatter a guest's vitals. 雙眉碎客膽
Two eyes pierce the lord's heart. 兩眼刺君心
But who could sell a single smile 誰能用一笑
So cheaply at a thousand in gold? 賤價賣千金

There was a broken copper clothes iron lying by the side of the dais. Shiniang suddenly composed a verse on it:

Heretofore its heart and belly burned—	舊來心肚熱
Pointlessly forced to iron the cloth of others.	無端強熨他
But now its form has turned cold—	即今形勢冷
Who is willing to rub it hot again?	誰肯重相磨

I responded:

Though its face has grown chilly,	若冷頭面在
It could not be worn away in a lifetime!	生平不熨空
Though now it seems cold and useless,	即今雖冷惡
We'll seek out the remaining metal.[1]	人自覓殘銅

Everyone laughed.

Notes

1. The iron poems are obscure. The second poem is probably meant to be a teasing description of Shiniang itself. For an interpretation, see Waley, "Colloquial in the *Yu-hsien K'u*," 563. He suggests that the phrase "remaining metal" is a pun for "come to an agreement."

13

Shiniang then summoned Fragrance 香兒 to bring out the musical instruments for me. Metal and stone were played together, flutes and pipes mingled their echoes. Storax 蘇合 plucked the lute, while Green Bamboo sounded the Tartar whistle. Fairy Maid 仙人 strummed the dulcimer, and Jade Girl 玉女 blew the mouth organ. Black cranes leaned forward to hear our zithers, and white fish leaped in answer to our rhythms. Clear notes echoed faster, and soon dust from the rafters went flying; elegant harmonies rang and tolled, and snow fell from the heavens. It was not false that Confucius lingered, forgetting the taste of meat; true indeed that Miss Han's 韓娥 echoing notes circled the rafters for three days.[1]

Shiniang said, "Since Your Lordship's arrival is a true rarity, we must taste the very limits of pleasure. Wusao is an extraordinary dancer; I hope she'll perform a piece."

Wusao did not shrink from this suggestion. She rose gracefully to her feet and strode coquettishly forward. A face that gnawed at men's hearts like vermin, she drove Yang city to the extremes of jealousy. Features as ruthless as devouring bandits, she bedazzled the heartstruck town of Xiacai 下蔡.[2] With a lift of the hand and a tap of the foot, she elegantly harmonized all the musical modes; with a peek behind and a glance before, how deeply she knew the rhythm of the piece. Like the twistings of a coiling serpent, like the soaring of a wild swan, as she turned her face the sun shone on the lotus blossoms; as she let fly her body, a breeze blew on fragile willows. Slanting brows, thievish glances, beauty beyond the commonplace, her languid steps grew ever swifter; carrying the amazing to its limit, creating ever new marvels. Her gauze robes shone and flashed, like turquoise phoenixes soaring through the clouds; brocade sleeves fluttering like a blue simurgh reflected on the water. Eyes of a thousand charms, as if the heavens lost their comets. A single slender waist, and Luo Stream grows ashamed of its whirling snow.[3] Brilliant advances and glamorous retreats so hard to find, so hard to meet. Back and forth, coming and going, so rarely heard of, so rarely seen.

The two of them rose to dance and urged me to do the same. I rose and demurred. "It's hard to be water in the midst of the vast sea, and hard to be thunder after the clash of the heavens. I dare not refuse, but I will certainly be ugly and clumsy." I then rose to dance. Cassia Heart lowered her head, convulsed with laughter. Shiniang asked, "What are you laughing at?"

"I laugh at the great skill of our music making," Cassia Heart replied.

"Why are we so skillful?"

"Because if we weren't, how could we lead all the animals to the dance?"[4]

I laughed. "It's not an animal you lead to the dance. Rather, the phoenix has come to your court."[5] Everyone burst into laughter.

Wusao said to Cassia Heart, "Don't make mistakes in your playing. Master Zhang keeps looking askance at you." Cassia Heart responded,

"I don't deny the grief of the singer—
Just regret that understanding listeners are few."[6]

"'If you meet a beauty on the road, no need to be formally acquainted,'" I replied. I then danced, and sang the following:

In the past I wandered through the world;	從來巡遶四邊
Suddenly met a pair of goddesses;	忽逢兩箇神仙

Their brows: 眉上冬天出柳
willow's green
in winter;
Their cheeks: 頰中旱地生蓮
lotus growth
in parched earth.
A thousand glances 千看千處嫵媚
in a thousand places
seductive;
A myriad glances, 萬看萬種㛹妍
myriad kinds
of charms.
This very night 今宵若其不得
if it happens
I don't obtain them—
Judge my life— 判命過與黃泉
let it pass
to the underworld below!

Again, everyone burst into laughter.

When the dance ended, I apologized. "Truly I have but a middling talent, yet I have had the opportunity to accompany the bright and the superior in ability. You have graced me with your lovely music. My shame at the gift is unbearable."

Shiniang chanted:

When satisfied, like mated ducks; 得意似鴛鴦
When thwarted, as far as steppe and jungle. 乖情若胡越
If I don't do my utmost for you, 不向君邊盡
Then where will I ever do so? 更知何處歇

She then said, "We are nothing special worth considering. You said we were 'willow's green in winter' and 'lotus growth in parched earth.' You're just teasing us."

I replied, "Though your face may not be spring, willow leaves flutter upon it."

Shiniang replied, "Since you've water on the brain, shouldn't it grow lotus flowers?"

I laughed. "You're clever with your word games. You can take advantage of any situation."

Shiniang quoted, "'If we don't take our pleasure now, who will it be next year?'"

Notes

1. *Analects* 7.14: The Master listened to the *shao* 韶 (the music of the sage emperor Shun) and for three months paid no attention to the taste of meat. He said, "I never imagined that the pleasures of music could reach to this!" For Miss Han, see Section 1, note 5.

2. From Song Yu's *Master Dengtu the Lecher* (see Chapter 2): "With a single captivating smile / She brings confusion to the city of Yang / Leads the state of Xiacai astray."

3. The Luo River goddess: see Section 3, note 5.

4. Allusion to the talents of the ancient legendary music master Yushi 予石.

5. The appearance of the dancing phoenix at court was a sign of sagely rule.

6. A couplet taken from the fourth of the Nineteen Old Poems.

14

There was an inkstone at the side of the dais. I therefore composed a verse on brush and inkstone:

Crush down the hairs, write as you wish—	摧毛任便點
If you crave hue, you must rub ever more.[1]	愛色轉須磨
The reason why it's hard to grind it enough	所以研難竟
Is because there's too much fluid!	良由水太多

Shiniang noticed a duck-headed vessel, and so composed upon it:

A beak that's long, but not for pecking—	觜長非爲嘲
A neck that's crooked, but not from pulling.	項曲不由攀
Just let the legs rise up straight—	但令腳直上
It looks like the eyes will roll away.[2]	他自眼雙翻

Wusao said, "This is really too perverse. You're getting in deeper and deeper!"

A pair of swallows came chasing each other and flew among the eaves. I recited:

Pair of swallows,	雙燕子
Fly wing to wing, a myriad turns—	聯翩幾萬迴
Well aware that I am a guest here,	強知人是客
They come only to stir me up.	方便惱他來

Shiniang replied:

Pair of swallows,	雙燕子
As it happens, are really making love.	可可事風流
Since I've obtained a partner,	即令人得伴
I need search no more for another.	更亦不相求

The drinking cup was passed to Shiniang. I then composed a verse upon the ladle:

When you move the end, move it swiftly—	尾動唯須急
When the front is low, it won't stay level.	頭低則不平
You should take up the beaker now,	渠今合把爵
Fill it deep or shallow, with my lord's mood.	深淺任君情

Shiniang chanted on the drinking cup:

When you first take it, put it to the mouth—	發初先向口
If you want to end, gradually make the tip rise.	欲竟漸昇頭
If you should pause midway,	從君中道歇
Then you'll have to stop altogether.	到底即須休

I rose up abruptly and apologized. "The lines of your verses truly approach the divine. Your ability must be innate, not a result of study."

Notes

1. One of the more obviously sexual of the poems. It plays on the meaning *se* 色, "hue" and "sex appeal."

2. I am unsure precisely of the sexual nuances here. The first two lines are probably references to the male organ. The last line may represent the woman's eyes rolling back into her head from pleasure.

15

Wusao said, "Since Master Zhang has only recently arrived, he still has no way to express himself completely. Let's take a walk in the rear garden and find a suitable background for our feelings."

Within the garden fresh fruit grew on myriad trees, filled with spring they put forth green; clumps of flowers glowed on all sides, spreading their violet and letting fly their crimson. Scoured stones and ringing waters, sparse cliffs and hollow crags. No seasonal differences could be found there: charming orioles fluttered about among brocade branches. No difference in old or new: brightly colored bream leaped in silver pools. Beguiling lush growth, clear cool zephyrs, geese and ducks drifted from their flocks, lotuses emerged in their midst. Large and small bamboo could be vaunted a superior to Qianmou 千畝, south of the Wei 渭; flowers filled with color opened, laughed to scorn all of Heyang's 河陽 beauty. Green, green the threadlike branches of the riverbank willows that brushed aside Wuchang's 武昌 trees; bright and glowing the arrow-like limbs of the mountain poplars, more lush than those of Dong Marsh 董澤.

I composed a verse on the flowers:

The breeze blows every tree to violet, 風吹遍樹紫
The sun shines the full pool red. 日照滿池丹
If you make me break a branch off for a time, 若爲交暫折
We'll hold it in our hands to see. 擎就掌中看

Shiniang responded:

Reflected in water, they all know to smile— 映水俱知笑
They form paths, though they never speak. 成蹊竟不言
Since now we have no restraint 即今無自在
You may pick as you wish, high or low. 高下任渠攀

I then rose and apologized. "A superior man does not speak lightly. I cannot repay your tremendous favor, Madam. Let's all compose verses now." I then began:

In the past I visited a little garden; 昔時過小苑
This morning I play in a rear courtyard. 今朝戲後園
Second year—the plum blossoms return; 兩歲梅花匝
Third month of spring—willow colors grow rich. 三春柳色繁
Water is bright, fish reflections calm; 水明魚影淨
The trees are emerald, bird song raucous. 林翠鳥歌喧
What need of an Apricot Tree Ridge? 何須杏樹嶺
This is our Peach Blossom Spring![1] 即是桃花源

Shiniang continued:

Plum spring commands the Taoist master; 梅溪命道士
Peach spring detains immortals. 桃澗佇神仙
The former fish has turned to a mighty sword—[2] 舊魚成大劍
The new tortoise resembles a small coin. 新龜類小錢

At the water's edge we see only willows;	水湄唯見柳
By the curve of the pool, lotus about to spring forth.	池曲且生蓮
If you want to know a place to cherish,	欲知賞心處
Peach blossoms fall before our eyes.	桃花落眼前

Wusao chanted:

We wander the fragrant garden as far as one can see;	極目遊芳苑
Hand in hand, we face the flowering forest.	相將對花林
Mist is calmed—mountain light emerges;	霧淨山光出
The pool is refreshed—tree reflections deep.	池鮮樹影沈
Falling flowers from time to time float in our ale;	落花時泛酒
Singing birds now and then sing to our lutes.	歌鳥或鳴琴
Just now, as dusk approaches,	是時日將夕
We take our cups up and retreat to cassia shadows.	攜樽就桂陰

Just then a plum fell into my lap. I composed the following:

I ask the plum trees	問李樹
Why are your hopes not mine?	如何意不同
You ought to drop plums in your mistress' hand	應來主手裏
But instead they fly to the robes of the guest.	翻入客懷中

Wusao replied to my verses:

Oh, plum tree,	李樹子
You've never been biased in your choice!	元來不是偏
You cleverly guessed the maid's desires	巧知娘子意
And threw your fruit at him.	擲果到渠邊

Suddenly a bee flew about Shiniang's face. She composed:

I tell the bee,	問蜂子
Bee, you're just too heartless!	蜂子太無情
You fly up and walk on my face,	飛來踏人面
As if you'd hold me in contempt!	欲似意相輕

I answered for the bee:

I go where there's a fragrant tree,	觸處尋芳樹
But few things are blooming now.	都盧少物華
And so I sought out scented places	試從香處覓
Till I met with this charming flower.	正植可憐花

Everyone applauded, laughing.

A pheasant entered the garden. I called for bow and arrow and shot at it. It dropped with a twang of my string. Wusao laughed. "You have a natural talent equal to Cao Zhi's. But now we see your military attainments are equal to Gongsun Chu's 公孫楚. If you make a match with My Lady here, then there won't be another such couple in all the world."

Shiniang composed the following upon my shooting of the pheasant:

A gentleman strolls in wheat fields, 大夫巡麥隴
The virtuous maid is accustomed to mulberries. 處子習桑間
If it weren't because of a single arrow, 若非由一箭
Who would break into a smile for you?[3] 誰能爲解顏

I replied:

If your feelings should coincide with mine, 心緒恰相當
Then who should keep to good or bad? 誰能護短長
Not both are beauties in the bed, 一床無兩好
But what matter if one is ugly? 半醜亦何妨

Wusao asked, "How are you at shooting a long target?" I replied, "I'm at your service." I then took my shots. All three arrows hit the mark, one after another. All praised my skill. Shiniang then composed verses on the bow:

All your life you're good at the bow. 平生好須弩
When you get to pull, then you lower your head. 得挽即低頭
I hear you're so quick with your thumbguard; 聞渠把提快
I beg you to hit a few more times! 更乞五三籌

I replied:

A shrunken shaft can never reach— 縮幹全不到
But when its head is raised, it goes quite beyond. 抬頭剩大過
If it should enter below the navel, 若令臍下入
A hundred times would I hit that mark. 百放故籌多

Notes

1. The first is a reference to an apricot tree grove planted by a Taoist immortal; the second possibly to Tao Qian's utopia or (again) to the story of Liu Chen and Ruan Zhao.

2. There are a number of legends about swords transforming into fish and vice versa. I am unsure if any further meaning is implied (or if there is any significance to the turtle in the following line).

3. Two allusions here: first to Pan Yue's rhapsody on hunting pheasants, which compares the motions of the hunted pheasants to that of a young girl hiding in mulberry trees. There is also an allusion to an anecdote in the *Zuo zhuan* 左傳 about a feudal lord whose newlywed wife refused to smile for three years until one day she saw her husband display his hunting skills.

16

The sun was sinking in the Western Pool,[1] and the moon rose in the eastern peaks. Wusao said, "All our jokes and games have been very fine, but now it's getting on toward dusk, and we ought to retire to our rooms. I hope that Master Zhang and our lady will rest together." Shiniang replied, "When two people meet for the first time, they enjoy dawdling a bit over drink and conversation. Besides, my room is small. What's the hurry?"

But Wusao was leading me to Shiniang's bedroom. Twelve folding screens, with four or five painted panels each; multicolored drapes on both sides of the room, and every corner hung with sachets filled with betel nut, cardamom, storax, and green aloeswood. Pillows and mats covered with patterned weave: in a riot of colors, chests filled with folded robes. We entered the room together, where everywhere gleamed gauze and painted silk. Lotus flowers grew from the mirror stand, kingfishers rose on golden slippers. At the bedcurtain's edge silver dragon ornaments; atop the bed, jade lions. Ten-layered carpets figured with mythical beasts; eightfold coverlets decorated with mandarin ducks. Several pairs of padded trousers, each extraordinary and seductive.

Her manner was innocent, though imbued with her romantic nature. A red chemise hugged her arms closely; green folds wrapped around her slender waist. At times she brushed herself with a kerchief, which emitted a fine blended incense. Sufficiently seductive and lovely, beauteous and charming, heaven-given; yet all along she knew how to adorn herself. Suppressing a smile, she straightened her gold hairpins; holding back her charm, she took up embroidered bedclothes. In comparison, the leisurely methods of adornment used by the Liang 梁 family have been falsely praised; and of what account are the curving brows drawn by the capital intendant?[2]

Shiniang then withdrew to a rear room. I stood brooding, but she did not return for a while. I asked Wusao, "Where did she go? Perhaps someone else is hindering her." She replied, "A girl is ashamed of marrying herself off.

It's a ruse—she's waiting for *you* to call *her*." Before she had finished speaking, Shiniang arrived.

I said to her, "When dawn came, it swept away the mists, and I sought flowers in a fragrant place. Then suddenly I met a mad gust of wind, and the lotus was deprived of its roots.[3] Where did you go off so carelessly?"

She turned her head and smiled. "The stars detained the Weaving Maid, who then returned to the human world. The moon kept Heng'e 恆娥 waiting, and she went back to heaven for a while.[4] Why must you bother yourself to blame me?"

The two of us sat facing each other, neither daring to touch. As the night deepened, our passions sharpened. In our excitement we felt oblivious of life and death. I chanted:

A thousand looks, a thousand thoughts grow intimate:	千看千意密
One glance, all my love grows deep.	一見一憐深
If I only were to take your hand	但當把手子
I'd be content, though torn asunder.	寸斬亦甘心

Shiniang's features hardened and she was about to leave. Wusao said,

She understands things well enough—	他家解事在
She's not yet willing, though *he* glares again and again.	未肯輒相嗔
You have no choice but to hold on tight—	徑須剛捉著
No matter how much spirit you put into it.	遮莫造精神

I then grasped her hand. No longer standing it, I said:

A thousand longings, a thousand innards burn;	千思千腸熱
One thought, all my heart is scorched.	一念一心燋
How can I succeed in my plea?	若爲求守得
And possess for a while your lovable waist?	暫借可憐腰

Shiniang was still unwilling. I pulled at her hands, and we struggled. Wusao chanted:

Cleverly she blocks her mouth with her robe	巧將衣障口
And shields her body with a blanket.	能用被遮身
I know she must be willing enough—	定知心肯在
Just blocks him as a ruse!	方便強邀人

Shiniang fell silent and broke into a smile, then came sidling into my arms. I felt a madness seize upon my vitals, and my heart beat wildly. I chanted:

Once her waist is brought under my power,	腰支一遇勒
My heart breaks in a hundred places.	心中百處傷
If I could now but win her mouth,	若爲得口子
I'd not wish for anything else.	餘事不承望

Shiniang angrily responded:

I've resigned my hand to him,	手子從君把
Have let my waist be turned.	腰支亦任迴
Though I'm still a worthless thing,	人家不中物
Slowly he forces his intentions.	漸漸逼他來

Though she continued to resist, she couldn't avoid surrendering to my mouth. Her lips were rich and fragrant, filling my nostrils with their scent. Her tongue was a scented herb, and I felt the inside of my cheeks bursting from its touch.

Wusao chanted:

Well aware a romance draws nigh,	自隱風流到
Yet she maintains propriety before you.	人前法用多
She finds that sometimes one must resist:	計時應拒得
She'll then feign a giving in.	佯作不禁他

Shiniang said, "In the past I've teased others, but now must bear their taunts."

I rose and asked: "There is one thing on my mind that I intend to discuss with you, but I still fear mentioning it directly. I ask for Wusao's help." Wusao answered, "Go ahead—don't feel any constraint." I replied:

Medicinal herbs—I've tasted many—	藥草俱嘗遍
But none are suitable now.	并悉不相宜
I only need a single thing:	唯須一箇物
You'll know without my naming it.	不道亦應知

Shiniang chanted:

My white hands have been grasped;	素手曾經捉
My slender waist has been captured.	纖腰又被將

And now I've lost to your mouth— 即今輸口子
For the rest, our troth is plighted! 餘事可平章

I kowtowed to Wusao. "Formerly I felt at a total loss, and truly my fear was exceptional. But Shiniang has taken pity on this guest and has saved him from death, as if she had restored flesh to white bone and brought flowers back to the withered tree. I prostrate myself on the ground and kowtow to you, earnestly begging the punishment of death."

Wusao arose and apologized. "I have heard that a thread goes through with the help of the needle, but it's not the needle that holds the garment together. If a girl marries with the help of a matchmaker, it's not the matchmaker's job if the couple gets along or not. I've been working hard to arrange the affair, but I won't predict how it will come out. Now that Madam has settled in, I'll be retiring to my room."

Notes

1. The legendary resting place of the sun god.

2. The family of Liang Ji 梁冀 of the Eastern Han was famous for its lovely ladies. Zhang Chang 張敞 of the Eastern Han (the capital intendant) became famous for applying his wife's cosmetics himself.

3. A pun: "My love was separated from its companion."

4. The Weaving Maid is the star in love with the Herd Boy star. Heng'e is another name for Chang'e 常娥, the goddess of the moon.

17

The night was growing late and our feelings grew more passionate and intimate. The fish-oil lamp shown bright all around; the wax candles illumined all sides. Shiniang then called for Cassia Heart and also for Peony 芍藥. They took off my slippers, folded up my robe, put aside my turban, and hung up my sash. I then helped Shiniang take off her damask cape, undo her gauze skirt, shed her red chemise, remove her green stockings. Her flower-like features filled my eyes, and a fragrant breeze assaulted my nostrils. My heart leapt uncontrollably; passion came irrepressibly. I slid my hand into her crimson trousers while we entwined our limbs under the turquoise coverlet. We held our mouths lip to lip, while I supported her head with my arm. I fondled and squeezed her breasts, rubbed and stroked her thighs. A nip brought elated feelings, an embrace brought a broken heart. My nostrils throbbed, and my heart was tied in knots. Before long my eyes were blurred and my ears burned, my veins bulged and my sinews grew slack. For the first time I knew how rare it had been to meet her, and how precious she was. In a short while we had joined several times.

Who knew that the hateful, ill-bearing magpie should wake us in the middle of the night and that that scampish, crazy rooster should cry dawn at midnight? We then donned our clothes and sat facing each other, looking at each other with tears streaming down. I wiped them away and said, "It is bitter that it is so easy to part but so hard to meet; one will stay, one leave—so perversely separated. But the king's affairs bring restraints, and I do not dare to tarry long. Whenever I think of it, I ache to the marrow of my bones."

"All our lives we never revealed ourselves to each other. And now, by chance, we have recently joined. Before we can taste our joys to the fill, we suddenly must part. A human life is full of partings and meetings. What else can we know?" She then recited:

At first we were unacquainted; 元來不相識
It was natural we should not be close. 判自斷知聞
The Lord of Heaven is meddlesome, 天公強多事
Now forcing us to such a parting. 今遣若爲分

I responded:

Grief on grief, heart already broken, 積愁腸已斷
My grace will still reach you from a great distance. 懸望眼應穿
Don't shut your door tonight, 今宵莫閉戶
For in dreams I'll come to your side. 夢裏向渠邊

18

Before long the sky grew light. We still wept, our hearts choked up within; we could not calm ourselves. Several serving maids entered sighing and could not raise their heads. Wusao said, "If there is meeting, there must be parting; and it has always been so. When joy reaches its peak, sadness comes: a certainty from ancient times. I hope Madam can control herself somewhat."

I then gave the sleeve of my robe to Shiniang to wipe away her tears. She composed a parting poem:

A parting is still a parting: 別時終是別
A spring heart finds no spring. 春心不値春

Ashamed to see a simurgh reflected alone,[1]	羞見孤鸞影
Grieve to behold one rider in the dust.	悲看一騎塵
Emerald willows open their eyebrow-colored leaves;	翠柳開眉色
Red peaches run riot with cheek-hues renewed.	紅桃亂臉新
But this time you will not be here—	此時君不在
The tender orioles will tease me horribly.	嬌鸎弄殺人

Wusao responded:

Now, after he has left us for good,	此時經一去
Who knows how many years will pass?	誰知隔幾年
A pair of ducks wounded by threads of separation,	雙鳧傷別緒
A lone crane grieving over parting strings.	獨鶴慘離絃
Resentment arises after intoxication fades;	怨起移醒後
Melancholy born before we are lost in our cups.	愁生落醉前
If you'll let your heart stay intimate,	若使人心密
You won't regret the wear on your horse's hooves.	莫惜馬蹄穿

I chanted:

So soon I hear you say goodbye—	忽然聞道別
Melancholy comes, cannot be repressed.	愁來不自禁
Below the eyes, a thousand streams of tears;	眼下千行淚
From my vitals hangs one inch of heart.	腸懸一寸心
Paired swords now slide into separate cases,	兩劍俄分匣
Paired ducks suddenly in different groves.	雙鳧忽異林
Do take care of your precious jade body—	懃懃惜玉體
Don't let outsiders trespass upon it.	勿使外人侵

Shiniang's childhood name was Jasper Bloom 瓊英. I therefore chanted:

A Bian He who never dug for jade;	卞和山未斲
A Yang Boyong who never tilled.[2]	羊雍地不耕
I regret that I am no jade at all;	自憐無玉子
When will I see Jasper Bloom again?	何日見瓊英

Shiniang echoed my verses:

A phoenix quilt I must give in parting,	鳳錦行須贈
Though my dragon shuttle has long fallen silent.	龍梭久絕聲
I regret that I have no loom—	自恨無機杼
When will I see cloth's pattern completed [Wencheng]?[3]	何日見文成

I was astounded by these lines. It cut through my melancholy, and I broke into a smile.

I then called for my servant Song Harp 曲琴 to take in a "love-longing" pillow and give it to Shiniang as a memento. I chanted:

In southern lands, they use coconut shells—	南國傳椰子
Eastern families offer pomegranate wood in tribute.	東家賦石榴
I hope that this, replacing my left arm	聊將代左腕
Will pillow you through the long night.	長夜枕渠頭

Shiniang repaid with a pair of slippers. She chanted:

Paired ducks—suddenly one mate is lost;	雙鳧乍失伴
But these two swallows still belong to you.	兩燕還相屬
I hope that they will match your heart	聊以當兒心
And fit your feet to the end of the day.	竟日承君足

I also sent Song Harp to present her with a bronze mirror from Yangzhou 揚州 and the following poem:

The Immortals appreciate Fuju;[4]	仙人好負局
Recluses often secretly watch themselves.[5]	隱士屢潛觀
Reflecting water, its caltrop light disperses,	映水菱光散
Facing the breeze, bamboo images chill.	臨風竹影寒
Under the moon, it will at times startle magpies;	月下時驚鵲
At pool's edge, a lone, dancing simurgh.	池邊獨舞鸞
If you say my heart will change,	若道人心變
Then reflect it yourself and see.	從渠照膽看

She then gave me one of her own fans and this poem:

Acacia flowers sporting by jade waters—[6]	合歡遊璧水
Linking hearts serving by flowered halls.	同心侍華闕
Whoosh! It rustles like a morning breeze;	颯颯似朝風
So round, like the evening moon.	團團如夜月
Simurgh shapes invade the mists and rise,	鸞姿侵霧起
Crane shadows ply the air and soar.	鶴影排空發
I pray my lord will hold it in his hand	希君掌中握
And not let your kindness to it fade away.	勿使恩情歇

After I had said goodbye, I sent the maids to fetch a new-style bolt of Yizhou 益州 brocade to give to Wusao, with the poem:

I now leave you a small memento　今留片子信
Which you can give at a romantic tryst.　可以贈佳期
Cut it to form an eight-layered coverlet　裁爲八幅被
At times bringing back thoughts of love.　時復一相思

Wusao then pulled out one of her hairpins and gave it to me, with this poem:

I give you this, as we part;　兒今贈君別
I know future meetings will never come.　情知後會難
Don't think this hairpin but a trifle:　莫言釵意小
You can pin it to your cap.　可以掛渠冠

Again, I took a bolt of small-weave damask from Huazhou 滑州 and presented it to Cassia Heart, Fragrance, and several other maids to distribute among themselves. Cassia Heart and the others removed silver hairpins, golden bracelets, and gauze and silk handkerchiefs to give to me.

"Bon voyage. Come visit us again, if your travels bring you near." Fragrance composed the following:

My lord, look after your travels.　大夫存行跡
Try hard to come again.　慇懃爲數來
Don't act like the floating duckweed grass:　莫作浮萍草
It drifts with the current, knows not to return.　逐浪不知迴

I wiped away tears and said, "Although dogs and horses know little, they still recognize the pain of parting. Birds and beasts have no feeling, but they still resent separation. My heart is not wood or stone—how can I forget your great favor to me?" Shiniang responded:

They say grief is still better than death,　他道愁勝死
But I say death is better than grief.　兒言死勝愁
When grief comes, you're wounded in a hundred places;　愁來百處痛
When death comes, all ceases at once.　死去一時休

And also composed:

They say grief is still better than death,　他道愁勝死
But I say death is better than grief.　兒言死勝愁
Day and night, tortured by thoughts,　日夜懸心憶
How many autumns till we meet again?　知隔幾年秋

I chanted:

Someone leaves (so far away), parted on two sides of the sky.	人去悠悠隔兩天
Nor can we yet tell (so long a time), how many years shall pass.	未審迢迢度幾年
Even if my body's sent beyond ten thousand miles,	縱使身遊萬裏外
Always will my thoughts of return come back to your dear side.	終歸意在十娘編

Shiniang responded:

Edge of the sky, corner of the earth, who knows where?	天崖地角知何處
Jade-white body and ruddy face will never come again.	玉體紅顏難再遇
If only I could make a pair of wings grow for me now—	但令翅羽爲人生
Surely I would fly high and away off to where you'll be.	會些高飛共君去

Notes

1. A simurgh, or love-bird, owned by a nobleman was pining away for its mate. When he attempted to cheer it up by placing a mirror in its cage, it was so struck by its own reflection it died on the spot (see Chapter 4).

2. Bian He discovered a piece of jade that was later made into a highly valued disk during the Warring States period. Yang Boyong obtained some magic pebbles that he planted like seeds and so obtained jade from them.

3. Wencheng was Zhang Zhuo's 張 鷟 polite name. This is the main piece of evidence that Zhang was the author.

4. An immortal who polished mirrors for a living.

5. Recluses would often employ mirrors for the purpose of meditation and magic.

6. Acacia flowers became a symbol for conjugal bliss because their petals shut during the night. Their abstract pattern often adorned fans.

19

I could not bear to look on her. I quickly grasped her hand and departed. After I had gone about a mile, I turned my head to look. Several of them still stood in the same place. As I went on, I gradually drew further and further away, and their voices and faces gradually faded. Finally I looked back and could see them no more. I went on with bitter sadness.

When I reached the entrance of the mountain, I departed by boat. As night came, I lay restless and sleepless; my heart was lonely and had no place to lodge. I grew sad at the cries of the gibbons, was stricken by the swan separated from its flock. Swallowing my breath and repressing my voice—it is the way of Heaven and of human feelings that if there are partings, then there must be resentment; and if there is resentment, then there must be a surfeit of it.

How short the day departing, and how long the night before us! The *bimu* 比目 fish is severed from its mate, and the paired ducks are parted. Daily my robes grew looser; every morning my sash was slacker. My lips parted on their own accord, and emotion-filled gasps filled my chest. Tears ran down my face in a thousand streams, and my grieving vitals broke into pieces. I sat straight in front of my harp, and tears of blood fell on my robe. A thousand thoughts vied for attention, and a hundred worries impinged upon me. Alone I knit my brows and brooded long, held my knees to my chest, and chanted long:

I gaze off to the immortals, but I cannot see them!	望神仙兮不可見
All of Heaven and Earth must know of my heart.	普天地兮知余心
I long for those immortals, but I cannot reach them!	思神仙兮不可得
I seek for Shiniang, but no news at all.	覓十娘兮斷知聞
Wishing to hear of this, my heart's core grows wild;	欲聞此兮腸亦亂
Wishing to see this troubles soul within.	更見此兮惱余心

REFERENCE MATTER

Notes

For complete author names, titles, and publication data for works cited here in short form, see the Works Cited, pp. 399–414.

Introduction

EPIGRAPH: Ban Gu, *Han shu*, 2346.

1. *Mengzi zhengyi*, 12.4b–5b.

2. I am, of course, ignoring the fact that philosophers like Mencius and Confucius themselves traveled from state to state. But I believe the analogy still holds, at least as a tentative moment in the passage: the failure of both Mencius and Confucius to find a lord willing to accept them was a source of anxiety and a reason for doubting one's own self-worth and independence. Mencius' "Great Man" is more what he would like himself to be than what he actually is.

3. Liu Xiang, *Gu lienü zhuan*, 1.11a.

4. Yang Xiong, *Fa yan zhu*, 32. What I translate as "excessive" (*yin* 淫) is the same word used in Mencius' description of the Great Man: "if you are incapable of excess in spite of wealth and rank."

5. For a detailed discussion of the distrust of rhetoric and especially of Yang Xiong's position, see Wai-yee Li, *Enchantment and Disenchantment*, 17–21. She translates *qiefu zhi dao* as "the way of the concubine" (although she does not mention the Mencian origins of the phrase), and it is clear in the context of her argument that this phrase indeed came to mean that in later attacks on rhetoric. However, in the context of Mencius' passage, it is clear he is referring to married women in general and that *qiefu* is an inclusive, not a limiting, term (Legge and D. C. Lau so translate it). Of course, in a different context, *qiefu* could indeed mean "concubine," when placed in relation to one's "proper" wife (see, e.g., *Chunqiu jing zhuan jijie*, 893).

6. My descriptions of the Han literati here bear some resemblance to Stephen W. Durrant's (*The Cloudy Mirror*, 1–10) recent description of the psychology of the Western Han historian Sima Qian. However, Durrant attempts to trace a psychological portrait of an individual, not the "zeitgeist" of a class—a task I find highly problematic for so early a time, when descriptions of seemingly "personal" traits are often stylized and represent traditions of biographical and autobiographical representation. For a critique of Durrant's reading, see Puett, Review of Durrant, *The Cloudy Mirror*.

7. *Chu ci buzhu*, 3–4.

8. This distaste for the literature has not prevented the creation of a detailed scholarship on the historical and social positions of Chinese women both in the West and in China. Almost all of it, unfortunately, deals with post-Song China. One exception is Patricia Ebrey's recent *The Inner Quarters* (1993). On the literary side, one can cite Maureen Robertson's germinal article "Voicing the Feminine: Constructions of the Gendered Subject in Lyric Poetry by Women of Medieval and Late Imperial China" (1992), as well as Jowen R. Tung's *Fables for the Patriarchs* (2000). Kang Zhengguo's *Feng sao yu yanqing* (1988) deals with the portrayal of gender in early literature from a somewhat simple feminist viewpoint.

9. For a discussion of some of these issues in regard to pre-modern English literature, see Guillory, *Cultural Capital*, 19–38.

10. Lest I be thought to be arguing that a legitimate "female voice" arising from male texts somehow compensates for female silence, let me emphasize here that I am only suggesting that the constitution of sexuality in the subject at times allows for its own occasional subversion, in which aporias in the gendered construction of the self reveal multiple voices. Barbara L. Estrin (*Laura*, 10–19) has argued for the occasional appearance of the subversive female voice in male-authored poetry in the Petrarchan tradition.

11. See Robertson's reading of later women poets in "Voicing the Feminine." Interestingly enough, her considerations of earlier woman-authored texts (Tang and pre-Tang) tend to focus on the degree to which they created models of "female authorship" that were influential upon the coherent community of female authors active in the Qing. Such communities may have existed earlier (as in Tang courtesan circles), but the surviving poems do not allow a complex reconstruction of such a circle.

12. Jameson, *Postmodernism*, 181–217.

13. Chandler, *England in 1819*, 155–85; see also the first half of his book *passim*. Chandler solves the problem of situating literature historically by dealing with one appropriate moment: he discusses the English romantics precisely at a point when they themselves were obsessed with "representation" and the anecdote. He thus suggests that "New Historicism" to some extent repeats a romantic English gesture.

14. Chandler, *England in 1819*, 165; K. Burke, *Grammar of Motives*, 59.

15. See, e.g., Laqueur, *Making Sex*. For a consideration of the "incorporeality" of the Chinese body, see Hay, "The Body Invisible in Chinese Art?"; for "other discourses" of sexuality, see Zito's stimulating comments in "Ritualizing *Li*."

16. My use of gender theory here also accounts for my avoidance of any psychoanalysis of the Chinese male subject. Although a number of scholars (e.g., Jowen R. Tung, Zheng Yuyu) have started to use the language of Freud and Lacan to discuss early Chinese literary texts, I believe this is seriously misguided; we are becoming increasingly aware of the dependence of psychoanalytic theory on the relatively late social and cultural developments of bourgeois Europe.

17. I draw some inspiration here from recent scholars of English Renaissance literature, who have come to see the intimate relationships between the erotic literary traditions of the sixteenth and seventeenth century and the political world of the courtier; as with the early Chinese aristocrat/literatus, the Elizabethan nobleman worked out his own anatomy of desire through the politically determined routes that guided him to the ruler. Such a perspective first emerged with Stephen Greenblatt's *Renaissance Self-Fashioning* (1980), although his work emerged concurrently with more "feminist" approaches—e.g., Ferguson et al., *Rewriting the Renaissance* (1986), and with the social historical work of Peter Burke and of Natalie Zemon Davis, *Society and Culture in Early Modern France*. More recent works have built on New Historicist reading techniques while remaining skeptical of their disregard of bigger social issues. Most useful for me in their employment of "close reading" techniques of literature have been Jonathan Goldberg, *Sodometries* (1992); and Estrin, *Laura* (1994).

18. E.g., Schafer, *The Divine Woman* (1973); and Cahill, *Transcendence and Divine Passion* (1993).

Chapter 1

EPIGRAPH: Quoted by Dumézil, *Mitra-Varuna*, p. 11.

1. Although I attempt a literal rendering here, independent of interpretation, I rely to some extent for nouns and verbs on Karlgren, *The Book of Odes*, 44.

2. E.g., by Van Zoeren, *Poetry and Personality*. I can also suggest one scenario: the incomprehensibility of the decontextualized songs allowed for the evolution of a hermeneutic apparatus whose very operations helped create the concept of what it meant to be "Confucian"—both in a doctrinal and pedagogical sense. A commentator on the songs could legislate morality in the guise of merely "explaining," and the young scholar would memorize and remember.

3. For such idealized readings of gendered (especially female-voiced) Odes, see McCraw, *How the Chinawoman Lost Her Voice*; and also Kang, *Feng sao yu yanqing*, 11–51. In this regard, consider the following comment in Judith Butler's *Gender Trouble*, 36:

Some feminists have found in the prejuridical past traces of a utopian future, a potential resource for subversion or insurrection that promises to lead to the destruction of the law and the instatement of a new order. But if the imaginary "before" is inevitably figured within the terms of a prehistorical narrative that serves to legitimate the present state of the law or, alternatively, the imaginary future beyond the law, then this "before" is always already imbued with the self-justificatory fabrications of present and future interests, whether feminist or antifeminist. The postulation of the "before" within feminist theory becomes politically problematic when it constrains the future to materialize an idealized notion of the past or when it supports, even inadvertently, the reification of a precultural sphere of the authentic feminine. This recourse to an original or genuine femininity is a nostalgic and parochial ideal that refuses the contemporary demand to formulate an account of gender as a complex cultural construction. This ideal tends not only to serve culturally conservative aims, but to constitute an exclusionary practice within feminism, precipitating precisely the kind of fragmentation that the ideal purports to overcome.

4. This "romantic" interpretation is followed by a wide variety of commentators and translators, English and Chinese: e.g., Waley, *The Book of Songs*, 54–55; Gao Heng, *Shijing jin zhu*, 94; Chen Zizhan, *Shijing zhijie*, 197–98; Zhu Shouliang, *Shijing pingshi*, 203.

5. Of course, such female silence is not necessarily typical in an anthropological reading of individual poems—e.g., Odes nos. 75, 94, and 106 (translated as "courtship" poems in the female voice by Waley). I mean to suggest only that when gender is unclear, certain unquestioned assumptions usually come into play.

6. Again, by calling this an anthropological reading, I mean more a reading tradition popular in the past century that tends to romanticize folk cultures and that makes some use (or misuse) of anthropological scholarship and theory. Anyone courageous enough to undertake a more substantial anthropological reading of the *Odes* would need to go much deeper than Waley's interpretations.

7. For the classic feminist reading of marriage and kinship, see Rubin, "The Traffic in Women." Rubin's account is rooted in a sensitive critique of Lévi-Straussian models of kinship; although much work has been done to refine her arguments, she remains a basic starting point for many discussions (including my own).

8. Habits of grammar force me here to make "Confucianism" a subject and consequently, a reified actor; I am aware of the dangers of this. The reader should understand that I have Foucauldian concepts of power in mind here, in which power, although for convenience' sake is said to act on the subject, is indistinguishable from the sum total of social activities carried out by those subjects.

9. *Mao shi zhengyi*, 140.

10. *Chunqiu jing zhuan jijie*, 223, 237.

11. *Mao shi zhengyi*, 140.

12. Such a reading was indeed hypothesized by the Qing dynasty commentator Fang Yurun (1811–83) in his *Shijing yuanshi*, 188; Fang accepted Mao's context but assumed the text was meant to satirize the pride and ingratitude of the people and ruler of Wei.

13. For a detailed discussion in English of this hermeneutic shift, see Van Zoeren, *Poetry and Personality*, 219–21, 230–49.

14. Zhu Xi, *Shi jizhuan*, 3.28b. Of "Gentle Girl" he says, "This is a poem about an illicit liaison" (2:24a). The reading of certain odes as "depraved" begins with Ouyang Xiu, and is elevated to a central tenet in Zhu Xi ("Gentle Girl" is a prime example); see Van Zoeren, *Poetry and Personality*, 169–72, 219–21.

15. Lu Qinli, *Quan xian Qin Han Wei Jin Nanbei chao shi*, 187, 683.

16. Zhu Shouliang, *Shijing pingshi*, 203.

17. Saussy, *The Problem of a Chinese Aesthetic*, 148–49. His graceful work is an important discussion of the implications of the often maligned Mao readings for Chinese aesthetics and ethics. Here I can merely point out that he traces the implications of "relativity of reading" to a much greater extent than I can here. Note, however, the following passage:

> Here the declared function of interpretation is to construct, for the examples, the thing they are to be examples of. Insofar as that thing is not given, but rather made through reading, the interpretations subject the texts to a qualitative change; they remake them according to their wish. But to the degree that the wish is realized, and culture is created by the will to read the poems as if they were (already) manifestos of the sage, the distance between wish and fact shrinks. It is the reader who adheres to the second, non-natural, model of exemplarity who most successfully counters the charge of "allegorization" leveled against the *Mao Odes*. But that reader has already "become a parable by following the parables," to borrow a phrase from Kafka. Tropes read in that way obliterate themselves as tropes. (Ibid., 149.)

18. Van Zoeren, *Poetry and Personality*, 35–44, 84–94; Kang, *Feng sao yu yanqing*, 51–52.

19. Saussy (*The Problem of a Chinese Aesthetic*, 56–66) discusses this point at great length. Readers outside Chinese literary studies and comfortable with general concepts underlying "reader-response criticism" and the like may find these points unnecessary. However, much scholarship in the study of Chinese literature, especially of the biographical kind, has been devoted to ascertaining the individual's motivation in the act of composition and to reconstructing the poet's life (whatever that is) from the texts he left behind. I mean here to recognize the significance of this need in understanding Chinese poetics itself (that is, that poets wrote quite often with

the realization that later readers would be attempting such a reconstructive move), while at the same time suggesting the futility of arriving at some absolute knowledge of the poet. I am also interested in the consequences the assumption brings to the performative act of poetry writing within society and culture.

20. Supporters of the "friendship" reading include Cui Shu, *Du feng ou shi*; Fang Yurun, *Shijing yuanshi*, 188 (who actually sees it as "originally" [*ben* 本] a friendship poem applied satirically to the Wei ruler—see note 12 to this chapter); and Qu Wanli, *Shijing quan shi*, 117.

21. *Li ji*, 1.2b; trans. Legge, *Li Chi: Book of Rites*, 1: 65.

22. Mauss, *The Gift*.

23. Qian Zhongshu, *Guan zhui bian*, 99–100. Of course, one problem with such a reading is the seemingly unchanging value of the gifts as we progress from stanza to stanza (unless the gems about which we know so little increase in value).

24. Takigawa, *Shiki kaichū kōshō*, 129.13.

25. Mauss, *The Gift*, 6.

26. I wish to emphasize here that I speak of male social ideals, and not necessarily of all actual cultural practices. For the most recent and reasoned consideration, see Ebrey, *The Inner Quarters*, especially "Separating the Sexes," 21–44.

27. Ibid., 24.

28. It is telling in this respect that Ebrey (*The Inner Quarters*) approaches the lives of Song dynasty women mostly through the various social practices surrounding and resulting from their marriages.

29. Barlow, "Theorizing Woman," 253–61.

30. Again, as I have noted above, this position was first laid out by Rubin ("The Traffic in Women," 169–98); she draws first of all on Lévi-Strauss's *The Elementary Structures of Kinship* and Jacques Lacan's discussion of the role of language in formulating sexual orientations; it is also Rubin who suggests using the term "sex/gender system" as opposed to "patriarchy" to prevent misleading associations. Rubin's scenario for woman's position in basic kinship systems remains essentially unchallenged, although there have been many fruitful revisions, notably a challenge to the inevitability of a Lacanian reading; see, e.g., Butler, *Gender Trouble*, 72–75. I am also not the first to begin a study of traditional Chinese literature with a consideration of the linkages of gender, desire, and clan/family. Note Keith McMahon (*Misers, Shrews, and Polygamists*, 5):

> Gender derives central meaning from the kinship structure. The laws of kinship define the symbolic order, which is primarily structured by the paternal family—its rules of descent, incest prohibitions, and the binaries of male and female, senior and junior, inner and outer. The symbolic function of this order is in effect to assign each subject a place and a role in the kin group even before he or she is born. A basic premise of my study is that the rules or laws of kinship—in-

cluding gender definition and hierarchy—are such because of historical and social construction, not innate or natural necessity.

31. Butler, *Gender Trouble*, 17.

32. Ibid., 22.

33. For an example of such assumptions, note the beginning of Donald Harper's ("The Sexual Arts of Ancient China," 539) discussion of an ancient hygienic text: "In addition to their interest in procreation and in satisfying mutual desires *between man and woman*, the ancient Chinese regarded sexual relations as a vital part of the therapeutic arts of physical cultivation" (my italics). One may also pause at the broad attribution of such interests to "the ancient Chinese."

34. Foucault, *The History of Sexuality*, 106.

35. Ibid., 101.

36. Bray, *Homosexuality in Renaissance England*, esp. chaps. 2 and 3. Goldberg (*Sodometries*, 18–22) has stressed at greater length the importance of Foucault and Bray in coming to an understanding of sodomy and of male/male relations.

37. For an elaboration of this point using a literary text, see Goldberg's discussion of Marlowe's *Edward II*, in *Sodometries*, 119–26.

38. I should note here in passing that in spite of my use of Foucault (*The History of Sexuality*) as an important tool in uncovering what goes unsaid in Chinese texts, Foucault himself (like many Western critical theorists) often displayed an embarrassing ignorance of other cultures. See, for example, his orientalizing descriptions of East Asian "sexual science" in ibid., 57–58, where he clearly locates himself within the camp of those drawn to Taoist sex manuals (cf. note 33 to this chapter). It should also be emphasized that in spite of the fact that Foucault's *History of Sexuality* has long been superseded in the details of its argument, it is still recognized as an important departure point. See, e.g., James Grantham Turner's (*Sexuality and Gender in Early Modern Europe*, xv) comments:

> Foucault's stimulating suggestions opened up an exciting possibility: he abolished, at a stroke, all existing methods of combining sex and history—the psychoanalytic speculations that were *ipso facto* impossible to prove, the "lascivious erudition" that seemed only one step removed from pornography, or the dour compilation of population studies. But he did not solve the problem of how to put these insights into practice. Most theorists and historians in the rapidly expanding fields of sexuality, family, and gender would now reject his findings and hypotheses, even if they respect the gadfly quality of his mind and explore the implications of his once-startling conjunction, "history" with "sexuality."

39. *Mao shi zhengyi*, 14.

40. *Li ji*, 15.4b–5a; trans. from Legge, *Li Chi: Book of Rites*, 2: 264. I have altered Legge's translation of *ren* 人 ("people," a non-gendered term) as "men" to the vaguer "others." However, I believe that the passage still conceives the act of governing as a

public one, and hence that the direct relation of "love" here is one between the ruler and his public "male" subjects; the female subjects are private and part of the domestic sphere of control.

41. In her essay "Women on Top," Natalie Zemon Davis (*Society and Culture in Early Modern France,* 128) notes similar connections between marriage and state/subject relations:

> Thus, Jean Calvin, himself a collapser of ecclesiastical hierarchies, saw the subjection of the wife to the husband as a guarantee of the subjection of both of them to the authority of the Lord. Kings and political theorists saw the increasing legal subjection of wives to their husbands (and of children to their parents) as a guarantee of the obedience of both men and women to the slowly centralizing state—a training for the loyal subject of seventeenth-century France or for the dutiful citizen of seventeenth-century England. "Marriages are the seminaries of States," began the preamble to the French ordinance strengthening paternal power within the family. For John Locke, opponent of despotic rule in commonwealth and in marriage, the wife's relinquishing her right of decision to her husband as "naturally . . . the abler and stronger" was analogous to the individual's relinquishing his natural liberties of decision and action to the legislative branch of government.

It would be tempting (though difficult to prove) to see the *Li ji,* a text essentially constructed in the Han, as representing similar moves toward a more centralized state. Characteristically, however, the Chinese version of this relationship places particular emphasis on the duties of the hierarchically superior member.

42. *Chu ci buzhu,* 14–15.

43. Ibid., 30–35.

44. Of course, it may very well be that these shamanic texts were much less ambiguous while being performed and that original context made the meaning clear (although I am not totally convinced of this). Regardless, once the text enters into the reading tradition, interpretation can no longer decide definitively how the text should be read.

45. Joan Kelly, "Did Women Have a Renaissance?" in idem, *Women, History and Theory,* 44. For a specific close reading that illustrates this, see Marguerite Waller's discussion of Wyatt's "Whoso list to hount" in "The Empire's New Clothes."

46. Of course, this represents the world of literature as a discourse of largely male-authored texts that are preserved and often canonized. Whether "Women" as such are silenced outside this ideologically constructed world is another matter altogether (and the same may be said of "popular" literary genres possibly created outside elite interference).

47. A point deftly made by Alison H. Black in "Gender and Cosmology in Chinese Correlative Thinking," 179–81. See also Tung, *Fables for the Patriarchs*, 30–40, for a nice discussion of these issues in the Tang.

48. See Butler, *Gender Trouble*, esp. 128–41; and idem, *Bodies That Matter*, esp. 1–23, 93–119.

49. Butler, *Bodies That Matter*, 94–95.

50. See especially Hay, "The Body Invisible in Chinese Art?," 62–70.

51. Ames, "Taoism and the Androgynous Ideal," 44.

52. I mean here of course the popular conception of "Taoism," as opposed to the more political reading of Laozi characteristic of the Huang-Lao school during the pre-Han and Han periods.

53. For work in this area, see Zito, "Ritualizing *Li*," 332–41; and idem, "Silk and Skin." Zito has already discussed at great length the degree to which gender and sexuality are defined by "non-medical" discourses in traditional China and elaborates on the "ritualization" of gender to a much greater extent than I do here.

54. It is the failure to deal with the consequences of a Foucauldian model of sexuality (and of the historically contingent nature of homosexuality) that renders the only full-length treatment of Chinese homosexual behavior so unsatisfying (even beyond its philological and historical problems). I mean of course Bret Hinsch's *Passions of the Cut Sleeve*. Not only does any identification of a "tradition" as a subjective enterprise contingent on modern attitudes go unrecognized, but there are moments where, in spite of his claims of cultural sensitivity and relativism, he implicitly recognizes modern "tolerant" Western attitudes as an ideal "goal" for homosexual activity. E.g.: "With a kinship-structured society has come the sometimes problematic combination of heterosexual marriage and homosexual romance. What this social organization *prevented*, and still *prevents*, was the emergence of a self-identified homosexual life-style independent of marriage, as with gays of the contemporary West" (ibid., 19; italics in original). We should, however, recognize Hinsch's attempt to at least deal with the material. We might also note his discussion of the relatively large number of male favorites entertained by the Han emperors and their treatment by historians: they are usually condemned not for their sexual identities vis-à-vis the ruler, but for the threat they present in undermining the privileges of ministers to gain access to the ruler's presence and body (ibid., 34–54).

55. E.g., *Li ji* 4.20a; Legge, *Li Chi: Book of Rites*, 1: 248: "The seven lessons [of morality] were:—[the duties between] father and son; elder brother and younger; husband and wife; ruler and minister; old and young; friend and friend; host and guest." Anecdotal literature does address friendship at greater length, but it still remains problematic.

56. E.g., *Li ji* 7.7a–b; Legge, *Li Chi: Book of Rites,* 1: 379–80: "What are 'the things which men consider right?' Kindness on the part of the father, and filial duty on that of the son; gentleness on the part of the elder brother, and obedience on that of the younger; righteousness on the part of the husband, and submission on that of the wife; kindness on the part of elders, and deference on that of juniors; with benevolence on the part of the ruler, and loyalty on that of the minister;—these ten are the things which men consider to be right."

57. For one intelligent consideration of friendship within the context of literary production, see Connery, "Jian'an Poetic Discourse," 266–304. However, although he sees friendship as a relation that posed problems to authority (i.e., "factionalism"), he does not talk about the ideological tensions present within the construction of friendship itself—the degree to which it covers for competition and enmity.

58. For discussions of the honor codes and socially disruptive nature of the knight-errant, see James J. Y. Liu, *The Chinese Knight-Errant,* 1–54.

Chapter 2

EPIGRAPH: Xu Ling, *Yutai xinyong jian zhu,* 253.

1. Zhang Xuecheng, *Wenshi tongyi jiaozhu,* 80.

2. Those interested in pursuing the literary historical backgrounds of the genre have much new scholarship to peruse: Gong Kechang (*Han fu yanjiu*) and Ma Jigao (*Fu shi*) among others. If I fail to cite them more, it is because their concerns often lie outside my own interests. Much of their work demonstrates a desire on the part of Chinese scholars to rehabilitate the often-denigrated form. However, the very action of rehabilitation in such work often precludes the distancing of the scholar from his own role as advocate: there are frequent attempts in these works to champion anachronistic judgments on the nature of literature itself, in particular through a categorizing that polarizes the supposed "moral" or "aesthetic" goals of a text. Thus, as David Knechtges notes of Gong Kechang's study of the rhapsody:

> Gong Kechang began his study of the *fu* with no preconceptions. If he found that the *fu* was indeed a "mass of dregs," then he would "ruthlessly sweep it away." On the other hand, if he found something of value in it, he was prepared to affirm its worth.

Knechtges goes on to quote from an autobiographical piece of Gong's:

> As I read and wrote, I gradually discovered that the Han *fu* was not as worthless as people had imagined. If one assesses it against the concrete historical background of the Han and judges it from its place in the early stage of the development of literary art, we can see that not only is the Han *fu* not a mass of dregs, but even has its own outstanding features. The Han *fu* in fact has made a beneficial contribution to the development of Chinese literary art. (Gong, *Studies,* 7;

Knechtges translates from "Wo yanjiu Han fu de qianqian houhou" 我研究漢賦的前前後後, in *Han fu yanjiu*, 454–55).

Although Gong's championing of *fu* may be gratifying for the *fu* fan, the quality of that championing has far more to do with the cultural politics of the past hundred years than with any sensitive understanding of the special conditions of Han society and politics. Gong notes that poets employ "romantic" imagery on the one hand (p. 80) or are unfortunately "superstitious" (p. 275); poets are denigrated for embracing the "nihilism" of Laozi and Zhuangzi (p. 108) or for exhibiting a "conservative Confucianism" (p. 227). Above all, Gong and many other *fu* critics treat rhapsody texts as if they were transparent windows onto the attitudes of their authors and onto Han society. If we are to relate poetry to its society, subtler models are required.

3. Wai-yee Li, *Enchantment and Disenchantment*, 29.

4. Chen Shou, *San guo zhi*, 973–74. For this quote and the general reputation of the *Zhanguo ce*, see Crump, *Chan-Kuo ts'e*, 2–11.

5. Li Wai-yee, *Enchantment and Disenchantment*, 11. I cannot discuss the erotic and "feminized" aspects of the Han rhapsody without acknowledging my debt to her incisive discussion in *Enchantment and Disenchantment*, chap. 1. However, much of her argument is directed toward her broad interests in the fiction and drama of the late imperial period. My engagement with Li's work in this chapter is meant more to refocus her conclusions.

6. Yang Xiong, *Fa yan zhu*, 25.

7. Hawkes, *Classical, Modern and Humane*, 139–40, describing Sima Xiangru's *Rhapsody on the Great Man* (*Daren fu* 大人賦).

8. Xiao Tong, *Wen xuan*, 26; I benefit from Knechtges's thoroughly annotated translation, *Wen Xuan*, 1: 92–179. See also his discussion of the importance of ritual in Ban Gu's poem in "To Praise the Han."

9. Four eminent men who served as imperial chancellors under various Western Han rulers.

10. Xiao Tong, *Wen xuan*, 26.

11. Ibid., 30.

12. Three royal structures associated with early, idealized Zhou kings.

13. Xiao Tong, *Wen xuan*, 31–32.

14. Individuals are mentioned only for their ritual roles, not for their talent or abilities. To take a few lines at random: "Ritual officers put right the ceremonies, and only then does the royal carriage depart" 禮官整儀乘輿乃出 (ibid., 32); "The court officers follow him like his shadow, with majesty spreading and flourishing show" 天官景從寖威盛容 (32); "Then he lets rites flourish and music prosper, sets out the tents at his cloud dragon court" 勝禮興樂供帳置於雲龍之庭 (33); "He sets out the hundred officers and has them guide in all the nobles" 陳百寮而

贊群侯 (33); "They strike the bell to announce dismissal; and the hundred officials all withdraw" 撞鐘告罷百寮遂退 (34). Knechtges ("To Praise the Han," 129, 139) believes that the supremacy of ritual in the Eastern Han is one of the main points of the poem.

15. Xiao Tong, *Wen xuan*, 35.

16. For the questionability of reading Sima Qian as a "Confucian," see Puett, Review of Durrant, *The Cloudy Mirror*. The same factors would apply to Sima Xiangru: the fact that he protested to the emperor on policies he did not agree with does not make him "Confucian." The issue here, of course, is not whether they were "Confucians" or not, but whether a coherent "Confucianism" can be said to have existed during their lives.

17. This perspective on Han intellectual and literary history is common in scholarship: For discussions in English, see Loewe, *Crisis and Conflict in Han China* (for the rise of the *ru* perspective); Ebrey, "The Economic and Social History of Later Han"; and Ch'en Ch'i-yun, "Confucian, Legalist, and Taoist Thought in Later Han." It is also frequently noted how Ban Gu's *Han History* attempts to Confucianize the historiographic tradition in opposition to Sima Qian's supposedly more "open" attitudes toward the didacticism of past events; see, e.g., Watson, *Early Chinese Literature*, 106–8.

There is still debate over how this increasing power of the *ru* scholar represents the increasing power of a social/economic class, since the sources of social history in the Han are so meager: for an overview, see Ebrey, "The Economic and Social History of Later Han," and T'ung-tsu Ch'u, *Han Social Structure*. Martin Powers (*Art and Political Expression in Early China*, 73–103) attempts to argue for a "middle class" meritocracy emerging in the Eastern Han that expresses itself through its "classical" monumental art. Michael Nylan ("Style, Patronage and Confucian Ideals in Han Dynasty Art"), on the other hand, sees instead a Confucian elite increasing their stranglehold over economic power and access to government positions; for her, Confucian rhetoric concerning social status and merit is an ideological statement that conceals more cynical motives on the part of the *ru* group. She thus suggests that Powers's enthusiasm for the artistic expression of the *ru* in the Eastern Han ends in making him fall victim to their own propaganda. Christopher Connery (*Empire of the Text*, 79–109) has made similar criticisms of Powers's model in the course of his discussion of the *shi* class; although he does not show any awareness of Nylan's critique, he comes to many of the same conclusions. He perhaps takes the denial of ideological consciousness on the part of Eastern Han literati a little further than the evidence suggests—but he is concerned mostly to demonstrate the fallacy of interpreting late Han literati as valiant dissident "intellectuals" acting in opposition to imperial policy.

18. In critiquing the standard view of factionalism between "old text" and "new text" adherents, Michael Nylan ("The *Chin Wen/Ku Wen* Controversy in Han

Times," 117–20) suggests that court competitions on the Classics were most important for providing a road by which a scholar could attain high rank for himself and his family and that exhaustive erudition in the Classics and their commentaries had become essential for such competitors.

19. Xiao Tong, *Wen xuan*, 129.

20. The close readings of this chapter vary considerably from recent readings of the same texts, e.g., Kang, *Feng sao yu yanqing*, 76–91; and Zheng Yuyu, "Shennü lunshu yu xingbie yanyi," 44–53. Zheng's in particular is the most critically sophisticated and intelligent discussion of the Song Yu poems to date, and she is sensitive to issues of power, class difference, and the representation of gender. Nevertheless, in the end I disagree with many of her assumptions. First, there is her use of Freudian psychoanalytic theory, an application I find unjustified in discussing premodern Chinese literature. Second, I find that she accepts many of the conventional preconceptions of traditional "Confucian" literary criticism. For example, she accepts the standard (though questionable) biographical accounts of Qu Yuan and Song Yu; she believes that rhapsody poems are for the most part deliberately produced political texts, whose main purpose is to criticize the vices of the ruler or to provide a medium for expressing the poet's emotions and beliefs; and that even a poem like *Master Dengtu* is implicated inevitably in the dynamics of ministerial employment and the representation of public virtues. Making the texts so "monologic" threatens to suppress their complexity and the way they exploit the representation of audiences both inside and outside the poem. It also threatens to deny the marvelous humor and wit of the texts themselves.

21. At least since the Qing scholar Cui Shu (*Du feng ou shi*), the authorship of the Song Yu rhapsodies has been suspected, not least because Song Yu is portrayed as a character in them. Most recent scholars East and West have tended to agree, beginning most prominently with Lu Kanru (*Song Yu*). See also Knechtges, *The Han Rhapsody*, 123*n*14; Asano, "So Gyoku no sakuhin no shigi ni tsuite"; and Yang Yinzong, "Song Yu fu kao." Hu Nianyi (*Zhongguo gudai wenxue lungao*, 135–51) virulently defends Song Yu's authorship of the rhapsodies attributed to Song Yu in the *Wenxuan*. Ma Jigao (*Fu shi*, 38–43) is more cautious but suggests that some of the more "reliable" rhapsodies (notably *The Wind*, *Gaotang*, and *The Goddess*) may be early enough to have come from Song Yu's hand. Jiang Shuge (*Xian Qin cifu yuan lun*, 119–55) strongly defends Song Yu's authorship of *Gaotang* and *The Goddess* and goes so far as to suggest that these poems "founded" the rhapsody form. Asano, the most extreme in his skepticism, is willing only to give *termini ad quem* for the various rhapsodies: beginning of the third century for *The Goddess* and fifth century for *Master Dengtu*. My own impressionistic guess (based on the language and the themes) is that the Song Yu rhapsodies dealt with here are most likely Han works. Modern scholars are more quick to dismiss the authenticity of the *Meiren fu*, since its earliest

sources are the *Chu xue ji* 初學記 (early Tang) and the *Guwen yuan* 古文苑 (early Song). Since we cannot know for sure the order of composition for any of these poems, I have tried not to place them in any sort of hierarchy of development or to assume that one poem is deliberately "playing off" another (except in the case of the *Shennü fu*, which explicitly mentions the *Gaotang fu*).

22. Xiao Tong, *Wen xuan*, 190–91.

23. See Watson's discussion of the poem, *Early Chinese Literature*, 260–64.

24. *Mengzi zhengyi* 4.11a–12a.

25. Liu Wu, who was granted the title Prince Xiao of the Liang; a patron of Sima Xiangru during the reign of Emperor Jing.

26. Zou Yang was a prominent literary contemporary of Xima Xiangru's; at the time this rhapsody supposedly was taking place, he had taken refuge with the Prince of Liang after his earlier patron, the Prince of Wu, had revolted against the emperor. See also the epigraph to the Introduction.

27. When a gift of female musicians from Qi interfered with the holding of court, Confucius left in disgust. Legend has it that the philosopher Mozi's dislike of music and its supposed immorality was so strong that he refused even to enter a town whose name was Zhaoge (Dawn song).

28. The states of Zheng and Wei were famous for their beautiful women. The valleys of the Zhen and Wei and Sangzhong are mentioned in the *Shi jing* as locations for romantic trysts.

29. Reading the variant *mo ding* 脈定 for the *Guwen yuan*'s *qi fu* 氣服.

30. Sima Xiangru, *Sima Xiangru ji jiaozhu*, 80–85.

31. See James Hightower's discussion of *ding qing* and his translation of surviving rhapsodies on the theme in "The *Fu* of T'ao Ch'ien," 169–96.

32. Xiao Tong, *Wen xuan*, 268–69; compare Knechtges's recent translation, *Wen Xuan*, 3: 349–53.

33. Voluntary symbolic emasculation in order to gain access to another man's woman (and thus to cuckold him) has been analyzed by Sedgwick (*Between Men*, 49–66) in her analysis of Wycherley's *The Country Wife*. See also Tung, *Fables for the Patriarchs*, 129–30; she fails to notice the advantage of emasculation, however.

34. Although they do not emphasize the reversal of sexual roles here, both Zheng, "Shennü lunshu yu xingbie yanyi," 51–53; and Tung, *Fables for the Patriarchs*, 130, 151–55, perceptively note the degree to which the minister must surrender his sexuality to imperial authority.

35. Jean-Pierre Diény (*Pastourelles et magnanarelles*, 27–61) traces the sexual nuances of mulberry picking to earlier rites of fertility; for his observations on *Master Dengtu* and its participation in the mulberry tradition, see ibid., 96–99. For a thorough discussion of the most famous character of the *pastourelle*, Luofu, and the poems about her, see Allen, "From Saint to Singing Girl."

36. Xiao Tong, *Wen xuan*, 264–65.

37. Wai-yee Li, *Enchantment and Disenchantment*, 28.

38. Hay, "The Body Invisible in Chinese Art?," 51–59.

39. Xiao Tong, *Wen xuan*, 267–68; compare Knechtges's recent translation, *Wen Xuan*, 3: 339–49.

40. For a brief overview of the debate on this issue, see Wai-yee Li, *Enchantment and Disenchantment*, 28*n*50; and Zheng, "Shennü lunshu yu xingbie yanyi," 45*n*28. Although the most rational reading of the controversy sides with reading the dream as Song Yu's, some respected scholars still follow the *Wenxuan* text—e.g., Hu Nianyi, *Zhongguo gudai wenxue lungao*, 131. Schafer (*The Divine Woman*, 196*n*108) believes that most readers from the Tang and before would have read the dream as the king's.

41. A point suggested by Wai-yee Li, *Enchantment and Disenchantment*, 28. Others have also noticed the use of the verb *bai* 白 for "to relate," a word generally used by one of lower social status to one of higher.

42. Historically, one might assume a male reader for the Han; a female reader who is caught up in the process of narrative desire would, however, complicate the play of gender in interesting ways.

43. E.g., *Li sao*, ll. 9–12 (*Chu ci buzhu*, 4): " Already I had inward beauty in great measure, / And I coupled it with a refined appearance. / I donned fragrant river grass and hidden angelica, / I twisted autumn orchids for my girdle."

44. This interpretation has been taken by Knechtges in his translation, *Wen Xuan*, 3: 347.

45. Xiao Tong, *Wen xuan*, 270.

46. For translations, see, e.g., Whitaker, "Tsau Jyr's 'Luoshern fuh'"; and Knechtges, *Wen Xuan*, 3: 355–65. For a discussion, see, e.g., Whitaker, "Tsau Jyr's 'Luoshern fuh'"; Cutter, "The Death of Empress Zhen"; and Wai-yee Li, *Enchantment and Disenchantment*, pp. 33–36.

47. Ibid., 269–70.

48. See Whitaker ("Tsaur Jyr's 'Luoshern fuh,'" 44–50) for a summary of these issues. For an excellent survey of the historical sources and what they say about the empress, see Cutter, "The Death of Empress Zhen."

49. Liu Yiqing, *Shishuo xinyu jiao jian*, 489; following the interpretation in Cutter, "The Death of Empress Zhen," 578.

Chapter 3

EPIGRAPH: Liu Yiqing, *Shishuo xinyu jiao jian*, 365. In translating from *Worldly Tales*, I have benefited considerably from Mekada, *Sesetsu shingo*; and of course Mather, *Shih Shuo Hsin Yü*. I have avoided to some extent the strict literalness of Mather's renderings, and I do not explain allusions or historical detail if they do not contribute to the point I am making. Interested readers are urged to consult the original, as well as the notes in Mekada and Mather. Henceforth, *Worldly Tales* citations will be identified in the text, first by chapter and item number and then followed by page number in Liu Yiqing, *Shishuo xinyu jiao jian*. I also include dates the first time the more important figures are mentioned, based on Mather's calculations; however, there is occasional disagreement among authorities on exact birth and death dates for public figures during this era.

1. The original title of the text seems simply to have been *Shishuo* (Worldy tales); the appended description was added later to distinguish it from the "Worldly Tales" of Liu Xiang (80–9 B.C.).

2. For this commonly held view, see, among others, Lin Wenyue, "The Decline and Revival of *Feng-ku*," 152–59; and Liu Dajie, *Zhongguo wenxue fazhan shi*, 207–14, 219–27. Typically, such condemnations often involve the attributions to the poets of "aestheticism" (*weimeizhuyi* 唯美主義) and a disregard for social injustice.

3. Historians still debate on what constituted "elite" membership during this era—how "aristocratic" it was, the degree of social mobility, the role of the emperor in promoting talent, etc. I have tried to construct my argument around this gap, partly by stressing the degree to which the upper classes still possessed a shared cultural knowledge that marked them off as "literati." Thus, I generally use the term "elite" or "literati" in describing this class, with no intention of conflating it with the literati classes of the later imperial period.

4. On a more sophisticated, post-Foucauldian method of analyzing ideology, see Eagleton, *Ideology*, 1–61. It is important in this respect to emphasize the importance of not "intellectualizing" ideology and divorcing it from social practice. There is a commonly held view of the post-Han era that a "liberation of the self" (even an "individualism") came about, accentuated by the disintegration of political and social institutions. While this view is vital for our understanding of the period, we must also note that such arguments are largely based on an emphasis on the importance of neo-Taoist philosophical texts (see, e.g., Ying-shih Yü, "Individualism and the Neo-Taoist Movement"; and Wei-ming Tu, "Profound Learning, Personal Knowledge, and Poetic Vision"). My perspective will consider instead the use-value of such philosophy: not whether it "explains" literati cultural positions, but what the social action of taking any philosophical position in the third and fourth century implies.

5. From the Chinese scholarly perspective, a sampling might include Zhu Ziqing, "Shi yan zhi bian," in idem, *Zhu Ziqing gudian wenxue lunwen ji*, 185–355; Qian Mu, *Liang Han jingxue jiguwen pingyi*; Pan Zhonggui, "Shixu mingbian"; Wei Peilan, "Maoshi xu zhuan weiyi kao"; and Wang Wensheng, "Liang Han de wenxue lilun piping." Western works include Owen, *Readings*, 37–56; Saussy, *The Problem of a Chinese Aesthetic*; Pauline Yu, *Reading of Imagery*, 44–83; Jullien, *La Valeur allusive*; and James J. Y. Liu, *Chinese Theories of Literature*.

6. *Mao shi zhengyi*, 15–17.

7. James J. Y. Liu, *Chinese Theories of Literature*, 67–70, 111–12.

8. Van Zoeren, *Poetry and Personality*, 114.

9. Ibid., 111.

10. This is seen most clearly in Confucius' brief autobiography:

> The Master said, "When I was fifteen I set my mind to study; when I was thirty I took a stand; at forty I no longer had any doubts; at fifty I knew what Heaven wanted of me; at sixty, my ear was in tune with things; and at seventy I could do what I wanted without going beyond what was right." (*Lunyu* 2.4; Yang Bojun, *Lunyu yizhu*, 12).

11. This argument has been made most cogently by Herbert Fingarette in *Confucius: The Secular as Sacred*. Although many have critiqued or revised his argument, his essential insight is still valid.

12. Owen, *Readings*, 45.

13. Pauline Yu, *Reading of Imagery*, 79–80. See also C. H. Wang, "Towards Defining a Chinese Heroism."

14. See the discussion in Chapter 2; for sources, see note 17 to that chapter.

15. For Confucian scholarship at this time, see Jian, *Jincun Nanbeichao jingxue yiji kao*; and Yang Donglin, "Lüe lun nanchao de jiazu yu wenxue," 5–7, for the social importance of family scholarly traditions. For the importance of scholarship in maintaining one's status as "elite," see Yan Zhitui's admonishment to his sons, *Yan shi jia xun jijie*, esp. 143–236.

16. Much later (from the Tang and especially the Song dynasties), the supposed "borrowing" of literary folk traditions from "the people" becomes a complex ideological discourse in its own right. Often perceived by its proponents as the "sweeping away" of decadent literary habits in the elite by the "fresh" products of the lower classes, such moves actually reveal in part the ambivalence later elites felt over creating specialized literatures for their own consumption in the light of Confucian claims that good literature is somehow spontaneous and ethically unmediated by rhetoric. In this early period, the "unrhetorical" is not associated with "the people," but as we have seen already, there is a "complexity vs. simplicity" discourse already partially in existence at this early date (cf. Yang Xiong's critique of rhapsodies).

17. *Yuefu* are named for the Western Han "music bureau," which in later legend was said to have collected folk songs from the populace. Later literati seem to have imitated the music of such poems and occasionally adopted themes characteristic of them. There is, of course, little textual evidence of a movement from "folk" or "popular" to "elite." Joseph Roe Allen (*In the Voice of Others*, 37–68) has demonstrated the difficulties of tracing literati *yuefu* back to folk antecedents. However, there is little doubt that early literati composition shows a shared use of topoi, images, and metaphors that derive from a world of formulaic oral composition,

although there is no need to see such a world as a "folk" world. It also seems evident that poems such as the "Nineteen Old Poems" reflect an urban popular culture; literati may have composed such poetry, but they did so while participating in such a culture. In the discussion that follows, I am more interested in showing how certain poets employed formulaic topoi to express historically defined situations.

18. See Frankel, "The Development of Han and Wei *Yueh-fu*," for an exhaustive catalogue of early literati themes.

19. Connery ("Jian'an Poetic Discourse") deals with this issue at much greater length. Although highly conscious of the social ideology of Jian'an poetry, he is also interested in how these poets constructed a peculiar "subjectivity." My point here is a simpler one and is limited only to the issues of poetry as a way of constructing class identity.

20. Lu Qinli, *Quan xian Qin Han Wei Jin Nanbei chao shi*, 332.

21. Thus, I believe Ying-shih Yü ("Individualism and the Neo-Taoist Movement," 127) is off the mark when he says that the poems "show in a highly personal way their authors' inner experiences with the fleetingness of life, the sadness of parting, the emptiness of fame, etc." Strong emotion in a poem is no guarantee that it is "personal" to its author, or that it (as Yü suggests) contributes to "individualism." In general, Yü fails to take into consideration the degree to which Jian'an and Wei literature share the same tropes and rhetorical structures.

22. Lu Qinli, *Quan xian Qin Han Wei Jin Nanbei chao shi*, 365. I have rendered the Chinese *chai* (dhole) as "jackal" to convey in English the force of Wang's meaning. Note Ronald Miao's sensitive reading of this poem in *Early Medieval Chinese Poetry*, 130–45.

23. Chen Shou, *San guo zhi*, 597–98. Everyone is in agreement on this as the occasion for writing. Miao (*Early Medieval Chinese Poetry*, 38) emphasizes the importance of historical contextualization for Wang's poem: "Indeed, the popularity of this poem in the native literature is due in part to its successful expression of the *zeitgeist* of the Chien-an [Jian'an] period."

24. Jean-Pierre Diény ("Lecture de Wang Can," 291–99) tends toward this self-consciously "literary" method of interpretation in stressing the intertextuality of Wang's poems. However, his brilliant examination of how poems can be constructed intertextually during this period is invaluable and a necessary corrective to excessive "individualism" in reading.

25. Ban Gu, *Han shu*, 1755.

26. *Mao shi zhengyi*, 271.

27. Although many anonymous "old poems" incorporate tag lines from the *Odes*, they make no reference to the historical contextualization of the *Odes* that would have been part of literati knowledge.

28. Zhang Keli, "Jian'an shige," 119–22, gives the standard criticism; for a defense of Cao Zhi's formal banquet poems, see Cutter, "Cao Zhi's Symposium Poems." For group composition during Jian'an, see Connery, "Jian'an Poetic Discourse," 281–304.

29. For Liu Xie's evocation of Jian'an social occasions, see Liu Xie, *Wenxin diaolong zhushi*, 49. For the *Literary Anthology*'s selection of Jian'an banquet poems, see Xiao Tong, *Wen xuan, juan* 20.

30. For an evaluation in the context of the late Han world and its literature, see Connery, "Jian'an Poetic Discourse," 243–66. For a review of the standard scholarship on the Nine Rank system, see Holcombe, *In the Shadow of the Han*, 77–81. Note also the odd resonance in the link between medieval Western literary and social hierarchies noted by Stallybrass and White (*The Politics and Poetics of Transgression*, 2); after observing Ernst Robert Curtius's discovery that the word "classic" was derived from a category of property qualification and taxation in the Roman empire, they comment: "From the first it seems that the ranking of types of author was modeled upon social rank according to property classifications and this interrelation was still being actively invoked in the nineteenth century."

31. Xiao Tong, *Wen xuan*, 720.

32. Ibid.

33. A similar argument has been made by Chi Xiao ("Lyric Archi-Occasion"), although he is addressing what he believes to be characteristics of Chinese poetics in general: he sees poetic individuality located in the ability of a poet to locate himself within the experiences of the poets of the past. However, the degree to which this operates differs from age to age. I am more interested here in poetry as a way of defining the coherence of social groups. Connery ("Jian'an Poetic Discourse") has suggested ways in which this construction occurred during the Jian'an.

34. The best overall discussion of *Worldly Tales* as literature is Qian Nanxiu, "Being One's Self," which focuses on the construction of typology as well as the narrative art of the text.

35. Liu Wenzhong ("*Shishuo xinyu zhong de wenlun gaishu*," pp. 102–3) points out that its compilation was part of a series of state-supported literary projects to inaugurate the new dynasty.

36. Of course, the movement from imperial bureaucracy to the more personal selection of entourages among local landowners and "warlords" contributes in part to the conception of this era as "feudal" (although this term is more misleading than helpful; see Ying-shih Yü, "Individualism and the Neo-Taoist Movement," 122–25). Regardless of how we analyze overall social change during this period, I believe we are not far off the mark if we point out the degree to which literati were forced to develop strategies of self-evaluation to replace imperial legitimization. Such activity can be interpreted as the development of "individualism," but it does not have to be.

For a standard account of literati attitudes during the period of crisis, see Balazs, "Political Philosophy and Social Crisis at the End of the Han Dynasty," in *Chinese Civilization and Bureaucracy*, 187–225.

37. Holzman (*Poetry and Politics*) demonstrates the importance of this factor in examining the works of Ruan Ji (210–63). See also Mather, "Individualist Expressions of the Outsiders."

38. The complexity of defining "aristocratic" status as opposed to "bureaucratic" status is reviewed by Holcombe, *In the Shadow of the Han*, 37–42; and Ebrey, *The Aristocratic Families of Early Imperial China*, 15–33. See also David Johnson's (*The Medieval Chinese Oligarchy*, 153n1) argument for terming the elite an "oligarchy."

39. Diény, *Portrait anecdotique d'un gentilhomme chinois*, 13–18.

40. On this issue, see Mather, "Intermarriage as a Gauge of Family Status."

41. I base my interpretation of the obscure conclusion to this item on Diény's reading in *Portrait anecdotique d'un gentilhomme chinois*, 17.

42. Holcombe, *In the Shadow of the Han*, 80–82.

43. Grafflin, "The Great Family in Medieval South China"; and idem, "Reinventing China."

44. Holcombe, "The Exemplar State." He draws much of his inspiration in this argument from Tanigawa, *Medieval Chinese Society*.

45. Two noted attempts to reconstruct *qingtan* from *Worldly Tales* evidence are Pu Meiling, *"Shishuo xinyu" zhong suo fanying de sixiang*; and Yiming Tang, "The Voices of Wei-Jin Scholars."

46. Such forms of cultural production are certainly not strange to other cultures; Pierre Bourdieu (see. e.g., *The Field of Cultural Production*, 29–61) for one has noted the creation within complex societies of autonomous "fields" wherein competence and expertise are granted special authority and prestige, in spite of their seeming independence from more transparent forms of power—e.g., economic, political, religious. I will not push the Bordieuian analysis here too far, because I believe that we still know too little of Six Dynasties society to be able to determine where power lay exactly; Holcombe's interesting analyses, for example, suggest that to some extent power could extend itself in nonmaterial ways. I think it indisputable, however, that the cultural competitions of the Jin era are portrayed in such a way to let us know that the parties feel they are fighting for significant prizes—however those prizes played out in precise societal terms.

47. Mei Jialing ("Yiwei yu fude yu caixing zhi jian") provides a thoughtful discussion of the "Worthy Beauties" (*xian yuan* 賢媛) chapter of *Worldly Tales*. Although her concern with the developing role of women in Chinese society and the discourse of "female virtue" overlaps mine in places, her focus is somewhat different.

48. There is some disagreement among commentators over the identity of each of these Xie cousins.

49. E.g., Van Gulik, *Sexual Life in Ancient China*, 92–93; and Hinsch, *Passions of the Cut Sleeve*, 68–69.

50. Some may see some resemblance between *yaliang* and the Castiglione-inspired idea of *sprezzatura* (see e.g., Chennault's brief mention of the term in conjunction with later literati culture, in "Palace-Style Ladies and Odes on Objects," 18*n*17).

51. In this respect, one should note that Mather's ("Individualist Expressions of the Outsiders," 200–204) characterization of the Seven Worthies as "outsiders" may hold true only for their own times and not for the period of their idealization enshrined in *Worldly Tales*. Although there was an explicit intellectual and moral debate over whether their position was correct, their actions came to participate fully within the "discourse" of literati action in the third century.

52. Note also Bourdieu's discussion of gift-giving and symbolic violence among the Kabyle tribesmen of Algeria in *Logic of Practice*, 98–111.

53. *Jin shu*, 1008.

54. Owen, *Mi-Lou*, 146.

55. Yang Donglin ("Lüe lun nanchao de jiazu yu wenxue," 5–7; 18–19) recognizes the similar roles of *qingtan* and poetry as marks of upper-class distinction.

56. My biography of Xie here is derived from Frodsham, *The Murmuring Stream*; and Obi, *Sha Reiun*.

57. Poem found in Lu Qinli, *Quan xian Qin Han Wei Jin Nanbei chao shi*, 935.

58. I follow previous precedent in calling the Song "Liu Song" after the surname of its founder and to distinguish it from the later Song dynasty (960–1279).

59. See, e.g., the comments of Zhong Rong, *Shi pin ji zhu*, 160–61; and Liu Xie, *Wenxin diaolong zhushi*, 49, 493.

60. E.g., Lin Wenyue, *Shanshui yu gudian*; Kang-i Sun Chang, *Six Dynasties Poetry*, 47–78; Westbrook, "Landscape Description"; and Frodsham, "The Origins of Chinese Nature Poetry." One of the most interesting recent discussions is Li Chunqing's ("Shiren yu ziran"), which relocates "landscape" issues within broader philosophical and literary critical issues. For a refreshingly skeptical view of Xie's landscape description, see Pauline Yu, *Reading of Imagery*, 148–59.

61. Kunlun was a great, semi-mythical mountain in China's far west.

62. Anqi was an early Taoist adept who had mastered the alchemical arts of immortality.

63. Lu Qinli, *Quan xian Qin Han Wei Jin Nanbei chao shi*, 1162; I have also consulted Huang Jie's commentary (see Xie Lingyun, *Xie Kangle shi zhu*, 92–93).

64. Shen Yue, *Song shu*, 1754. "Noble and low-born" here would not necessarily represent "literati and commoners," but rather high- and low-level gentry. Later in the passage, Shen explains that his literary fame made Xie generally famous, so that when he returned to the capital, both literati and commoners thronged to see him.

Of course, it is difficult to determine whether these categories have any meaning when the text seems intent on merely rhetorically demonstrating that Xie's poems were extremely popular. Surely the erudition of Xie's poetry and the difficulty of their language (a difficulty recognized in criticisms of Xie less than a hundred years after their composition) would make his works largely incomprehensible to all but the thoroughly literate.

65. Lu Qinli, *Quan xian Qin Han Wei Jin Nanbei chao shi*, 1167; Xie Lingyun, *Xie Kangle shi zhu*, 128–29.

66. *Nan shi*, 539.

67. On this, see especially Grafflin, "The Great Family in Medieval South China." Both Yang Donglin ("Lüe lun nanchao de jiazu yu wenxue") and Yan Caiping ("Shi shu guanxi") discuss shifting relations between aristocratic writer and ruler during the Liu Song; however, they disagree on the full significance of this relationship and the degree of imperial power it implies.

68. Wang Yunxi and Yang Ming (*Wei Jin Nanbei chao wenxue piping shi*, 495–96) discuss some of these early critical works, now lost.

Chapter 4

EPIGRAPH: Xu Ling, *Yutai xinyong jian zhu*, xiii.

1. Ban Gu, *Han shu*, 3983–84; Lady Fan was a consort of the King of Chu. She refused to eat meat when she felt her husband was wasting his time hunting.

2. Ibid., 3984; these are probably the names of female conduct books.

3. Ibid., 3984–85.

4. A "woman at dawn" is (metaphorically) a hen that crows instead of a rooster—i.e., a domineering woman. Baosi helped bring about the fall of the Western Zhou.

5. Ban Gu, *Han shu*, 3985–86.

6. Ibid., 3987. "Green Robe," Ode no. 27, is said to describe a wife replaced by concubines. "White Bloom," Ode no. 229, is said to criticize King You 幽 for rejecting a virtuous queen in favor of Baosi.

7. Text in Xiao Tong, *Wen xuan*, 198–200.

8. In this respect, one may point to two figures we have already seen as forerunners of Eastern Han scholarship. Yang Xiong was the author of a rhapsodic critique of Qu Yuan, the *Fan Sao* 反騷 or "Against the *Li sao*" (reprinted in *Chu ci jizhu*, 236–43); Ban Gu objected explicitly to Qu Yuan's lack of self-restraint in his biography of the poet in his "Preface" to the *Li sao*. For a brief survey of early attitudes toward Qu Yuan, see Schneider, *A Madman of Ch'u*, 17–86.

9. Lu Qinli, *Quan xian Qin Han Wei Jin Nanbei chao shi*, 116–17.

10. See, e.g., notes on the text's authenticity in *Liang Han wenxueshi cankao ziliao*, 533–34; and Yu Guanying, *Han Wei Liuchao shi xuan*, 39. The best recent discussions

of the poem's provenance and of its literary merit are You Shihong, "Shi lun 'Yuan ge xing,'"; and Knechtges, "The Poetry of an Imperial Concubine."

11. Zhong Rong, *Shi pin ji zhu*, 94.

12. For a review of the roots of the courtier / court lady reading, see Miao, "Palace-Style Poetry," 1–5. Although he is concerned mostly with a Tang allegorical reading, he points out the beginnings of the reading with Qu Yuan and surveys the roots of "palace poetry" thematics in the Lady Ban poems (ibid., 6*n*7). See also Chennault, "Palace-Style Ladies and Odes on Objects," for such readings in the Six Dynasties.

13. Xiao Tong, *Wen xuan*, 227.

14. Knechtges, "Ssu-ma Hsiang-ju's Tall Gate Palace Rhapsody," 51–54.

15. These two perspectives on representing women guide Kang's (*Feng sao yu yanqing*) history of gender in literature; he terms the two approaches the "*Odes* and *Li sao*" (*feng sao* 風騷) perspective, on the one hand, and "erotic passion" (*yanqing* 艷情), on the other. Although we might acknowledge the appropriateness of his model for some situations, we can only admit that divisions were never mapped that clearly in poetic practice. Such dialectical give-and-take (as well as the temptation to judge a work or poet as falling on one side or the other) is much more typical of modern literary histories.

16. Zhong Rong (*Shi pin ji zhu*) here quotes the *Analects*, 17.8:

> The Master said, "Little ones, why do you not study the *Odes*? With the *Odes* you can express your mood, you can observe, *you can use it to be social, and you can use it to express your resentment.* Nearby you can use them to serve your father; far away, to serve your lord. And you can learn the names of many birds, beasts, plants, and trees." (*Analects* 17.9; Yang Bojun, *Lunyu yizhu*, 185)

Wang Yunxi and Yang Ming (*Wei Jin Nanbei chao wenxue piping shi*, 503–4) point out that Zhong quotes only this passage because he wishes to emphasize the roots of the *Odes* in emotion and de-emphasize its explicit moral and educative aims. I also believe he does this because he wishes to emphasize the functions of poetry as social exchange.

17. Zhong Rong, *Shi pin ji zhu*, 47, 54.

18. The "tumbleweed" (*peng* 蓬) is a species of artemisia that comes loose from its roots when it dies, allowing its dried remains to be driven about by the wind. Although not exclusively a frontier image, poets use the tumbleweed as an image for the self when they are forced to travel long distances, often to frontier locations.

19. Zhong Rong, *Shi pin ji zhu*, 50. There is another opinion on the Han lady that associates her with Wang Zhaojun; for Wang's importance in the "self-expressive" tradition, see Chapter 5.

20. Lu Qinli, *Quan xian Qin Han Wei Jin Nanbei chao shi*, 330.

21. Xiao Tong, *Wen xuan*, 410.

22. However, it may very well be true for late imperial commentators living in a "neo-Confucian" world.

23. Note in particular nos. 1, 6, and 9. Later commentators try to be very specific about the scenario, but an honest reading reveals the impossibility of proving the gender of speaker and of addressee. Diény (*Les dix-neufs poèmes anciens*, 54–59) points this fact out in the case of the first poem, and I believe this holds true for six and nine as well, although I am less certain of the "real" meaning in these cases than he is.

24. Zhong Rong (*Shi pin ji zhu*, 97–98) is especially of this opinion. Note also the highly positive comments garnered from Xie Lingyun, Shen Yue, and Liu Xie (ibid., 109). Ge Xiaoyin ("Lun Nanbei chao Sui Tang wenren," 3–7) notes that fifth- and sixth-century opinion on Cao and on the Jian'an period in general is almost universally positive but often varies somewhat based on the personal stance of the critic. The propensity to see in Cao the best qualities (no matter what they might be) suggests a general consensus over his importance.

25. For a detailed discussion of this term in relation to Cao and the Jian'an poets, see Lin Wen-yueh, "The Decline and Revival of *Feng-ku*." Lin chooses to make value judgments on poets, reading *fenggu* as a quality sadly lacking in later poets until the Tang. Although she uses Zhong Rong as a support for this position, she makes no attempt to discuss how Zhong's position itself is rooted in his own ideological presuppositions.

26. Fong, "Persona and Mask in the Song Lyric," 459–65. More specifically:

> The female persona in *ci* [song lyric] disengages the writer's direct investment of self; its use creates an aesthetic distance between the poet and his work, and makes his work an art of presentation rather than an embodiment of self-expression. (462)

Also:

> One important aspect of these songs [i.e., those that employ "masks"] is the assumption of "pose" or "role," and I use the term "mask" to stress how these poses call attention to the difference between the actual person and the assumed role, a difference that is not in question in the use of persona. (465*n*11)

Obviously, my use (and then dismissal) of these terms to clarify Cao Zhi's use of voice does not in any way undermine Fong's use of the terms in her own discussion; I merely employ them as a helpful tool of analysis.

27. Lu Qinli, *Quan xian Qin Han Wei Jin Nanbei chao shi*, 457.

28. Ibid., 425.

29. The most famous example is the "Mingdu pian" 明都篇; see Lu Qinli, *Quan xian Qin Han Wei Jin Nanbei chao shi*, 431.

30. Ibid., 455–56.

31. Hans Frankel's famous dismissal of the autobiographical readings (in "Fifteen Poems by Ts'ao Chih") in trying to locate a universally valid work of art succeeds

ultimately only in divorcing the poet from the context that made him important and interesting. David Roy, in a slightly older essay ("The Theme of the Neglected Wife"), has no problem in assuming the poems are autobiographical, and other scholars join him. A fairly early tradition links this poem of Cao's with a specific case of an abandoned wife that elicited the sympathy of a number of the Jian'an poets (see Cao Zhi, *Cao Zijian shi zhu*, 103, for the account—now preserved in the Ming commentary on *Jade Terrace*); it may well be that in this case a "real" abandoned woman triggered Cao's own response because of his empathy for her position. Zhang Lei on the other hand suggests that an attempt to associate the poems with specific historical circumstances be abandoned in favor of an acknowledgment that such poems are a product of a certain psychological state of mind on the part of the author. I agree with Zhang here, except that the "psychological state" of the *reader* may be more important in the long run.

32. As will become clearer later on, I am suggesting that the possibility of "male" self-expression in "female"-voice sixth-century poetry is very real, in spite of most critics' dismissal; note, e.g., Miao's ("Palace-Style Poetry," 30–31) suggestion that such readings become viable only in the Tang, in spite of the previous existence of an allegorical tradition for female-voiced poetry. To a detailed extent, Cynthia Chennault ("Palace-Style Ladies and Odes on Objects") *has* suggested the possibility: she reads *yongwu* 詠物 (poems on things) as well as many female-voiced "palace poems" as pleas for advancement on the part of courtiers united to some extent in a court culture under the increasingly "autocratic" regimes of the Qi and Liang. I am indebted to her individual readings of poems and acknowledge her percipience in recognizing the importance of a social context for female-voiced poetry—her realization that such "self-expressive" readings (what she terms the creation of a "public persona") are always possible when literati are writing within a "courtly" social context. For a further discussion of how to interpret the social context for these issues, see the section "Aside: Court Poet or Poet at Court?" below in this chapter.

33. Some might claim that Zhong's admiration for the past, his mild dislike for Xie Tiao 謝朓, and his attack on "tonal propriety" make him essentially a conservative, and consequently hostile to court society. Although it is true that one can create "oppositional camps," placing Zhong on one side, Shen and Tiao on the other, such a view tends to ignore the way in which *Shi pin* itself participates in literati society through its rhetoric of evaluation and high-minded connoisseurship.

34. *Liang shu*, 109.

35. Kang, *Feng sao yu yanqing*, 133–34.

36. Miao ("Palace-Style Poetry," 12) suggests "poetry of glamour."

37. *Liang shu*, 446–47. Generally speaking, the *qing* element in *qingyan* suggests that "palace poetry" is not limited to boudoir subjects alone but could also include insufficiently "serious" poetry. The Xu Chi biography, on the other hand, empha

sizes the novelty and originality of the verse, no doubt referring in part to its increasing emphasis on metrical proprieties. With these different accounts in mind, Zhou Zhenfu (*Shiwen qianshu*, 141–48) suggests that "palace poetry" should apply in general to the poetry of Xiao Gang's salon and not specifically to the "erotic" verse. Such an interpretation emphasizes that later "moral" readers of the Sui and Tang condemned the poetry not necessarily because of its preoccupation with "sex" (the privileging of "sex" as the first target of premodern moralists is often a twentieth-century invention), but rather because of its betrayal of orthodox poetic practice. See also Yang Ming, "Gongti shi pingjia wenti"; and for an interesting discussion on the impact of Buddhism on court verse, Ma Jigao, "Lun gongtishi yu fojiao."

38. Zhang Pu, *Liang Jianwen Di ji*, in idem, *Han Wei Liu Chao bai san ming jia ji*, 1.62a (4: 210); see also Hightower's (*Topics in Chinese Literature*, 46) interpretations of Xiao Gang's attitudes.

39. Guo Shaoyu, *Zhongguo gudian wenxue lilun piping shi*, 81. In his more recent revisions of this work, he seems less hostile to Xiao Gang. On the other hand, John Marney (*Liang Chien-wen Ti*, 93–97) anachronistically suggests that Xiao Gang's comment implies an incipient "art for art's sake" attitude.

40. Wang Yunxi and Yang Ming, *Wei Jin Nanbei chao wenxue piping shi*, 299; Deng Shiliang, "Shi 'fandang.'"

41. Zhang Pu, *Liang Jianwen Di ji*, in idem, *Han Wei Liu Chao bai san ming jia ji*, 1.55–56 (4: 207).

42. Liu Su, *Da Tang xin yu*, 1: 28.

43. The use of *da* 大 in the anecdote (translated here as "magnify") has exercised a number of recent scholars, e.g., Mu Kehong, "Shi lun *Yutai xinyong*"; and Liu Yuejin, *Jie wang man lu*. Most of them assume that the use of the term *da* can be attributed to Xiao Gang himself and not to the author of the ninth-century miscellany, and so try to see it as a crucial moment in Xiao Gang's poetics.

44. Text for preface from Xu Ling, *Yutai xingyong jian zhi*, xi–xiii. Other translations may be found in Birrell, *New Songs* (337–41); and Hightower, "Some Characteristics of Parallel Prose," 77–87. My own translation owes much to the latter, as well as to Uchida Sennosuke's Japanese translation and commentary, *Gyokudai shin'ei*.

As will become evident from the translation, the preface is written in a highly rhetorical style in which parallel phrases abound; this form is commonly known as *pianti wen* 駢體文, or "parallel prose." This rhetorical structure (combined with the consistent use of rhyme) has led the other translators to render the text in verse. I have avoided emulating them chiefly because I believe Xu's "argument" becomes clearer when read as prose. I have also occasionally repressed the literal translation of allusions, preferring to give the meaning Xu is aiming for. I can only apologize for this practice by claiming that many of the allusions are introduced for the sake of ornament and have little to do with the meaning Xu is intending to convey; that many of the allusions would take a great deal of space to explain and would distract the reader unnecessarily; and that a number of the allusions are still inexplicable—either through a corruption of the text or through the subsequent loss of cultural knowledge. Those interested in reading a more "precise" rendering as well as an explanation of the allusions are urged to consult Hightower, "Some Characteristics of Parallel Prose"; Birrell, *New Songs*; and Uchida, *Gyokudai shin'ei*.

45. Not all these place-names have been identified; some of them are known as the original homes of a number of parvenu empresses. The source of "Lady Charming" is unknown, but "Skillful Smile" was a favorite of Cao Pi's credited with inventing the application of rouge to the face.

46. The Eastern Neighbor as a forward, promiscuous woman is mentioned in *Master Dengtu the Lecher*: see Chapter 2. Xi Shi was the peasant girl trained as a femme fatale by the King of Yue 越 and sent as a present to the King of Wu 吳.

47. For a discussion of this legend, see Chapter 5.

48. This is an important point. Although not all "palace poetry" was written at court under the eyes of the emperor (the competitive games that it plays are not always "court" games), the act of canonization attempted by Xu requires that the poetry be tied to significant political action in relation to the ruler.

49. The peony is possibly an allusion to Ode no. 95, in which the flower is exchanged between lovers. The grape vine has not been satisfactorily explained.

50. Note Chennault's ("Palace-Style Ladies and Odes on Objects," 5) suggestive comment on the link between female ornament and male talent: "Perhaps it is by the application of rouge or face powder, by the addition of an ornament to their coiffures, that we may understand the process of nature's conversion into things both alluring and serviceable."

51. Ban Gu, *Han shu*, 1755.

52. I base my reading here on Hightower's speculative note, "Some Characteristics of Parallel Prose," 83*n*7.

53. "Arrow-toss" is a game in which contestants attempt to throw arrows into a narrow-mouthed pot from a distance. An early myth states that when the goddess Jade Girl fails to hit the mark in her game, Heaven laughs, thus producing lightning.

54. *Liubo* was an early Chinese board game in which draughts were moved about on a board based on the throw of different colored tallies. See Lien-sheng Yang, "A Note on the So-called TLV Mirrors and the Game *Liu-po*"; and idem, "An Additional Note on the Ancient Game *Liu-Po*" for details.

55. *Mao shi zhengyi*, 104.

56. See Kang-i Sun Chang, *Six Dynasties Poetry*, 51–54; Lin Wenyue, *Shanshui yu gudian*, 41. Yu-kung Kao ("The Aesthetics of Regulated Verse," 348–51) approaches Xie's experiments from a slightly different direction, but stresses many of the same qualities.

57. See Mair and Mei, "Sanskrit Origins," for a thorough investigation into the origins of tonality.

58. Kang-i Sun Chang, *Six Dynasties Poetry*, 115–25.

59. Stephen Owen (*Poetry of the Early T'ang*, 425–28) has termed this the "tripartite form," which became fully standardized by the "Early Tang" period (seventh century).

60. It is the relative decline in laments from palace women specifically that leads Miao ("Palace-Style Poetry," 30–31) to suggest that they are not an "important" element in palace poetry. However, there are still quite a substantial number that express the despair of the harem favorite—even if the theme is less prevalent than in the Tang.

61. These issues have been discussed in part in Rouzer, *Writing Another's Dream*, 71–77.

62. Xu Ling, *Yutai xingyong jian zhi*, 330–31; Uchida, *Gyokudai shin'ei*, 502–3.

63. *Mao shi zhengyi*, 251.

64. Xu Ling, *Yutai xingyong jian zhi*, 238; Uchida, *Gyokudai shin'ei*, 385.

65. Owen, *Traditional Chinese Poetry and Poetics*, 63–73.

66. A dynamics observed by Rouzer, *Writing Another's Dream*, 71–77; Fong, "Persona and Mask in the Song Lyric," 460; and Robertson, "Voicing the Feminine," 68–72.

67. Girard, *Deceit, Desire, and the Novel*, 1–52. Girard's later work continues his theory of triangular desire (which he then terms "mimetic desire").

68. Stressing this multivalent quality perhaps somewhat more than Chennault ("Palace-Style Ladies and Odes on Objects") does, who sees the poems more as conscious appeals to imperial largess on the part of the courtier poet.

69. The poetry composition thus bears a limited resemblance to the "fair" as discussed by Stallybrass and White in *The Politics and Poetics of Transgression* (basing themselves on work of Bakhtin). We are not dealing here primarily with class divisions, however, and the mixture of an ideological "high" and "low" that is important to Bakhtin's work. The poetry salon is still "elite." The indeterminacy of gender, however, does create the sort of confusion between participant and observer that Stallybrass and White describe. For a more "Bakhtinian" text, see Chapter 7.

70. Chennault, "Palace-Style Ladies and Odes on Objects," 2.

71. Ibid., 7.

72. Chartier, *Cultural History*, 81–82.

73. Ibid., 83–84.

74. For *houjin*, see Morino, "Ryōsho no bungaku shūdan," 94.

75. Holcombe, *In the Shadow of the Han*, 136–37.

76. Grafflin, "The Great Family in Medieval South China," 73.

77. Yan Caiping, "Shi shuo guanxi"; Yang Donglin, "Lüe lun nanchao de jiazu yu wenxue."

78. Xu Ling, *Yutai xingyong jian zhi*, 301; Uchida, *Gyokudai shin'ei*, 479–80. This poem is also discussed in Rouzer, *Writing Another's Dream*, 78–79.

79. Xu Ling, *Yutai xingyong jian zhi*, 337; Uchida, *Gyokudai shin'ei*, 511–12. This poem may not have been written specifically to match Xiao's poem, because the rhyme is different. But obviously the situation is the same.

80. The siege is recorded in *Shi ji* (Takigawa, *Shiki kaichū kōshō*, 56.14); for a detailed account of how the story is represented, see Chapter 5 below.

81. For a thorough examination of the cultural and literary connotations of mirrors in *Yutai* verse, see Birrell, "Erotic Decor," 193–208.

82. Xu Ling, *Yutai xingyong jian zhi*, 251; Uchida *Gyokudai shin'ei*, 407–8.

83. Xu Ling, *Yutai xingyong jian zhi*, 205; Uchida, *Gyokudai shin'ei*, 339.

84. Xu Ling, *Yutai xingyong jian zhi*, 513; Uchida, *Gyokudai shin'ei*, 699.

85. For similar "subversive" strategies in the female use of the mirror, see La Belle, *Herself Beheld*.

86. Xu Ling, *Yutai xingyong jian zhi*, 357–58; Uchida, *Gyokudai shin'ei*, 541–42.

87. From the preface to Fan Tai's 范泰 (355–422) praise poem on the *luan* bird, in Lu Qinli, *Quan xian Qin Han Wei Jin Nanbei chao shi*, 1144.

88. Xu Ling, *Yutai xingyong jian zhi*, 255–56; Uchida, *Gyokudai shin'ei*, 416–17.

Chapter 5

EPIGRAPH: Lu Qinli, *Quan xian Qin Han Wei Jin Nanbei chao shi*, 199.

1. I use "khan" throughout this chapter to translate the Xiongnu title *shanyu* 單于; although this is historically anachronistic, I prefer it stylistically.

2. I.e., the *yanzhi* 閼氏, the official title of a khan's wife or mother.

3. Citations in this chapter are to Takigawa, *Shiki kaichū kōshō*; *juan* number is followed by pages (each juan is paginated separately).

4. For basic survey accounts of Xiongnu history in English, see Ying-shih Yü, "Han Foreign Relations," 383–405; and for an account from the Xiongnu point of view, Barfield, *The Perilous Frontier*, 32–84.

5. When I speak of "Chinese" identity throughout this chapter, I do not have modern conceptions of nationalism or ethnic identity in mind. However, by the Han, it is quite obvious that members of the Chinese cultural sphere are creating a language of cultural difference that separates themselves from non-Chinese peoples. It is in fact ethnographic discourse that largely defines what it is to be "Chinese" at this period. See Hartog, *The Mirror of Herodotus*, for a quite similar analysis of ancient ethnography in Greece. Schaberg ("Travel, Geography, and the Imperial

Imagination") goes into several theoretical issues more thoroughly than I have here and has paid particular attention to the foundation of imperial discourse in pre-Han texts.

6. For the accuracy of historical accounts of army sizes, see Loewe, "The Campaigns of Han Wu-ti," 90–96.

7. References in the text to the *Han shu* are to the Zhonghua edition, *juan* followed by page number.

8. Dissenting hostility to this policy during the Han is well-documented; see, e.g., Kwong, *Wang Zhaojun*, 43–55.

9. The account of how Danfu refused to fight with the barbarians and preferred instead to flee (thus earning the respect of his peaceloving subjects) was a popular story in pre-Han China, as attested by the *Zhuangzi* and the *Mencius*.

10. As we shall see, You Yu anticipates the figure of Zhonghang Shuo: a native Han Chinese who becomes spokesman for a non-Chinese culture.

11. Legge, *Shoo King*, 142–49.

12. Yang Lien-sheng, "Historical Notes on the Chinese World Order," 21–22; Ban Gu, *Han shu*, 3830–34.

13. Barfield, *The Perilous Frontier*, 1–31.

14. Ibid., 39.

15. For a review of racism and ethnocentrism in China before the modern period, see Dikötter, 7–30.

16. This becomes explicit in the *History of the Han* at one point. In 84 B.C., a Xiongnu courtier contemptuously alludes to the chaos of Chinese internal politics (somewhat earlier, the crown prince had revolted and started a brief civil war). The emissary replies, "How does that compare with the khan Modu, who personally killed his own father and took the throne. . . . That is the behavior of birds and beasts" (Ban Gu, *Han shu*, 3780).

17. Lattimore, *Inner Asian Frontiers*, 429–68.

18. Waldron, *Great Wall of China*, 30–51.

19. For example, see the exchange of letters between Han and Xiongnu rulers during Emperor Wen's reign, *Shi ji* 110.38–41.

20. Ying-shih Yü, *Trade and Expansion in Han China*, 37. One is reminded of Julius Caesar's reason for the strength and bellicosity of the Belgians: "Of all these people, the Belgae are the most courageous because they are farthest removed from the culture and the civilization of the Province, and least often visited by merchants introducing the commodities that make for effeminacy" (Julius Caesar, *The Gallic War*, 5).

21. One might compare this speech to the shorter one made by You Yu to Duke Mu of Qin in defense of barbarian customs (*Shi ji* 5.7).

22. David Schaberg ("Travel, Geography, and the Imperial Imagination," 178–83) gives a detailed analysis of the Zhonghang Shuo story and makes some of the same points as I have here.

23. Although the historical Cai Yan lived two hundred years after the historical Wang Zhaojun, I shall treat her first because the Wang Zhaojun *legend* was somewhat slow to develop, and I do not believe that the poem attributed to Cai that I discuss shows any influence from it. Moreover, I see Cai Yan as a sort of intermediate stage in the development of the alienated Chinese living among the barbarians: from the generals forced to defect and betray the emperor on the one hand, and the court lady forced to leave the emperor in order to be loyal to him.

24. Lu Qinli, *Quan xian Qin Han Wei Jin Nanbei chao shi*, 199–200.

25. Frankel, "Cai Yan," 146–54, gives his reasons for not accepting her authorship for either poem. Although I do not find his proof conclusive, I remain uncommitted to one side or the other and do not find a stand necessary for my argument here. I will refer to the persona of the poem as "the narrator."

26. Frankel, "Cai Yan," 149–50.

27. Lu Qinli, *Quan xian Qin Han Wei Jin Nanbei chao shi*, 335–36.

28. In reviewing the early accounts of Wang Zhaojun, I am indebted to the late Kwong Ying Foon (*Wang Zhaojun*), whose discussion of the historical and literary materials related to Wang is impressively thorough and perceptive.

29. Fan Ye, *Hou Han shu*, 89.2941.

30. Ibid.

31. Liu Xin, *Xijing zaji jiaozhu*, 6.

32. Lu Qinli, *Quan xian Qin Han Wei Jin Nanbei chao shi*, 315–16.

33. Ibid., 642–43.

34. For this movement from narrative to lyric, see Allen, "From Saint to Singing Girl."

35. Lu Qinli, *Quan xian Qin Han Wei Jin Nanbei chao shi*, 1614.

36. Ibid., 2129–30.

37. See, e.g., Xiao Ji 蕭紀, "Mingjun ci" 明君詞, in Lu Qinli, *Quan xian Qin Han Wei Jin Nanbei chao shi*, 1900. Bai Juyi made the same witty association in a piece of juvenilia, "Wang Zhaojun" (*Quan Tang shi*, 437.4858).

38. Lu Qinli, *Quan xian Qin Han Wei Jin Nanbei chao shi*, 2132–33.

39. My evaluation of Wang Zhaojun poems from the Tang derives from a reading of over eighty poems that take her as their main subject. I have not examined poems that use her merely as a passing allusion. I have also not dealt with the Wang Zhaojun *bianwen* 變文, since the social and cultural origins of this genre and its authors are still open to debate. For a discussion, see Eoyang, "The Wang Chaochün Legend."

40. *Quan Tang shi*, 139.1419. However, I have used the texts and the order as found in Huo Songlin, *Wan shou Tang ren jueju*, which puts first the poem that occurs last in *Quan Tang shi*. For annotations, see ibid., 2.76–77.

41. For a discussion of quatrains and the music tradition, see Rouzer, *Writing Another's Dream*, 29–39.

42. *Quan Tang shi*, 472.5361.

43. Ibid., 505.5743.

44. Ibid., 602.6960–61.

45. For the crisis in historical belief, see Rouzer, *Writing Another's Dream*, 95–118.

46. *Quan Tang shi*, 230.2511; I have consulted the text and commentary of Chou Zhao'ao, *Du shi xiang zhu*, 1502–5.

47. This "possession" may have been more poignant for the Tang reader because of the associations of the lute with shamanic possession rites; see Dudbridge, *Religious Experience*, 105–6.

48. *Quan Tang shi*, 425.4688.

49. Ibid., 434.4797.

50. Ibid., 435.4818.

51. One may note an analogy with Western seamen active in the south seas during the nineteenth century: for the narrator of Melville's *Typee*, tattooing symbolized the final abandonment of civilization, an irreversible "going native."

52. Here I disagree somewhat with Wai-yee Li ("The Idea of Authority in the *Shih Chi*," 379–80), who sees this anecdote as illustrating Sima Qian's willingness to accept a plurality of cultures. Although not denying Sima's capacity for tolerance, I find it more likely that he is trying to sinicize a disturbing alien custom and engage in the general Chinese habit of tracing the origins of non-Chinese peoples back to Chinese origins.

53. Waley, *Po Chü-i*, 130.

Chapter 6

1. Hence the title of this chapter, which I have cheerfully stolen from Jessie Weston's early anthropological examination of the Arthurian legends, *From Ritual to Romance*.

2. See Stephen Owen, *Poetry of the Early T'ang*, 325–80.

3. For a summary of pertinent facts concerning Zhang Zhuo's authorship, see Yagisawa, *Yū senkutsu zenkō*, 14–16.

4. References in the text are to the section divisions of my translation of *You xianku* in the Appendix.

5. For an overview of *zhiguai* narrative conventions, see Campany, *Strange Writing*, 222–33.

6. For a translation, commentary, and textual study of *Mu tianzi zhuan*, see Mathieu, *Le Mu Tianzi zhuan*; for the Queen Mother of the West, see Cahill, *Transcendence and Divine Passion*.

7. *Taiping guangji* citations in this chapter are to the four-volume Shanghai Guji edition.

8. For Chen Yinke (see "Du 'Yingying zhuan'" 讀鶯鶯傳, in idem, *Yuan Bai shi*, 106–16), this bit of genealogy was important in determining the "truth" behind the text. For him, the fact that Lady Cui and the heroine of Yuan Zhen's "Tale of Yingying" share the same surname is proof that Cui was a *nom de guerre* employed by courtesans. Zhou Zhenfu ("'Du Yingying zhuan' xian yi") has refuted much of the detail of his argument. However, the shape of the argument itself is questionable, both in discussing "Yingying" and *You xianku*. It assumes that Zhang's and Yuan's tales are rooted in autobiographical experience and that each author wanted us to read his encounter as one between courtesan and literatus. In Zhang's case, it also assumes that a highly cultivated class of courtesans capable of sophisticated literary exchange with literati was already in existence by the 670s, when most of our evidence for such a class is post–An Lushan. Rather, Lady Cui is an inhabitant of a privileged space: she shares traits with a prostitute only because both are outside the matrix of Confucian family relations. Such women become important because they can supposedly be courted or seduced through a man's "intrinsic" worth; their choice of sexual partner is a free one, and by making such a choice they guarantee the desirability of the man. Lady Cui is thus more likely to be a creation of male fantasy and speaks to the need to create such a class of courtesans that can validate the existence of educated literati as a whole. We will turn to such a class in Chapter 7; there is no substantial evidence in *Immortals' Den* that such a relationship is a social reality at this point.

9. See Zito, "Ritualizing *Li*"; and idem, "Silk and Skin"; for an overview of ritual studies by an anthropologist familiar with China, see Bell, *Ritual*.

10. There have been twentieth-century anxieties over the "immorality" of the text; Wang Chung-han ("The Authorship of the *Yu-hsien k'u*"), for example, assumed that a proper Chinese literatus could not write such a "pornographic" piece and that an enemy must have attributed it to Zhang Zhuo. Obviously, such standards of morality do not need to be applied to Tang poets. There is no evidence that Zhang (or whoever wrote *Immortals' Den*) would not have freely circulated it among friends and acquaintances.

11. It is the character of this exchange that should make us wary of reading Liu Lingxian's skeptical response to her husband as a poem that "subverts" objectification. See Chapter 4 above.

12. *Analects* 11.15; Yang Bojun, *Lunyu yizhu*, 114.

13. Again, if one were to follow Chen Yinke's ("Du 'Yingying zhuan'" 讀鶯鶯傳, in idem, *Yuan Bai shi*, 106–16) reading of the tale, she is obviously the madam or procuress of the establishment, whose task is to make sure that the gentleman is properly bedded by the end. Wusao's role within the fantasy of the text hardly demands this reading, although it contributes to our view of her importance.

14. See Chapter 1 above, 31–32.

15. This is in keeping with the more explicit comparison of arrows to the male organ found later in the text.

16. *Analects* 12.8; Yang Bojun, *Lunyu yizhu*, 126.

17. Although *TPGJ* reads Han Yi 翊, other versions of the story have Han Hong 翃, and various arguments make him the more likely figure. Sanders ("Poetry in Narrative," 197) concurs.

18. See Chapter 3, 99–101.

19. Most famously, in *Huo Xiaoyu zhuan* 霍小玉傳 (*TPGJ* 4: 550–55).

20. Ren Bantang, *Dunhuang geci*, 320–21.

21. Graham Sanders ("Poetry in Narrative," 194) reads this line differently: "How sad it has been given so many years of separation." This is surely possible, and it actually makes better sense. However, rarely in Chinese poetic usage does one see an abstraction like "separation" given as a gift. I choose provisionally to read the line as I have it, while leaving the other possibility open.

22. Readers would surely have assumed that what they were reading was based on fact (Han Hong was a real poet of the period). However, comparison with the version told in the *Ben shi shi* (Meng Qi, 234–36) suggests that different writers could tell the story different ways depending on certain emphases and interests (see Sanders, "Poetry in Narrative," 193–96, for an excellent translation of the *Ben shi shi* version).

23. The *Ben shi shi* version has neither memorial nor judgment and goes on to relate another anecdote surrounding Han Hong. This suggests that Meng Qi, the compiler of the *Ben shi shi*, was interested more in the drama of poetic exchange than in the wider consequences of the plot.

24. For an intelligent discussion of the subtleties of Hu Yinglin's analyses (more sophisticated than most modern interpreters have granted), see Wu, "From *Xiaoshuo* to Fiction," 352–65.

25. See Lu Hsun, *Brief History of Chinese Fiction*, 85–122.

26. Campany, *Strange Writing*; Dudbridge, *Religious Experience*.

27. Chen Yinke, "Han Yü and the T'ang Novel"; for circulation of scrolls, see Mair, "Scroll Presentation"; and also Cheng, *Tang dai jinshi xingjuan*.

28. Owen, *The End of the Chinese "Middle Ages,"* 130–48.

29. Ibid., 141.

30. In China itself we have the *caizi jiaren* 才子佳人 romances of the late imperial period, but one could also point to ancient Greek romances or to the chivalry

romances of medieval Europe. Even in one of the undisputed romances, "The Story of Huo Xiaoyu," the romance ends with a not unsubstantial account of how the ghost of the betrayed woman exacts vengeance on her faithless lover. Here the tone is factual and not unlike standard *zhiguai*.

31. *TPGJ* 4: 555–56.

32. Cf., e.g., Sima Qian's biography of the Lord of Xinling, the Prince of Wei (*Shi* ji, juan 77); there, his most gifted retainer and strategist, Hou Ying, concocts a scheme that will enable the prince to come to the assistance of the state of Zhao. The scheme is risky, but Hou is too old to accompany the prince and share his risks. As a result, he slits his own throat to demonstrate his loyalty. In the "Biographies of Assassin-Retainers" (ibid., *juan* 86), Tian Guang kills himself because he is too old to do his best for his master, the prince of Yan.

33. This variety is caused in part by the rich sources of traditional and religious folklore that contributed to it. See Dudbridge, "The Tale of Liu Yi and Its Analogues"; and Bai, "Longnü bao'en gushi de lai long qu mai."

34. Bai Huawen ("Longnü bao'en gushi de lai long qu mai," 82) argues that the meeting of failed returning examination candidates and goddesses is in fact a recurring motif in the Tang.

35. A variant makes the rhetoric even stronger: "But now because you love your son, I entrust him to you as a worthless pawn."

36. At one point Ren involves Zheng in a horse-selling scheme that could reap a substantial profit. Through her own (supernatural?) foreknowledge, she knows that Zheng will be able to sell a relatively worthless horse in the marketplace for an immense sum because the buyer (a man in charge of an imperial stable) is anxious to replace a horse of similar description that had died under his care. Although Zheng withstands the increasingly large offers from the buyer, he eventually caves in to pressure from bystanders and family before the sum Ren had promised him is reached. Perhaps Ren had hoped that this scheme would help her lover become more independent in character as well as financially. Yet he fails to trust her in the end.

37. Dudbridge, *The Tale of Li Wa*, 52–57.

38. Ibid., 56–57.

39. Ibid., 169.

Chapter 7

EPIGRAPH: Sun Qi, *Bei li zhi*, 41. Subsequent citations from this work are given in the text, keyed to the 1959 Zhonghua edition.

1. As is generally known, Chang'an was divided into wards (*fang* or more popularly, *li*), defined by the intersection of the large avenues, walled about and accessed by gates at the four cardinal points that were closed at curfew and opened at dawn;

most wards were several hundred yards on a side. For a standard account and illustration of Tang Chang'an, see Hiraoka, *Chōan to Rakuyō chizu.*

2. There are to my knowledge only two translations of *Northern Wards* (*Bei li zhi*). Howard Levy's "Record of the Gay Quarters"; and Robert des Rotours, *Courtisanes chinoises.* I owe a particular debt to the latter, and only occasionally disagree with his readings. *Northern Wards* is full of distinctive Tang idioms and word usages, which, combined with the absence of a more detailed social background, makes it often obscure for the modern reader. I apologize in advance for any errors in my own translation. I have also avoided giving detailed notes on the biographical circumstances of its figures, on historical allusions, etc., unless I felt that such information was significant for my argument; those interested in filling in such gaps should consult des Rotours's thorough annotations.

3. See Ishida, "Chōan no kagi" 長安の歌妓, reprinted in *Zōtei Chōan no haru,* 100–125. More recent considerations of the role of courtesans include Gao Shiyu, *Tangdai funü,* 56–74; and Song Dexi, "Tangdai de jinü."

4. Cass, "*Pei-li chih,*" 650.

5. See, e.g., the discussion of the relations of late Tang *jinshi* and courtesans raised by Chen Yinke in his *Tangdai zhengzhishi shulun kao.* For a more recent discussion of their literary interaction and its role in the rise of *ci,* see Wagner, *The Lotus Boat.* Tung (*Fables for the Patriarchs,* 101–22) gives an analysis that touches on some of the issues I discuss here.

6. Van Gulik, *Sexual Life in Ancient China,* 181.

7. *Jiao fang ji,* 25. For Ishida's discussion, see "Chōan no kagi" in *Zōtei Chōan no haru,* 111–12. See also Tung's comments, *Fables for the Patriarchs,* 115–22.

8. Stallybrass and White, *The Politics and Poetics of Transgression,* 27–43.

9. I follow des Rotours's (*Courtisanes chinoises,* 51) speculation that since the term *duwei* 都尉 occurs only in the Tang as a term for the emperor's married kin, some idiomatic expression for imperial kinsmen is meant here.

10. The "emissaries in search of flowers" were supposed to pick celebratory flowers from the most famous gardens of the city. Interesting here (although I do not have space to elaborate on it) is the fact that *tanhua* 探花 (seeking the flowers) became a specific term for third-rank examination laureates in the Song and that "seeking the flowers" was a euphemism for cruising for courtesans. (I am indebted to Professor Pei-yi Wu for clarifying this relation; see also des Rotours, *Courtisanes chinoises,* 52n1.)

11. Such gossip about imperial visits was already common and dates back at least to Han popular stories about Han Wudi.

12. As in the set term *xia ji* 狎妓, "to consort with prostitutes"; note also a term for courtesans still in use, *xiaxie* 狎邪.

13. The discussion of this issue begins of course with Chen Yinke; more recent contributions include Twitchett, "The Composition of the T'ang Ruling Class"; Johnson, *The Medieval Chinese Oligarchy;* Ebrey, *The Aristocratic Families of Early Imperial China;* and Bol, *"This Culture of Ours."*

14. The mutual validation of literatus and courtesan is noted by Tung, *Fables for the Patriarchs,* 108–11.

15. Sun does not give a surname for this person; des Rotours (*Courtisanes chinoises,* 109) for various reasons suggests Zhao 趙.

16. The rosy mists of Heaven are the food of immortals; the ceremonial music of Mount Yunhe is referred to in the *Zhou li* (here it is a compliment paid to Jiangzhen).

17. A reference to Han Gaozu's success in beating Xiang Yu to the Qin capital of Xianyang; an image of the physical conquest of a woman is implied by this image as well.

18. On the theme of recognition, see Henry, "The Motif of Recognition in Early China."

19. See, e.g., Owen, *Great Age of Chinese Poetry,* 5.

20. Our own standards of judgment in poetry tend to be Song and post-Song creations; an examination of the surviving fragments of Tang anthologies and poetry criticism suggests radically different standards. Figures later seen as significant are practically ignored (e.g., Du Fu) or seem to be appreciated for different reasons (e.g., Li Bai). We should recognize this not as a sign of Tang "blindness" but of a different aesthetic in operation. The return in the Late Tang to the regulated occasional verse form is a sign that poetry's most important role was still its contribution to social graciousness.

21. Although much has been said on the role of the courtesan in the rising popularity of the *ci* form, it is interesting to see just how small a role dance and song play in *Northern Wards* overall; this suggests the continuing primacy placed ideologically on *shi*. For the role of music, though, see the discussion of Yan Lingbin below.

22. "Lad Liu" can be a romantic and slightly disrespectful nickname for Han Wudi (the family name of the Han royal house was Liu). This term is usually applied in conjunction with Wudi's amorous adventures or his dabbling in Taoist arts. Some annotators, including des Rotours (*Courtisanes chinoises*), believe that "Lad Liu" is Liu Ling, a famous drunkard and one of the Seven Worthies of the Bamboo Grove. Since Liu Ling is an unromantic figure and because "Lad Liu" very obviously refers to Han Wudi in other Tang poems, I believe this identification is possible but less likely.

23. Des Rotours (*Courtisanes chinoises,* 82*n*3) tentatively suggests this translation for this office; he cannot find this term in any other Tang work.

24. Confucius' disciple Zilu was famous for his reputedly great physical strength.

25. I.e., become a nun.

26. There may also be an implied snobbish critique of Duan in this vignette. Since he is a petty official, not a romantic examination candidate, he comes off in this passage as distinctly "low class" in his violence and jealousy.

27. Most common terms for this relation include *zhiji* 知己 (he who knows oneself) and *zhiyin* 知音 (he who knows the tone; a term drawn from an anecdote about the great lute player Boya 伯牙 and his intimate friend, Zhongzi Qi 鐘子期).

28. See Sun Qi, *Bei li zhi*, 32, for another case in which a "foster mother" fails to appreciate one of her daughters sufficiently. For an example of a condemnation of greed among "mothers" and "daughters," see ibid., 36.

29. Referring to the anecdote in the *Zhuangzi* in which the philosopher beats on a pot and sings in mourning for the death of his wife. Although his gesture was meant to indicate his transcendence of the sorrows of death, poets tended to use it as a cliché indicating mourning.

30. In Han times, it was believed that departed souls went to Mount Tai; the mountain thus figures frequently in the imagery of mourning songs. The following line blends two images from the *Analects*: Confucius asking after the location of the ford (*wenjin*—read metaphorically as "seeking one's way in life"; *Analects* 18.6) and his sorrow over the endless flowing of the water (*Analects* 9.17).

31. An allusion to Song Yu's *Master Dengtu the Lecher*; the "Eastern Neighbor" became a very common allusion to girls of easy virtue (for both, see Chapter 2).

32. See Ren Bantang, *Tang sheng shi*; and Rouzer, *Writing Another's Dream*, chap. 2.

33. Xiao Tong, *Wen xuan, juan* 23.

34. In this respect, it becomes a literati counterpart to the "popular" dirge-singing contest described in the *Li Wa zhuan*; see Chapter 6.

35. No surname is supplied for this person. For a discussion on who this might be, see des Rotours, *Courtisanes chinoises*, 120–21.

36. It was a custom for examinees who had just passed to inscribe their names on the wall of the pagoda at Mercy Temple (*Ci'en si*) in the Jinchang Ward. Here, two pagodas are implied: Xiaorun's buttocks.

37. A reference to the writing style of Ouyang Xun (557–647), a noted Tang calligrapher. "Golden Hill" could refer to Jinling (sight of present-day Nanjing), although, as des Rotours (*Courtisanes chinoises*, 122) suggests, the place-name is probably not implied here.

38. This rather poor and egotistical joke is based on the Tang legend that Dongfang Shuo, Han Wudi's court wit, managed to steal some peaches from the garden of the Queen Mother of the West. Funiang is being compared to a peach tree, whose "fruit" Cui Zhizhi "steals."

39. Tung also (*Fables for the Patriarchs*, 117–18) discusses the power of this image and its relations to the examinations; she sees the possession of the woman as more important than the passing of the examinations.

40. See note 22 to this chapter. Again, the allusion here is unclear, since drinking is mentioned, but the romantic context again points to Han Wudi.

41. Xi Shi is the famous beauty who, when sent as a gift from the state of Yue 越 to the state of Wu 吳, brought about the downfall of the latter.

42. Literature of the time makes reference to a game of *dou cao* 鬥草, a competitive dueling game with blades of grass or straw.

43. A proverbial cure for dry or disfigured skin.

44. Sima Xiangru's verse was said to be worth such an extravagant sum; however, I suspect that Yizhi is mildly teasing Sun Qi here.

45. Evidently, to be on the official rolls of the *jiaofang* (and hence eligible to be called up for imperial or government duty) made it more difficult and expensive to be bought out of one's courtesan status.

46. See Chen Yinke's discussion in *Tangdai zhengzhishi shulun kao*; Dudbridge has weighed the evidence at greater length in *The Tale of Li Wa*, 72–80. In the few cases mentioned in *Northern Wards*, courtesans are always acquired by wealthy merchants or low-ranking clerks.

47. The image of transplantation is also played on by the courtesan poet Yu Xuanji in her famous "Planting Wilted Peonies": there, too, it expresses the hope of social mobility:

When, at last, their roots are transplanted to the Imperial Park,	及至移根上林苑
Then, my dear prince, you will regret that you are no longer able to buy them.	王孫方恨買無因

(Cited from Chen Wenhua, *Tang nüshiren*, 101.)

48. The commentator probably introduces this comment in order to explain why Yizhi continued to live with her "foster mother."

49. I am a little unsure of the translation here; des Rotours (*Courtisanes chinoises*, 137) thinks that the text is corrupt.

50. *Taiping guangji*, 4: 559.

51. See Chen Yinke, "Du 'Yingying zhuan,'" in idem, *Yuan Bai shi*, 106–16; and Chapter 6, note 8.

52. Master Dengtu has come to be synonymous with "lecher" in the Chinese tradition, due to his representation in the poem *Master Dengtu the Lecher*.

53. *Taiping guangji*, 4: 555.

54. Ibid., 4: 559.

Afterword

EPIGRAPH: Li Shangyin, *Li Shangyin shige jijie*, 79.

1. For an institutional examination of these changes, see McMullen, *State and Scholars in T'ang China*; for a detailed intellectual examination, see Bol, *"This Culture of Ours,"* 76–147. Owen, *The End of the Chinese "Middle Ages,"* discusses some reflections of this change in literature. See Rouzer, *Writing Another's Dream*, 172–85, for a discussion of "safe spaces."

2. Li Shangyin, *Li Shangyin shige jijie*, 79.

3. See Su Xuelin, *Yuxi shimi* (and other collections of her criticism).

4. Miao, "Palace-Style Poetry," 40–41.

5. The text erroneously has Huanzhi for Zhihuan; I have changed it in all cases.

6. A reference to the An Lushan rebellion, which broke out in 755.

7. The Pear Garden was the name for the imperial music academy that employed musicians for imperial outings and other events; musicians could be hired out as well for private parties. The current anecdote gives the impression that the musicians were holding a banquet for their own pleasure.

8. The hermit is Yang Xiong, often employed as a symbol of scholarly rectitude.

9. The subject of this quatrain is Lady Ban (see Chapter 4); she was one of the most popular subjects of Tang song quatrains.

10. The poet Song Yu boasted of his musical connoisseurship by saying that anyone could appreciate the vulgar songs of the peasants of Ba, but that only he and a few others could value songs like *Spring Light* or *White Snow*.

11. "Resent the Willow Branches" was the name of a popular song of the time; Jade Gate Pass led to the vast wastelands of northwest China. The poet argues that since "Resent the Willow Branches" describes the appearance of the newly green willows, it is pointless to play the piece in a land where spring never comes.

12. Xue, *Ji yi ji*, 8–9.

13. Owen, *The Great Age of Chinese Poetry*, 92–94. For another, even more detailed discussion of this anecdote and its relations to "performativity" in the Tang, see Ashmore, "Hearing Things," 14–30.

14. *Quan Tang shi*, 184.1882.

15. Ibid., 164.1701. For a quite different reading of these Li Bai quatrains, see Tung, *Fables for the Patriarchs*, 158–60.

16. *Quan Tang shi*, 128.1301.

17. S. Chou, "Beginning with Images in the Nature Poetry of Wang Wei," 128–36.

18. For example, his imitation of the "Xu Ling style," in Li Shangyin, *Li Shangyin shige jijie*, 1751.

19. For a "close reading" of these poems, see Ye Jiaying, *Jialing lun shi cong kao*, 147–209.

20. Li Shangyin, *Li Shangyin shige jijie*, 389.

21. Kang, *Feng sao yu yanqing*, 236–37.

22. Graham, *Poems of the Late Tang*, 148.

23. Li Shangyin, *Li Shangyin shige jijie*, 1451.

24. A meaning found, e.g., in Cen Shen's 岑參 "Sent to Reminder Du Fu of the Department of State Affairs" 寄左省杜拾遺, *Quan Tang shi*, 200.2064.

25. As we have seen, these conventions probably did not fully solidify until the age of palace poetry. Connery ("Jian'an Poetic Discourse," 227–43) suggests that they are in place by the Jian'an period, but he employs anachronistic assumptions in doing so.

26. Tung (*Fables for the Patriarchs*, 169–75) makes some astute comments here on the androgyny of Li in many of these poems.

27. Li Shangyin, *Li Shangyin shige jijie*, 1467.

28. Xiao Tong, *Wen xuan*, 228.

29. *Quan Tang shi*, 571.6626–27.

30. Story found in Sima Guang, *Zizhi tongjian*, 5508–9.

31. Story told in Liu Yiqing, *Shishuo xinyu jiao jian*, 491–92.

32. Li Shangyin, *Li Shangyin shige jijie*, 1420.

33. Ibid., 1452.

34. Kao and Mei, "Meaning, Metaphor, and Allusion in T'ang Poetry," 335–55.

35. Li Shangyin, *Li Shangyin shige jijie*, 1467.

36. James J. Y. Liu, *The Poetry of Li Shang-yin*, 63.

37. An alternative interpretation has been suggested: that Master Liu is Liu Chen, the friend of Ruan Zhao who encountered goddesses in the Tiantai Mountains (for a translation and discussion, see Chapter 6). In this case, the "search for immortality" still remains a theme in the poem.

38. Li Shangyin, *Li Shangyin shige jijie*, 1461.

39. Ibid., 1769–70.

40. "Night Rain: Sent to the North" 夜與寄北, ibid., 1230.

41. Throughout my discussion I have used the term "subjectivity"; I realize that this introduces complex questions that I cannot begin to answer: for example, how one precisely envisions a Chinese "subjectivity" in the Tang dynasty, and how that might differ from Western senses of the same idea, bound as they are to post-Cartesian philosophical issues. I do not wish to use Li's poetry as a tool for defining subjectivity as such, but merely as a way of articulating, difficult as that is, what Li is doing in poetic form that makes these poems so different from their predecessors.

42. There are of course, exceptions: e.g., Wu Wenying 吳文英.

43. Interest in issues of "voice" and self-expression can be seen to emerge almost obsessively in the essays in Pauline Yu, *Voices of the Song Lyric in China*.

44. Bryant, xxvii–xxx.

45. Tang Guizhang, *Quan Song ci*, 34–35; for an alternative translation, see Hightower, "The Songwriter Liu Yung," 355–56.

46. Tang Guizhang, *Quan Song ci*, 41; for an alternative translation, see Hightower, "The Songwriter Liu Yung," 345–46.

47. Of course, the Qing critic Zhang Huiyan (1761–1802) attempted to reverse this trend through a political rereading of Wen Tingyun's and other poets' *ci* (see Chia-ying Yeh Chao [Ye Jiaying], "The Ch'ang-chou School of *Tz'u* Criticism").

Works Cited

Allen, Joseph Roe, III. "From Saint to Singing Girl: The Rewriting of the Lo-fu Narrative in Chinese Literati Poetry." *Harvard Journal of Asiatic Studies* 48, no. 2 (1988): 321–61.

———. *In the Voice of Others: Chinese Music Bureau Poetry*. Ann Arbor: University of Michigan, Center for Chinese Studies, 1992.

Ames, Roger T. "Taoism and the Androgynous Ideal." In *Women in China: Current Directions in Historical Scholarship*, ed. Richard W. Guisso and Stanley Johannesen, pp. 21–45. Youngstown, N.Y.: Philo Press, 1981.

Asano Michiari 淺野通有. "Sō Gyoku no sakuhin no shigi ni tsuite" 宋玉の作品の眞僞について. *Kambun gakkai kaihō* 漢文學會會報 12 (1961): 3–12.

Ashmore, Robert. "Hearing Things: Performance and Lyric Imagination in Chinese Literature of the Early Ninth Century." Ph.D. diss., Harvard University, 1997.

Bai Huawen 白化文. "Longnü bao'en gushi de lai long qu mai: 'Liu Yi zhuan' yu 'Zhu she zhuan' bijiao guan" 龍女報恩故事的來龍去脈—《柳毅傳》與《朱蛇傳》比較觀. *Wenxue yichan* 文學遺產 1992, no. 3: 78–84.

Balasz, Etienne. *Chinese Civilization and Bureaucracy: Variations on a Theme*. New Haven: Yale University Press, 1964.

Ban Gu 班固. *Han shu* 漢書. Beijing: Zhonghua, 1962.

Bao Jialin 鮑家麟, ed. *Zhongguo funü lunji xuji* 中國婦女論集續集. Taibei: Daoxiang, 1991.

Barfield, Thomas J. *The Perilous Frontier: Nomadic Empires and China, 221 BC to AD 1757*. Cambridge, Mass.: Blackwell, 1989.

Barlow, Tani E. "Theorizing Woman: *Funü, Guojia, Jiating* (Chinese Woman, Chinese State, Chinese Family)." In *Body, Subject and Power in China*, ed. Angela Zito and Tani E. Barlow, pp. 253–89. Chicago: Chicago University Press, 1994.

Bell, Catherine. *Ritual: Perspectives and Dimensions*. Oxford: Oxford University Press, 1997.

Birrell, Anne M. "Erotic Décor: A Study of Love Imagery in the Sixth Century A.D. Anthology, *Yü-t'ai hsin-yung,* New Poems from a Jade Terrace." Ph.D. diss., Columbia University, 1979.

Birrell, Anne M., trans. *New Songs from a Jade Terrace.* London: Allen & Unwin, 1982.

Black, Alison H. "Gender and Cosmology in Chinese Correlative Thinking." In *Gender and Religion: On the Complexity of Symbols,* ed. Caroline Walker Bynum, Stevan Harrell, and Paula Richman, pp. 166–95. Boston: Beacon Press, 1986.

Bol, Peter K. *"This Culture of Ours": Intellectual Transition in T'ang and Sung China.* Stanford: Stanford University Press, 1992.

Bourdieu, Pierre. *The Field of Cultural Production: Essays on Art and Literature.* New York: Columbia University Press, 1993.

———. *The Logic of Practice.* Stanford: Stanford University Press, 1990.

Bray, Alan. *Homosexuality in Renaissance England.* Rev. ed. New York: Columbia University Press, 1995.

Bryant, Daniel. *Lyric Poets of the Southern T'ang: Feng Yen-ssu, 903–960, and Li Yü, 937–978.* Vancouver: University of British Columbia Press, 1982.

Burke, Kenneth. *A Grammar of Motives.* Berkeley: University of California Press, 1945.

Butler, Judith. *Bodies That Matter: On the Discursive Limits of "Sex."* New York: Routledge, 1993.

———. *Gender Trouble: Feminism and the Subversion of Identity.* New York: Routledge, 1990.

Cahill, Suzanne E. *Transcendence and Divine Passion: The Queen Mother of the West in Medieval China.* Stanford: Stanford University Press, 1993.

Campany, Robert Ford. *Strange Writing: Anomaly Accounts in Early Medieval China.* Albany: State University of New York Press, 1996.

Cao Zhi 曹植. *Cao Zijian shi zhu* 曹子建詩注. Commentary by Huang Jie 黃節. Taibei: Yiwen yinshu guan, 1971.

Cass, Victoria. *"Pei-li chih."* In *The Indiana Companion to Traditional Chinese Literature,* ed. William H. Nienhauser, p. 650. Bloomington: Indiana University Press, 1986.

Chandler, James. *England in 1819: The Politics of Literary Culture and the Case of Romantic Historicism.* Chicago: University of Chicago Press, 1998.

Chang, Kang-i Sun. *Six Dynasties Poetry.* Princeton: Princeton University Press, 1986.

Chartier, Roger. *Cultural History: Between Practices and Representations.* Ithaca, N.Y.: Cornell University Press, 1988.

Ch'en, Ch'i-yün. "Confucian, Legalist, and Taoist Thought in Later Han." In *The Cambridge History of China,* vol. 1, *The Ch'in and Han Empires, 221 B.C.–A.D. 220,* ed. Denis Twitchett and Michael Loewe, pp. 766–807. Cambridge, Eng.: Cambridge University Press, 1986.

Chen Shou 陳壽. *San guo zhi* 三國志. Beijing: Zhonghua, 1959.

Chen Wenhua 陳文華, ed. *Tang nüshiren ji san zhong* 唐女詩人集三種. Shanghai: Guji, 1984.

Chen Yinke 陳寅恪 (Tschen Yinkoh). "Han Yü and the T'ang Novel." *Harvard Journal of Asiatic Studies* 1 (1936): 39–43.

———. *Tangdai zhengzhishi shulun gao* 唐代政治史述論稿. Shanghai: Guji, 1982.

———. *Yuan Bai shi jianzheng gao* 元白詩箋証稿. Shanghai: Guji, 1978.

Chen Zizhan 陳子展. *Shijing zhijie* 詩經直解. Shanghai: Fudan daxue, 1983.

Cheng Qianfan 程千帆. *Tang dai jinshi xingjuan yu wenxue* 唐代進士行卷與文學. Shanghai: Guji, 1980.

Chennault, Cynthia L. "Palace-Style Ladies and Odes on Objects." Paper presented at the annual meeting of the Association for Asian Studies, Boston, March 1994.

Chi Xiao. "Lyric Archi-Occasion: Coexistence of 'Now and 'Then.'" *Chinese Literature: Essays, Articles, Reviews* 15 (1993): 17–35.

Chou, Shan. "Beginning with Images in the Nature Poetry of Wang Wei." *Harvard Journal of Asiatic Studies* 42, no. 1 (1982): 117–37.

Chou Zhao'ao 仇兆鰲. *Du shi xiang zhu* 杜詩詳注. Beijing: Zhonghua, 1979.

Chu ci buzhu 楚辭補注. Commentary by Hong Xingzu 洪興祖. Beijing: Zhonghua, 1983.

Chu ci jizhu 楚辭集注. Commentary by Zhu Xi 朱熹. Beijing: Zhonghua, 1987.

Ch'ü, T'ung-tsu. *Han Social Structure*. Seattle: University of Washington Press, 1972.

Chunqiu jing zhuan jijie 春秋經傳集解. Shanghai: Guji, 1978.

Connery, Christopher Leigh. *The Empire of the Text*. Lanham, Md.: Rowman and Littlefield, 1998.

———. "Jian'an Poetic Discourse." Ph.D. diss., Princeton University, 1991.

Crump, J. I., Jr., trans. *Chan-Kuo Ts'e*. Oxford: Clarendon, 1970.

Cui Shu 崔述. *Du feng ou shi* 讀風偶識. Congshu jicheng, 1st series. Beijing: Zhonghua, 1985.

Cutter, Robert Joe. "Cao Zhi's (192–232) Symposium Poems." *Chinese Literature: Essays, Articles, Reviews* 6 (1984): 1–32.

———. "The Death of Empress Zhen: Fiction and Historiography in Early Medieval China." *Journal of the American Oriental Society* 112, no. 4 (1992): 577–83.

Davis, Natalie Zemon. *Society and Culture in Early Modern France: Eight Essays*. Stanford: Stanford University Press, 1975.

Deng Shiliang 鄧仕樑. "Shi 'fangdang': jian lun Liu chao wen feng 釋「放蕩」— 兼論六朝文風." *Chūgoku bungaku hō* 中國文學報 35 (1983): 37–53.

des Rotours, Robert, trans. *Courtisanes chinoises à la fin des T'ang entre circa 789 et le 8 janvier 881:* Pei-li tche *(Anecdotes du quartier du Nord) par Souen K'i*. Bibliothèque de L'Institut des hautes études chinoises, 22. Paris: Presses universitaires de France, 1968.

Diény, Jean-Pierre. *Les Dix-neuf poèmes anciens*. Paris: Presses universitaires de France, 1963.

———. "Lecture de Wang Can (177–217)." *T'oung Pao* 73 (1987): 286–312.

———. *Pastourelles et magnanarelles: essai sur un thème littéraire chinois*. Geneva: Librairie Droz, 1977.

———. *Portrait anecdotique d'un gentilhomme chinois: Xie An (320–385) d'après le "Shishuo xinyu."* Bibliothèque de L'Institut des hautes études chinoises, 28. Paris: Collège de France, Institut des hautes études chinoises, 1993.

Dikötter, Frank. *The Discourse of Race in Modern China*. Stanford: Stanford University Press, 1992.

Dudbridge, Glen. *Religious Experience and Lay Society in T'ang China: A Reading of Tai Fu's "Kuang-i chi."* Cambridge, Eng.: Cambridge University Press, 1995.

———. *The Tale of Li Wa: Study and Critical Edition of a Chinese Story from the Ninth Century*. London: Oxford University, Board of the Faculty of Oriental Studies, 1983.

———. "The Tale of Liu Yi and Its Analogues." In *Paradoxes of Traditional Chinese Literature*, ed. Eva Hung, pp. 61–88. Hong Kong: Hong Kong University Press, 1994.

Dumézil, Georges. *Mitra-Varuna: An Essay on Two European Representations of Sovereignty*. New York: Zone Books, 1988.

Durrant, Stephen W. *The Cloudy Mirror: Tension and Conflict in the Writings of Sima Qian*. Albany: State University of New York Press, 1995.

Eagleton, Terry. *Ideology: An Introduction*. London: Verso, 1991.

Ebrey, Patricia. *The Aristocratic Families of Early Imperial China: A Case Study of the Po-ling Ts'ui Family*. Cambridge, Eng.: Cambridge University Press, 1978.

———. "The Economic and Social History of Later Han." In *The Cambridge History of China*, vol. 1, *The Ch'in and Han Empires, 221 B.C.–A.D. 220*, ed. Denis Twitchett and Michael Loewe, pp. 608–48. Cambridge: Cambridge University Press, 1986.

———. *The Inner Quarters: Marriage and the Lives of Chinese Women in the Sung Period*. Berkeley: University of California Press, 1993.

Eoyang, Eugene. "The Wang Chao-chün Legend: Configurations of the Classic." *Chinese Literature: Essays, Articles, Reviews* 4 (1982): 3–22.

Estrin, Barbara L. *Laura: Uncovering Gender and Genre in Wyatt, Donne, and Marvell*. Durham: Duke University Press, 1994.

Fan Ye 范曄. *Hou Han shu* 後漢書. Beijing: Zhonghua, 1965.

Fang Yurun 方玉潤. *Shijing yuanshi* 詩經原始. Beijing: Zhonghua, 1986.

Ferguson, Margaret W., Maureen Quilligan, and Nancy J. Vickers, eds. *Rewriting the Renaissance: The Discourse of Sexual Difference in Early Modern Europe*. Chicago: University of Chicago Press, 1986.

Fingarette, Herbert. *Confucius: The Secular as Sacred*. New York: Harper & Row, 1972.

Fong, Grace S. "Persona and Mask in the Song Lyric (*Ci*)." *Harvard Journal of Asiatic Studies* 50, no. 2 (1990): 459–84.

Foucault, Michel. *The History of Sexuality*, vol. 1, *An Introduction*. New York: Random House, 1978.

Frankel, Hans H. "Cai Yan and the Poems Attributed to Her." *Chinese Literature: Essays, Articles, Reviews* 5 (1983): 133–56.

———. "The Development of Han and Wei *Yüeh-fu* as a High Literary Genre." In *The Vitality of the Lyric Voice:* Shih *Poetry from the Late Han to the T'ang*, ed. Shuen-fu Lin and Stephen Owen, pp. 255–86. Princeton: Princeton University Press, 1986.

———. "Fifteen Poems by Ts'ao Chih: An Attempt at a New Approach." *Journal of the American Oriental Society* 84 (1964): 1–14.

Frodsham, J. D. *The Murmuring Stream: The Life and Works of the Chinese Nature Poet Hsieh Ling-yün (385–433), Duke of K'ang-lo*. 2 vols. Kuala Lampur: University of Malaya Press, 1967.

———. "The Origins of Chinese Nature Poetry." *Asia Major*, n.s. 8, no. 1 (1960): 68–104.

Gao Heng 高亨. *Shijing jin zhu* 詩經今注. Shanghai: Guji, 1980.

Gao Shiyu 高世瑜. *Tangdai funü* 唐代婦女. Xi'an: San Qin, 1988.

Ge Xiaoyin 葛曉音. "Lun Nanbei chao Sui Tang wenren dui Jian'an qianhou wen-feng yanbiande butong pingjia" 論南北朝隋唐文人對建安前後文風演變的不同評價. *Wenxue pinglun congkan* 文學評論叢刊 30 (1988): 1–19.

Girard, René. *Deceit, Desire, and the Novel: Self and Other in Literary Structure*. Baltimore: Johns Hopkins University Press, 1965.

Goldberg, Jonathan. *Sodometries: Renaissance Texts, Modern Sexualities*. Stanford: Stanford University Press, 1992.

Gong Kechang 龔克昌. *Han fu yanjiu* 漢賦研究. Ji'nan: Wenyi chubanshe, 1990.

———. *Studies on the Han* Fu. New Haven: American Oriental Society, 1997.

Grafflin, Dennis. "The Great Family in Medieval South China." *Harvard Journal of Asiatic Studies* 41, no. 1 (1981): 65–74.

———. "Reinventing China: Pseudobureaucracy in the Early Southern Dynasties." In *State and Society in Early Medieval China*, ed. Albert E. Dien, pp. 139–70. Stanford: Stanford University Press, 1990.

Graham, A. C. *Poems of the Late T'ang*. Harmondsworth, Eng.: Penguin Books, 1965.

Greenblatt, Stephen. *Renaissance Self-Fashioning: From More to Shakespeare*. Chicago: University of Chicago Press, 1980.

Guillory, John. *Cultural Capital: The Problem of Literary Canon Formation*. Chicago: University of Chicago Press, 1993.

Guo Shaoyu 郭紹虞. *Zhongguo gudian wenxue lilun piping shi* 中國古典文學理論批評史. Beijing: Renmin wenxue, 1959.

Harper, Donald. "The Sexual Arts of Ancient China as Described in a Manuscript of the Second Century B.C." *Harvard Journal of Asiatic Studies* 47, no. 2 (1987): 539–93.

Hartog, François. *The Mirror of Herodotus: The Representation of the Other in the Writing of History.* Berkeley: University of California Press, 1988.

Hawkes, David. *Classical, Modern and Humane: Essays in Chinese Literature.* Ed. John Minford and Siu-kit Wong. Hong Kong: Chinese University Press, 1989.

Hay, John. "The Body Invisible in Chinese Art?" In *Body, Subject and Power in China,* ed. Angela Zito and Tani E. Barlow, pp. 42–77. Chicago: Chicago University Press, 1994.

Henry, Eric. "The Motif of Recognition in Early China." *Harvard Journal of Asiatic Studies* 47, no. 1 (1987): 5–30.

Hightower, James Robert. "The *Fu* of T'ao Ch'ien." *Harvard Journal of Asiatic Studies* 17 (1954): 169–230.

———. "Some Characteristics of Parallel Prose." In *Studia Serica Bernhard Karlgren Dedicata,* ed. Søren Egerod and Else Glahn, pp. 60–91. Copenhagen: Ejnar Munksgaard, 1959.

———. "The Songwriter Liu Yung, Part I." *Harvard Journal of Asiatic Studies* 41, no. 2 (1981): 323–76.

———. *Topics in Chinese Literature: Outlines and Bibliographies.* Rev. ed. Cambridge, Mass.: Harvard University Press, 1953.

Hinsch, Bret. *Passions of the Cut Sleeve: The Male Homosexual Tradition in China.* Berkeley: University of California Press, 1990.

Hiraoka Takeo 平岡武夫. *Chōan to Rakuyō chizu* 長安と洛陽地圖. Tōdai kenkyū no shiori, 7. Kyoto: Kyōto daigaku, Bunjin kagaku kenkyūsho, 1956.

Holcombe, Charles. "The Exemplar State: Ideology, Self-Cultivation, and Power in Fourth-Century China." *Harvard Journal of Asiatic Studies* 49, no. 1 (1989): 93–139.

———. *In the Shadow of the Han: Literati Thought and Society at the Beginning of the Southern Dynasties.* Honolulu: University of Hawaii Press, 1994.

Holzman, Donald. *Poetry and Politics: The Life and Works of Juan Chi, A.D. 210–263.* Cambridge, Eng.: Cambridge University Press, 1976.

Hu Nianyi 胡念貽. *Zhongguo gudai wenxue lungao* 中國古代文學論稿. Shanghai: Guji, 1987.

Huo Songlin 霍松林, ed. *Wan shou Tang ren jueju jiaozhu jiping* 萬首唐人絕句校注集評. Taiyuan: Shanxi renmin, 1991.

Ishida Mikinosuke 石田幹之助. *Zōtei Chōan no haru* 增訂長安の春. Tokyo: Heibonsha, 1967.

Jameson, Fredric. *Postmodernism: or, the Cultural Logic of Late Capitalism.* Durham, N.C.: Duke University Press, 1991.

Jian Boxian 簡博賢. *Jincun Nanbeichao jingxue yiji kao* 今存南北朝經學遺籍考. Taibei: Sanmin, 1986.

Jiang Shuge 姜書閣. *Xian Qin cifu yuan lun* 先秦辭賦原論. Ji'nan: Qi Lu shushe, 1983.

Jiao fang ji jianding 教坊記箋訂. Commentary by Ren Bantang 任半塘. Beijing: Zhonghua, 1962.

Jin shu 晉書. Beijing: Zhonghua, 1974.

Johnson, David G. *The Medieval Chinese Oligarchy*. Boulder, Colo.: Westview, 1977.

Julius Caesar. *The Gallic War*. Loeb Classical Library. Cambridge, Mass.: Harvard University Press, 1917.

Jullien, François. *La Valeur allusive: des catégories originales d'interprétation poétique dans la tradition chinoise (Contribution à une réflexion sur l'altérité interculturelle)*. Paris: Ecole française d'Extrême-Orient, 1985.

Kang Zhengguo 康正果. *Feng sao yu yanqing: Zhongguo gudian shici de nüxing yanjiu* 風騷與艷情—中國古典詩詞的女性研究. Zhengzhou: Henan renmin, 1988.

Kao, Yu-kung. "The Aesthetics of Regulated Verse." In *The Vitality of the Lyric Voice:* Shih *Poetry from the Late Han to the T'ang*, ed. Shuen-fu Lin and Stephen Owen, pp. 332–85. Princeton: Princeton University Press, 1986.

Kao, Yu-kung, and Tsu-lin Mei. "Meaning, Metaphor, and Allusion in T'ang Poetry." *Harvard Journal of Asiatic Studies* 38, no. 2 (1978): 281–356.

Karlgren, Bernhard. *The Book of Odes*. Stockholm: Museum of Far Eastern Antiquities, 1950.

Kelly, Joan. *Women, History, and Theory: The Essays of Joan Kelly*. Chicago: University of Chicago Press, 1984.

Knechtges, David R. *The Han Rhapsody: A Study of the* Fu *of Yang Hsiung (53 B.C.–A.D. 18)*. Cambridge, Eng.: Cambridge University Press, 1976.

———. "The Poetry of an Imperial Concubine: The Favorite Beauty Ban." *Oriens Extremus* 36, no. 2 (1993): 127–38.

———. "Ssu-ma Hsiang-ju's 'Tall Gate Palace Rhapsody.'" *Harvard Journal of Asiatic Studies* 41, no. 1 (1981): 47–64.

———. "To Praise the Han: The Eastern Capital *Fu* of Pan Ku and His Contemporaries." In *Thought and Law in Qin and Han China: Studies Dedicated to Anthony Hulsewé on His Eightieth Birthday*, ed. W. L. Idema and E. Zürcher, pp. 118–39. Leiden: E. J. Brill, 1990.

Knechtges, David R., trans. *Wen Xuan, or Selections of Refined Literature*, vol. 1, *Rhapsodies on Metropolises and Capitals*. Princeton: Princeton University Press, 1982.

———. *Wen Xuan, or Selections of Refined Literature*, vol. 3, *Rhapsodies on Natural Phenomena, Birds and Animals, Aspirations and Feelings, Sorrowful Laments, Literature, Music, and Passions*. Princeton: Princeton University Press, 1996.

Kwong Hing Foon. *Wang Zhaojun: une héroïne chinoise de l'histoire à la légende.* Mémoires de l'Institut des hautes études chinoises, 27. Paris: Collège de France, Institut des hautes études chinoises, 1986.

La Belle, Jennijoy. *Herself Beheld: The Literature of the Looking Glass.* Ithaca, N.Y.: Cornell University Press, 1988.

Laqueur, Thomas. *Making Sex: Body and Gender from the Greeks to Freud.* Cambridge, Mass.: Harvard University Press, 1990.

Lattimore, Owen. *Inner Asian Frontiers of China.* 2d ed. New York: American Geographical Society, 1951.

Legge, James, trans. *The Chinese Classics,* vol. 3, *The Shoo King.* Hong Kong: Hong Kong University Press, 1960.

———. *Li Chi: Book of Rites, An Encyclopedia of Ancient Ceremonial Usages, Religious Creeds and Social Institutions.* 2 vols. Reprinted—New Hyde Park, N.Y.: University Books, 1967.

Levy, Howard S., trans. *China's First Novellette: The Dwelling of Playful Goddesses by Chang Wen-ch'eng (ca. 657–730).* Tokyo: Dai Nippon Insatsu, 1965.

———. "Record of the Gay Quarters." 3 pts. *Orient/West* 7, no. 10 (1962): 121–28; 8, no. 6 (1963): 115–22; 9, no. 1 (1964): 103–10.

Li Chunqing 李春青. "Shiren yu ziran: Zhongguo gudai shanshui wenxue jiazhiguan zhi wenhua diyun" 士人與自然—中國古代山水文學價值觀之文化底蘊. *Wenxue pinglun* 文學評論 1995, no. 3: 141–51.

Li ji 禮記. SBBY ed.

Li Shangyin 李商隱. *Li Shangyin shige jijie* 李商隱詩歌集解. Commentary by Liu Xuekai 劉學鍇 and Yu Shucheng 余恕誠. Beijing: Zhonghua, 1988.

Li, Wai-yee. *Enchantment and Disenchantment: Love and Illusion in Chinese Literature.* Princeton: Princeton University Press, 1993.

———. "The Idea of Authority in the *Shih chi* (Records of the Historian)." *Harvard Journal of Asiatic Studies* 54, no. 2 (Dec. 1994): 345–405.

Liang Han wenxueshi cankao ziliao 兩漢文學史參考資料. Ed. Beijing Daxue, Zhongguo wenxueshi jiaoyanshi 北京大學中國文學史教研室. Beijing: Zhonghua, 1986.

Liang shu 梁書. Beijing: Zhonghua, 1973.

Lin Wenyue 林文月 (Lin Wen-yüeh). "The Decline and Revival of *Feng-ku* (Wind and Bone): On the Changing Poetic Styles from the Chien-an Era Through the High T'ang Period." In *The Vitality of the Lyric Voice:* Shih *Poetry from the Late Han to the T'ang,* ed. Shuen-fu Lin and Stephen Owen, pp. 130–66. Princeton: Princeton University Press, 1986.

———. *Shanshui yu gudian* 山水與古典. Taibei: Chun wenxue, 1976.

Liu Dajie 劉大杰. *Zhongguo wenxue fazhan shi* 中國文學發展史. Minguo congshu, 2d ser., 58. Shanghai: Shanghai shudian, 1990.

Liu, James J. Y. *The Chinese Knight-Errant*. Chicago: University of Chicago Press, 1967.

———. *Chinese Theories of Literature*. Chicago: University of Chicago Press, 1975.

———. *The Poetry of Li Shang-yin, Ninth-Century Baroque Chinese Poet*. Chicago: University of Chicago Press, 1969.

Liu Su 劉肅. *Da Tang xin yu* 大唐新語. Congshu jicheng, 1st ser. Beijing: Zhonghua, 1985.

Liu Wenzhong 劉文中. "*Shishuo xinyu* zhong de wenlun gaishu" 《世說新語》中的文論概述. *Gudai wenxue lilun yanjiu* 古代文學理論研究, no. 3 (1981): 102–12.

Liu Xiang 劉向. *Liu Xiang Gu lienü zhuan* 劉向古烈女傳. SBCK ed.

Liu Xie 劉勰. *Wenxin diaolong zhushi* 文心雕龍注釋. Commentary by Zhou Zhenfu 周振甫. Beijing: Renmin wenxue, 1981.

Liu Xin 劉歆. *Xijing zaji jiaozhu* 西京雜記校註. Commentary by Xiang Xinyang 向新陽 and Liu Keren 劉克任. Shanghai: Guji, 1991.

Liu Yiqing 劉義慶. *Shishuo xinyu jiao jian* 世說新語校箋. Commentary by Xu Zhen'e 徐震堮. Beijing: Zhonghua 中華, 1987.

Liu Yuejin 劉躍進. *Jie wang man lu: Zhongguo gudian wenxue zuopin tanyou* 結網漫錄：中國古典文學作品探幽. Beijing: Xueyuan, 1996.

Loewe, Michael. "The Campaigns of Han Wu-ti." In *Chinese Ways in Warfare*, ed. Frank A. Kierman Jr. and John K. Fairbank, pp. 67–122. Cambridge, Mass.: Harvard University Press, 1974.

———. *Crisis and Conflict in Han China*. London: George Allen & Unwin, 1974.

Lu Hsun (Xun). *A Brief History of Chinese Fiction*. Beijing: Foreign Languages Press, 1959.

Lu Kanru 陸侃如. *Song Yu* 宋玉. Shanghai: Yatong tushuguan, 1929.

Lu Qinli 逯欽立, ed. *Quan xian Qin Han Wei Jin Nanbei chao shi* 全先秦漢魏晉南北朝詩. Rev. ed. Beijing: Zhonghua, 1993.

Ma Jigao 馬積高. *Fu shi* 賦史. Shanghai: Guji, 1987.

———. "Lun gongtishi yu fojiao" 論宮體詩與佛教. *Qiusuo* 求索 1990, no. 6: 86–92.

Mair, Victor H. "Scroll Presentation in the T'ang Dynasty." *Harvard Journal of Asiatic Studies* 38, no. 1 (1978): 35–60.

Mair, Victor H., and Tsu-lin Mei. "The Sanskrit Origins of Recent Style Prosody." *Harvard Journal of Asiatic Studies* 51, no. 2 (1991): 375–470.

Mao shi zhengyi 毛詩正義. Shanghai: Guji, 1990.

Marney, John. *Liang Chien-wen Ti*. Boston: Twayne, 1976.

Mather, Richard B. "Individualist Expressions of the Outsiders During the Six Dynasties." In *Individualism and Holism: Studies in Confucian and Taoist Values*, ed. Donald Munro, pp. 199–214. Ann Arbor: University of Michigan, Center for Chinese Studies, 1985.

———. "Intermarriage as a Gauge of Family Status in the Southern Dynasties." In *State and Society in Early Medieval China*, ed. Albert E. Dien, pp. 211–28. Stanford: Stanford University Press, 1990.

Mather, Richard B., trans. *Shih Shuo Hsin Yü: A New Account of Tales of the World*. Minneapolis: University of Minnesota Press, 1976.

Mathieu, Rémi. *Le Mu Tianzi zhuan: traduction annotée, étude critique*. Mémoires de l'Institut des hautes études chinoises, 9. Paris: Collège de France, Institut des hautes études chinoises, 1978.

Mauss, Marcel. *The Gift: The Form and Reason for Exchange in Archaic Societies*. New York: Norton, 1990.

McCraw, David. *How the Chinawoman Lost Her Voice*. Sino-Platonic Papers, 32. August 1992.

McMahon, Keith. *Misers, Shrews, and Polygamists: Sexuality and Male-Female Relations in Eighteenth-Century Chinese Fiction*. Durham, N.C.: Duke University Press, 1995.

McMullen, David. *State and Scholars in T'ang China*. Cambridge, Eng.: Cambridge University Press, 1988.

Mei Jialing 梅家玲. "Yiwei yu fude yu caixing zhi jian: *Shishuo xinyu* 'Xianyuan pian' de nüxing" fengmao 依違於婦德與才性之間：《世說新語》〈賢媛篇〉的女性風貌. In *Gudian wenxue yu xingbie yanjiu* 古典文學與性別研究, ed. Hong Shuling 洪淑苓 et al., pp. 129–66. Taibei: Liren shuju 里仁書局, 1996.

Mekada Makoto 目加田誠, trans. *Sesetsu shingo* 世說新語. 3 vols. Shinshaku kambun taikei, 76–78. Tokyo: Meiji shoin, 1975–78.

Meng Qi 孟棨. *Ben shi shi* 本事詩. In *Qinding siku quan shu* 欽定四庫全書, vol. 1478, pp. 231–45.

Mengzi zhengyi 孟子正義. SBBY ed.

Miao, Ronald C. *Early Medieval Chinese Poetry: The Life and Verse of Wang Ts'an (A.D. 177–217)*. Wiesbaden: Franz Steiner Verlag, 1982.

———. "Palace-Style Poetry: The Courtly Treatment of Glamour and Love." In *Studies in Chinese Poetry and Poetics*, vol. 1, ed. Ronald C. Miao. San Francisco: Chinese Materials Center, 1978.

Morino Shigeo 森野繁夫. "Ryōsho no bungaku shūdan" 梁朝の文學集團. *Chūgoku bungaku hō* 中國文學報 21 (1966): 83–108.

Mu Kehong 穆克宏. "Shi lun *Yutai xinyong*" 試論〈玉臺新詠〉. *Wenxue pinglun* 文學評論 1985, no. 6: 107–15.

Nan shi 南史. Beijing: Zhonghua, 1975.

Nylan, Michael. "The *Chin Wen / Ku Wen* Controversy in Han Times." *T'oung Pao* 80 (1994): 83–145.

———. "Style, Patronage and Confucian Ideals in Han Dynasty Art: Problems in Interpretation." *Early China* 18 (1993): 227–47.

Obi Kōichi 小尾郊一. *Sha Reiun: Kodoku no shijin* 謝靈運—孤獨の詩人. Tokyo: Kyūkō shoin, 1983.

Owen, Stephen. *The End of the Chinese "Middle Ages": Essays in Mid-Tang Literary Culture*. Stanford: Stanford University Press, 1996.

———. *The Great Age of Chinese Poetry: The High T'ang*. New Haven: Yale University Press, 1981.

———. *Mi-lou: Poetry and the Labyrinth of Desire*. Cambridge, Mass.: Harvard University Press, 1989.

———. *The Poetry of the Early T'ang*. New Haven: Yale University Press, 1977.

———. *Readings in Chinese Literary Thought*. Cambridge, Mass.: Harvard University, Council on East Asian Studies, 1992.

———. *Traditional Chinese Poetry and Poetics: Omen of the World*. Madison: University of Wisconsin Press, 1985.

Pan Zhonggui 潘重規. "Shixu mingbian" 詩序明辨. *Xueshu jikan* 學術季刊 4, no. 4 (June 1956): 20–25.

Powers, Martin. *Art and Political Expression in Early China*. New Haven: Yale University Press, 1991.

Pu Meiling 朴美齡. *"Shishuo xinyu" zhong suo fanying de sixiang* 世說新語中所反映的思想. Taibei: Wenjin, 1990.

Puett, Michael. Review of Stephen W. Durrant, *The Cloudy Mirror: Tension and Conflict in the Writings of Sima Qian*. *Harvard Journal of Asiatic Studies* 57, no. 1 (1997): 290–301.

Qian Mu 錢穆. *Liang Han jingxue jinguwen pingyi* 兩漢經學今古文平議. Hong Kong: Xinya yanjiusuo, 1958.

Qian Nanxiu. "Being One's Self: Narrative Art and Taxonomy of Human Nature in the *Shih-shuo hsin-yü*." Ph.D. diss., Yale University, 1994.

Qian Zhongshu 錢鍾書. *Guan zhui bian* 管錐編. Beijing: Zhonghua, 1979.

Qu Wanli 屈萬里. *Shijing quan shi* 詩經詮釋. Taibei: Lianjing, 1983.

Quan Tang shi 全唐詩. Beijing: Zhonghua, 1960.

Ren Bantang 任半塘, ed. *Dunhuang geci zongbian* 敦煌歌辭總編. Shanghai: Guji, 1987.

———. *Tang sheng shi* 唐聲詩. 2 vols. Shanghai: Guji, 1982.

Robertson, Maureen. "Voicing the Feminine: Constructions of the Gendered Subject in Lyric Poetry by Women of Medieval and Late Imperial China." *Late Imperial China* 13, no. 1 (1992): 63–110.

Rouzer, Paul F. *Writing Another's Dream: The Poetry of Wen Tingyun*. Stanford: Stanford University Press, 1993.

Roy, David T. "The Theme of the Neglected Wife in the Poetry of Ts'ao Chih." *Journal of Asian Studies* 19, no. 1 (1959): 25–31.

Rubin, Gayle. "The Traffic in Women: Notes on the 'Political Economy' of Sex." In *Toward an Anthropology of Women*, ed. Rayna R. Reitner, pp. 157–210. New York: Monthly Review Press, 1975.

Sanders, Graham. "Poetry in Narrative: Meng Ch'i (fl. 841–866) and *True Stories of Poems* (*Pen-shih shih*)." Ph.D. diss., Harvard University, 1996.

Saussy, Haun. *The Problem of a Chinese Aesthetic*. Stanford: Stanford University Press, 1993.

Schaberg, David. "Travel, Geography, and the Imperial Imagination in Fifth-Century Athens and Han China." *Comparative Literature* 51, no. 2 (1999): 152–91.

Schafer, Edward H. *The Divine Woman: Dragon Ladies and Rain Maidens*. Berkeley: University of California Press, 1973.

Schneider, Laurence A. *A Madman of Ch'u: The Chinese Myth of Loyalty and Dissent*. Berkeley: University of California Press, 1980.

Sedgwick, Eve. *Between Men: English Literature and Male Homosocial Desire*. New York: Columbia University Press, 1985.

Shen Yue 沈約. *Song shu* 宋書. Beijing: Zhonghua, 1974.

Sima Guang 司馬光. *Zizhi tongjian* 資治通鑑. Beijing: Zhonghua, 1956.

Sima Xiangru 司馬相如. *Sima Xiangru ji jiaozhu* 司馬相如集校注. Commentary by Zhu Yiqing 朱一清 and Sun Yizhao 孫以昭. Beijing: Renmin wenxue, 1996.

Song Dexi 宋德熹. "Tangdai de jinü" 唐代的妓女. In *Zhongguo funü lunji xuji* 中國婦女論集續集, ed. Bao Jialin 鮑家麟, pp. 67–121. Taibei: Daoxiang, 1991.

Stallybrass, Peter, and Allon White. *The Politics and Poetics of Transgression*. Ithaca, N.Y.: Cornell University Press, 1986.

Su Xuelin 蘇雪林. *Yuxi shimi* 玉溪詩謎. Ren ren wen ku, 1057. Taibei: Taiwan shangwu, 1969.

Sun Qi 孫棨. *Bei li zhi* 北里誌. In *Jiaofang ji, Beili zhi, Qinglou ji* 教坊記，北里誌，青樓集. Beijing: Zhonghua, 1959.

Taiping guangji 太平廣記. 4 vols. Shanghai: Guji, 1990.

Takigawa Kametarō 瀧川龜太郎. *Shiki kaichū kōshō* 史記會注考證. Tokyo, 1934.

Tang Guizhang 唐圭璋, ed. *Quan Song ci* 全宋詞. Beijing: Zhonghua, 1965.

Tang, Yiming. "The Voices of Wei-Jin Scholars: A Study of Qingtan." Ph.D. diss., Columbia University, 1991.

Tanigawa, Michio. *Medieval Chinese Society and the Local "Community."* Berkeley: University of California Press, 1985.

Tu, Wei-ming. "Profound Learning, Personal Knowledge, and Poetic Vision." In *The Vitality of the Lyric Voice:* Shih *Poetry from the Late Han to the T'ang*, ed. Shuen-fu Lin and Stephen Owen, pp. 3–31. Princeton: Princeton University Press, 1986.

Tung, Jowen R. *Fables for the Patriarchs: Gender Politics in Tang Discourse*. Lanham, Md.: Rowman and Littlefield, 2000.

Turner, James Grantham, ed. *Sexuality and Gender in Early Modern Europe: Institutions, Texts, Images*. Cambridge, Eng.: Cambridge University Press, 1993.

Twitchett, Denis. "The Composition of the T'ang Ruling Class: New Evidence from Tun-huang." In *Perspectives on the T'ang*, ed. Denis Twitchett and Arthur F. Wright, pp. 47–85. New Haven: Yale University Press, 1973.

Uchida Sennosuke 内田泉之助. *Gyokudai shin'ei* 玉台新詠. 2 vols. Shinshaku kambun taikei, 60–61. Tokyo: Meijin shoin, 1974–75.

Van Gulik, R. H. *Sexual Life in Ancient China*. Leiden: Brill, 1961.

Van Zoeren, Steven. *Poetry and Personality: Reading, Exegesis, and Hermeneutics in Traditional China*. Stanford: Stanford University Press, 1991.

Wagner, Marsha. *The Lotus Boat: The Origins of Chinese* Tz'u *Poetry in T'ang Popular Culture*. New York: Columbia University Press, 1984.

Waldron, Arthur. *The Great Wall of China: From History to Myth*. Cambridge, Eng.: Cambridge University Press, 1990.

Waley, Arthur. "Colloquial in the *Yu-Hsien K'u*." *Bulletin of the School of Oriental and African Studies* 29, no. 3 (1966): 559–65.

———. *The Life and Times of Po Chü-i, 772–846 A.D.* London: George Allen & Unwin, 1949.

Waley, Arthur, trans. *The Book of Songs*. Edited with additional translations by Joseph R. Allen. New York: Grove Press, 1996.

Waller, Marguerite. "The Empire's New Clothes: Refashioning the Renaissance." In *Seeking the Woman in Late Medieval and Renaissance Writings: Essays in Feminist Contextual Criticism*, ed. Sheila Fisher and Janet E. Halley, pp. 160–83. Knoxville: University of Tennessee Press, 1989.

Wang, C. H. "Towards Defining a Chinese Heroism." *Journal of the American Oriental Society* 95, no. 1 (1975): 25–35.

Wang Chung-han. "The Authorship of the *Yu-hsien k'u*." *Harvard Journal of Asiatic Studies* 11 (1948): 153–62.

Wang Wensheng 王文生. "Liang Han de wenxue lilun piping" 兩漢的文學理論批評. *Gudai wenxue lilun yanjiu* 古代文學理論研究 3 (1981): 84–101.

Wang Yunxi 王運熙 and Yang Ming 楊明. *Wei Jin Nanbei chao wenxue piping shi* 魏晋南北朝文學批評史. Shanghai: Guji, 1989.

Watson, Burton. *Early Chinese Literature*. New York: Columbia University Press, 1962.

Wei Peilan 魏佩蘭. "Maoshi xu zhuan weiyi kao 《毛詩序傳》違異考." *Dalu zazhi* 大陸雜誌 33, no. 8 (1966): 15–21.

Westbrook, Francis A. "Landscape Description in the Lyric Poetry and 'Fuh on Dwelling in the Mountains' of Shieh Ling-yunn.'" Ph.D. diss., Yale University, 1972.

Weston, Jessie L. *From Ritual to Romance*. Cambridge, Eng.: Cambridge University Press, 1920.

Whitaker, K. P. K. "Tsaur Jyr's 'Luohshern fuh.'" *Asia Major*, n.s. 4, no. 1 (1954): 36–56.

Wu, Laura Hua. "From *Xiaoshuo* to Fiction: Hu Yinglin's Genre Study of *Xiaoshuo*." *Harvard Journal of Asiatic Studies* 55, no. 2 (1995): 339–71.

Xiao Tong 蕭統, ed. *Wen xuan* 文選. Commentary by 李善. Bejing: Zhonghua, 1977.

Xie Lingyun 謝靈運. *Xie Kangle shi zhu* 謝康樂詩註. Commentary by Huang Jie 黃節. Taibei: Yiwen yinshu guan, 1975.

Xu Ling 徐陵, ed. *Yutai xinyong jian zhu* 玉臺新詠箋注. Commentary by Wu Zhaoyi 吳兆宜, Cheng Yanshan 程琰刪, and Mu Kehong 穆克宏. Beijing: Zhonghua, 1985.

Xue Yongruo 薛用弱. *Ji yi ji* 集異記. Chongshu jicheng, 1st ser. Beijing: Zhonghua, 1985.

Yagisawa Hajime 八木澤元. *"Yū senkutsu" zenkō* 遊仙窟全講. Rev. ed. Tokyo: Meiji shoin, 1976.

Yan Caiping 閻采平. "Shi shu guanxi yu Qi Liang wenxue jituan" 士庶關系與齊梁文學集團. *Wenxue yichan* 文學遺產 1994, no. 3: 22–28.

Yan Zhitui 顏之推. *Yan shi jia xun jijie* 顏氏家訓集解. Commentary by Wang Liqi 王利器. Beijing: Zhonghua, 1982.

Yang Bojun 楊伯峻, ed. *Lunyu yizhu* 論語譯注. Beijing: Zhonghua, 1980.

Yang Donglin 楊東林. "Lüe lun nanchao de jiazu yu wenxue" 略論南朝的家族與文學. *Wenxue pinglun* 文學評論 1994, no. 3: 5–19.

Yang, Lien-sheng. "An Additional Note on the Ancient Game *Liu-po*." *Harvard Journal of Asiatic Studies* 15 (1952): 124–39.

———. "Historical Notes on the Chinese World Order." In *The Chinese World Order: Traditional China's Foreign Relations*, ed. John K. Fairbank, pp. 20–33. Cambridge, Mass.: Harvard University Press, 1968.

———. "A Note on the So-Called TLV Mirrors and the Game *Liu-po* 六博." *Harvard Journal of Asiatic Studies* 9 (1945–47): 202–6.

Yang Ming 楊明. "Gongti shi pingjia wenti" 宮體詩評價問題. *Fudan xuebao* 复旦學報 (*shehui kexue ban* 社會科學版) 1988, no. 5: 46–52.

Yang Xiong 揚雄. *Fa yan zhu* 法言注. Commentary by Han Jing 韓敬. Beijing: Zhonghua, 1992.

Yang Yinzong 楊胤宗. "Song Yu fu kao" 宋玉賦考. 2 pts. *Dalu zazhi* 大陸雜誌 27, no. 3 (1963): 85–90; 27, no. 4 (1963): 126–32.

Ye Jiaying 葉嘉瑩 (Chia-ying Yeh Chao). "The Ch'ang-chou School of *Tz'u* Criticism." *Harvard Journal of Asiatic Studies* 35 (1975): 101–32.

———. *Jialing lun shi cong gao* 迦陵論詩叢稿. Shanghai: Guji, 1984.

You Shihong 游適宏. "Shi lun 'Yuan ge xing'" 識論〈怨歌行〉. *Zhonghua xueyuan* 中華學苑 42 (1992): 117–45.

Yu Guanying 余冠英, ed. *Han Wei Liuchao shi xuan* 漢魏六朝詩選. Beijing: Renmin wenxue, 1958.

Yu, Pauline. *The Reading of Imagery in the Chinese Poetic Tradition*. Princeton University Press, 1987.

Yu, Pauline, ed. *Voices of the Song Lyric in China*. Berkeley: University of California Press, 1994.

Yü, Ying-shih. "Han Foreign Relations." In *The Cambridge History of China*, vol. 1, *The Ch'in and Han Empires, 221 B.C.–A.D. 220*, ed. Denis Twitchett and Michael Loewe, pp. 377–462. Cambridge, Eng.: Cambridge University Press, 1986.

———. "Individualism and the Neo-Taoist Movement in Wei-Chin China." In *Individualism and Holism: Studies in Confucian and Taoist Values*, ed. Donald Munro, pp. 121–56. Ann Arbor: University of Michigan, Center for Chinese Studies, 1985.

———. *Trade and Expansion in Han China: A Study in the Structure of Sino-Barbarian Economic Relations*. Berkeley: University of California Press, 1967.

Zhang Keli 張可禮. "Jian'an shige de san zhong chansheng fangshi ji qi tedian" 建安詩歌的三種產生方式及其特點. *Wenxue pinglun congkan* 文學評論叢刊, no. 30 (1989): 111–24.

Zhang Lei 張蕾. "Cao Zhi funü ticai shizuo jianshang yishuo" 曹植婦女題材詩作鑒賞異說. *Hebei shifan daxue xuebao* 河北師範大學學報 1992, no. 2: 32–39.

Zhang Pu 張溥. *Han Wei Liu Chao bai san ming jia ji* 漢魏六朝百三名家集. 5 vols. Yangzhou: Yangzhou guji, 1990.

Zhang Xuecheng 章學誠. *Wenshi tongyi jiaozhu* 文史通義校注. Commentary by Ye Ying 葉瑛. Beijing: Zhonghua, 1983.

Zheng Yuyu 鄭毓瑜. "Shennü lunshu yu xingbie yanyi: yi Qu Yuan, Song Yu fu wei zhu de taolun" 神女論述與性別演義—以屈原、宋玉賦爲主的討論. In *Gudian wenxue yu xingbie yanjiu* 古典文學與性別研究, ed. Hong Shuling 洪淑苓 et al., pp. 29–56. Taibei: Liren shuju, 1996.

Zhong Rong 鍾嶸. *Shi pin ji zhu* 詩品集注. Commentary by Cao Xu 曹旭. Shanghai: Guji, 1994.

Zhou Zhenfu 周振甫. "'Du Yingying zhuan' xian yi" 〈讀鶯鶯傳〉獻疑. *Wenxue yichan* 文學遺產 1992, no. 6: 60–65.

———. *Shiwen qian shi* 詩文淺釋. Beijing: Shifan xueyuan, 1986.

Zhu Shouliang 朱守亮. *Shijing pingshi* 詩經評釋. Taibei: Taiwan xuesheng, 1984.

Zhu Xi 朱熹. *Shi jizhuan* 詩集傳. SBCK ed.

Zhu Ziqing 朱自青. *Zhu Ziqing gudian wenxue lunwen ji* 朱自青古典文學論文集. Shanghai: Guji, 1981.

Zito, Angela. "Ritualizing *Li*." *positions: east asia cultures critique* 1, no. 2 (1993): 321–48.
———. "Silk and Skin: Significant Boundaries." In *Body, Subject and Power in China*, ed. Angela Zito and Tani E. Barlow, pp. 103–30. Chicago: Chicago University Press, 1994.

Index

Harvard-Yenching Institute Monograph Series *(titles now in print)*

11. *Han Shi Wai Chuan: Han Ying's Illustrations of the Didactic Application of the Classic of Songs,* translated and annotated by James Robert Hightower
21. *The Chinese Short Story: Studies in Dating, Authorship, and Composition,* by Patrick Hanan
22. *Songs of Flying Dragons: A Critical Reading,* by Peter H. Lee
23. *Early Chinese Civilization: Anthropological Perspectives,* by K. C. Chang
24. *Population, Disease, and Land in Early Japan, 645-900,* by William Wayne Farris
25. *Shikitei Sanba and the Comic Tradition in Edo Fiction,* by Robert W. Leutner
26. *Washing Silk: The Life and Selected Poetry of Wei Chuang (834?-910),* by Robin D. S. Yates
27. *National Polity and Local Power: The Transformation of Late Imperial China,* by Min Tu-ki
28. *Tang Transformation Texts: A Study of the Buddhist Contribution to the Rise of Vernacular Fiction and Drama in China,* by Victor H. Mair
29. *Mongolian Rule in China: Local Administration in the Yuan Dynasty,* by Elizabeth Endicott-West
30. *Readings in Chinese Literary Thought,* by Stephen Owen
31. *Remembering Paradise: Nativism and Nostalgia in Eighteenth-Century Japan,* by Peter Nosco
32. *Taxing Heaven's Storehouse: Horses, Bureaucrats, and the Destruction of the Sichuan Tea Industry, 1074-1224,* by Paul J. Smith
33. *Escape from the Wasteland: Romanticism and Realism in the Fiction of Mishima Yukio and Oe Kenzaburo,* by Susan Jolliffe Napier
34. *Inside a Service Trade: Studies in Contemporary Chinese Prose,* by Rudolf G. Wagner
35. *The Willow in Autumn: Ryutei Tanehiko, 1783-1842,* by Andrew Lawrence Markus
36. *The Confucian Transformation of Korea: A Study of Society and Ideology,* by Martina Deuchler
37. *The Korean Singer of Tales,* by Marshall R. Pihl
38. *Praying for Power: Buddhism and the Formation of Gentry Society in Late-Ming China,* by Timothy Brook
39. *Word, Image, and Deed in the Life of Su Shi,* by Ronald C. Egan
40. *The Chinese Virago: A Literary Theme,* by Yenna Wu

Harvard-Yenching Institute Monograph Series

41. *Studies in the Comic Spirit in Modern Japanese Fiction,* by Joel R. Cohn
42. *Wind Against the Mountain: The Crisis of Politics and Culture in Thirteenth-Century China,* by Richard L. Davis
43. *Powerful Relations: Kinship, Status, and the State in Sung China (960-1279),* by Beverly Bossler
44. *Limited Views: Essays on Ideas and Letters,* by Qian Zhongshu; selected and translated by Ronald Egan
45. *Sugar and Society in China: Peasants, Technology, and the World Market,* by Sucheta Mazumdar
46. *Chinese History: A Manual,* by Endymion Wilkinson
47. *Studies in Chinese Poetry,* by James R. Hightower and Florence Chia-Ying Yeh
48. *Crazy Ji: Chinese Religion and Popular Literature,* by Meir Shahar
49. *Precious Volumes: An Introduction to Chinese Sectarian Scriptures from the Sixteenth and Seventeenth Centuries,* by Daniel L. Overmyer
50. *Poetry and Painting in Song China: The Subtle Art of Dissent,* by Alfreda Murck
51. *Evil and/or/as the Good: Omnicentrism, Intersubjectivity, and Value Paradox in Tiantai Buddhist Thought,* by Brook Ziporyn
52. *Chinese History: A Manual, Revised and Enlarged Edition,* by Endymion Wilkinson
53. *Articulated Ladies: Gender and the Male Community in Early Chinese Texts,* by Paul Rouzer